NORTH AMERICAN WILDLAND PLANTS

Third Edition

James Stubbendieck, Stephan L. Hatch,
Neal M. Bryan, and Cheryl D. Dunn

North American

UNIVERSITY OF NEBRASKA PRESS ⬇ LINCOLN AND LONDON

Wildland Plants

THIRD EDITION | A FIELD GUIDE

Illustrated by *Neal M. Bryan,*
Angie Fox, Kelly L. Rhodes Hays,
Bellamy Parks Jansen, and *Debra Meier*

Maps by *Kathleen Lonergan-Orr*
and *Neal M. Bryan*

Library of Congress Cataloging-in-Publication Data
Names: Stubbendieck, James L., author. | Hatch, Stephan L., 1945– author. | Bryan, Neal M., author. | Dunn, Cheryl D., author.
Title: North American wildland plants: a field guide / James Stubbendieck, Stephan L. Hatch, Neal M. Bryan, and Cheryl D. Dunn; illustrated by Neal M. Bryan, Angie Fox, Kelly L. Rhodes Hays, Bellamy Parks Jansen, and Debra Meier; maps by Kathleen Lonergan-Orr and Neal M. Bryan.
Description: Third edition. | Lincoln: University of Nebraska Press, [2017] | Includes bibliographical references and index.
Identifiers: LCCN 2016047607
ISBN 9780803299658 (pbk.: alk. paper)
ISBN 9781496200914 (epub)
ISBN 9781496200921 (mobi)
ISBN 9781496200938 (pdf)
Subjects: LCSH: Range plants—North America—Identification. | Forage plants—North America—Identification. | Range plants—North America. | Forage plants—North America.
Classification: LCC SB193.3.N67 S88 2017 | DDC 633.2/02—dc23 LC record available at https://lccn.loc.gov/2016047607

Set in Arno Pro by L. Auten.

To students striving to understand the complex nature of the plant world.

Contents

GRASSLIKE PLANTS

FORBS AND WOODY PLANTS

ASTERACEAE

Acknowledgments

The authors wish to acknowledge Kathie J. Diller, Charles H. Butterfield, and Lori M. Landholt for their contributions to earlier editions of this book. Angie Fox, Kelly L. Rhodes Hays, Bellamy Parks Jansen, Debra Meier, and Neal M. Bryan prepared the illustrations. Kathleen Lonergan-Orr and Neal M. Bryan prepared the maps.

Juan M. Martínez-Reyna, Jesús Valdés-Reyna, Juvenal Gutiérrez-Castillo, and Juan A. Encina-Dominguez are recognized for furnishing the common names used in Mexico. Others contributing to this and earlier editions include Kelly Allred, Val Anderson, William T. Barker, Kylie Faulk, Jessica Garlick, T. Mark Hart, Marshall Hervey, Barry Irving, Rixey Jenkins, Stanley D. Jones, Robert B. Kaul, Gary E. Larson, Elizabeth Manrique, Jacob Meador, Catherine Mills, Daniel Nosal, Linda L. Rader, Jesús Valdés-Reyna, Walter H. Schacht, Susan Schuckert, Karen Spath, and J. K. Wipff.

We recognize the work and dedication of the range plant team coaches and thank Rixey Jenkins for providing the names: Eddie Alford, Gary Baird, Fee Busby, Gustavo Carillo, Craig Carr, Rob Cox, Shawn Dekeyser, Charlene Duncan, Cheryl Dunn, Amy Ganguli, Todd Golder, Ricardo Mata-Gonzalez, Laura Goodman, Mitch Greer, Micha Humphreys, Barry Irving, Toutcha Lebgue-Keleng, Keith Klement, Bob Knight, Gary Larson, Karen Launchbaugh, Casey Matney, Sarah Nowlle, Jennifer Obrigewitch, Steve Petersen, Tami Plechaty, Gustavo Quintata, Juan Martínez-Reyna, Zac Roehers, Dan Rogers, Alfonso Ortega-Santos, Ally Searle, Randall Violett, and Brett Wolk.

NORTH AMERICAN WILDLAND PLANTS

Introduction

A comprehensive reference containing the important characteristics of the most important wildland plants of North America is critical for ecologists, range managers, land managers, and other natural resource professionals. In addition, university students and range plant identification teams needed a single, primary resource for learning about important wildland plant species. *North American Range Plants* was developed to meet these needs and was first published in 1981. Subsequent editions (1982, 1986, 1992, 1997) included changes in nomenclature, refinement of distributions, additional information on each of the species, and new illustrations. The illustrations were prepared to highlight general and specific characteristics to aid identification of the featured range plants. The fifth edition (1997) reflected changing attitudes toward riparian areas and wetlands. Reflecting this increased concern and interest, about 10 percent of the species included in this book occur on these sites. *North American Wildland Plants* (2003, 2011) included many nomenclatural changes, and the illustrations were labeled to accentuate specific characteristics. The title change reflected the importance of plants across ecosystems and the multiple uses of the plant resources within ecosystems. This edition contains additional refinements in the nomenclature, distribution, illustrations, and descriptions of plants. This reference will help both individuals with limited botanical knowledge and natural resource professionals to identify wildland plants.

The two hundred species in this book were selected because of their abundance, desirability, or noxious properties; in short, they are important wildland species. The list of plant species was developed over the course of nearly sixty years by faculty from the universities and colleges with rangeland management and ecology programs and by coaches of range plant identification teams. The formal list is now the Master Plant List for the International Range Plant Identification Contest sponsored by the Society for Range Management (6901 South Pierce Street, Littleton CO 80128; www .rangelands.org).

Plant species descriptions in this book include characteristics for their identification, a labeled illustration of a typical plant (with enlarged plant parts), and a general distribution map for North America. Each species description includes nomenclature; life span; origin; season of growth; inflorescence, flower or spikelet or other reproductive parts; vegetative parts; and growth characteristics. Forage values for wildlife and livestock are estimated. Brief notes are included on habitat; livestock losses; and historical, food, and medicinal uses. Information on historical, food, and medicinal uses was gathered from numerous sources and is presented as a point of interest and to broaden readers' appreciation of these plants. It is strongly emphasized that these plant species should not be used for these purposes.

Grasses (POACEAE family) are described first and are aligned by tribe, genus, and specific epithet in alphabetical order by rank. Grasslike plants (CYPERACEAE and JUN-CACEAE families) are next. All other families follow in alphabetical order by rank for family, genus, and specific epithet with the exception of members of the ASTERACEAE family, which are aligned as the POACEAE family.

The grass (POACEAE) and composite (ASTERACEAE) families are treated by tribe to help the reader relate to smaller groups within these large, complex families. Recognition of species within tribes builds a concept of tribal characteristics. When an unknown species of either family is encountered, knowledge of tribal alignments below family may reduce the time required for making an identification using a diagnostic key.

Classification generally follows Tropicos (www.tropicos.org), sponsored by the Missouri Botanical Garden. Tropicos is the world's largest botanical database.

Numerous authoritative floristic treatments from the wildland areas of North America were consulted for species names and authorities. Selected synonyms, noting other names for the same species, are included on the illustration page for each species to help clarify the species concept used in this text. The synonyms will help in finding additional information in other floristic treatments.

Common and alternative common names are given for the plants, but they may not include the common name used in a particular area. Common names were restricted to two words, sometimes resulting in long and cumbersome words. Common names used in Mexico are listed for the appropriate taxa and may exceed two words.

The origin of each species is given as native or introduced. Origins of introduced taxa are given parenthetically. Many species are known to be introduced, while others are thought to have been introduced. *Poa pratensis* L. is an example of a species that is listed as introduced but may be both native and introduced to North America.

Season of growth is listed as cool, warm, or evergreen. Cool-season plants complete most of their growth in the autumn, winter, and spring. Warm-season plants grow most in the summer when temperatures are the highest. The evergreen plants retain their ability to grow whenever climatic conditions are suitable. A summary of this information may be found in the Checklist of Wildland Plants included in this book.

Plant characteristics for each species are separated into categories to help in making comparisons between species. Bold type is used in this edition to emphasize important characteristics that separate species. These characteristics are intended to be useful to students preparing to compete in the International Range Plant Identification Contest and to amateur botanists. Conservative characteristics, those that are not greatly influenced by the environment, should be the basis for identification. These may include floral, spikelet, leaf, and inflorescence type but may vary with the species. Pubescence, ligule lengths, and awn lengths are highly variable characteristics, and primary importance should not be placed on these when identifying grasses. Presence or absence of rhizomes is another variable characteristic that is somewhat dependent upon moisture, other features of the habitat, and techniques used by the collector.

Forage values of the plants discussed in this book are relative values that vary with the type of animal utilizing the particular plant species. Values are determined on the basis of palatability, nutrient content, and the amount of forage produced by the plant species. These values may vary with the climatic conditions, the part of North America where the plant is growing, when the forage is consumed, associated plant species, and the age and class of each animal species utilizing the forage.

Losses due to poisonous plants, one of the major problems facing the livestock industry in some regions, are included in these plant descriptions. Annual losses on wildlands amount to hundreds of millions of dollars, with the effects of poisonous plants varying from slightly reduced rate of gain to deformities or death of the animal. Losses that are easy to document, such as death, are not as economically important as the losses wherein growth rate or milk production is reduced. The brief mention of livestock losses in this book include the animals affected and the type of poison, commonly referred to as the poisonous principle, contained in the plant species.

This book includes a glossary, list of nomenclature authorities, list of selected references, and a checklist of the species. This supplementary information will give the student, professional natural resource manager, and anyone else interested in plants a more complete knowledge of plants and a starting place in the literature to seek additional information. The index is comprehensive, including all scientific and common names used in the text.

The information contained in *North American Wildland Plants* is by no means complete. The authors have opted for brevity with the expectation that this book will be a starting point for those interested in wildland plant identification. Plant taxonomists and extension personnel in each locality can provide additional information on plant species of interest.

Wildland Plants

LIFE SPAN

Most wildland plant species are classified as perennials or annuals. Perennials generally live three or more years, while annuals complete their life cycle in one growing season. Herbaceous perennials have aerial stems that die back to the soil level each year while the underground parts remain alive. Perennial grasses, grasslike plants, and forbs are in this category. Woody perennials have aerial stems that remain alive throughout the year, although they may become dormant for part of the year. Shrubs and trees are in this category. Biennial is a third lifespan category. Biennials require two growing seasons to complete their life cycle. Growth during the first year is generally vegetative, and seed is produced during the second growing season. Relatively few plants fit into this category.

ORIGIN

Wildland plants originating in North America are classified as "native." The term "introduced" refers to plants that have been brought into North America from another continent and adapted to conditions here. Several introduced species are valuable forage plants that were intentionally introduced for that purpose. Some introduced species were brought in for various reasons (e.g., landscaping) and then escaped; of these, some are now troublesome weeds. Some species were accidentally introduced through contaminated crop seed, packing material, or ballast.

CLASSIFICATION

Botanical nomenclature refers to a system of naming plants. Plants are described and grouped according to their structure, particularly structure of the flowering or other reproductive parts. One classification system from general to specific is:

KINGDOM (PLANT)
DIVISION (PHYLUM)
CLASS
ORDER
FAMILY
TRIBE
GENUS
SPECIFIC EPITHET

The genus and specific epithet are combined to form the species name which is written in italics or underlined.

While plants included in this book come from various divisions, classes, and orders, we will be concerned only with the last four parts of the classification system:

A. FAMILY

A plant family is the basic division of plant orders. Morphological characteristics or similarities determine the family to which a plant belongs. However, molecular genetics is becoming increasingly important in plant classification. Flowering characteristics are extremely important in the classification of families. All grasses have similar flowers in spikelets and belong to the same family, POACEAE. For the other plants, numbers of petals, sepals, stamens, pistils, and other flowering parts are used for assignment to family. All family names of vascular plants used in this book have a standard ending-ACEAE.

B. TRIBE

A plant family may be divided into tribes. In this book, POACEAE and ASTERACEAE are the only families for which tribes are recognized. An example is the ANDRO-POGONEAE tribe of the POACEAE family. All tribe names of vascular plants have the standard ending-EAE.

C. SCIENTIFIC NAME

There is only one correct scientific name for each species. The scientific name, or binomial, consists of two main parts. The first part is the genus, and the second is the specific epithet. The binomial is unique to each species. The authority is added for completeness and accuracy.

1. GENUS

Classification of plants into genera (plural of genus) is based on similarities in flowering and/or morphological and nonmorphological characteristics, although with more specific divisions. An example is the genus *Schizachyrium*, which is part of the ANDROPOGONEAE tribe of the POACEAE family. The first letter of the genus is capitalized, and the word is underlined or italicized.

2. SPECIFIC EPITHET

The second part of the scientific name is the specific epithet. It is italicized or underlined but not capitalized. A species is the kind of plant and is named by the combination of the genus and specific epithet. This classification is based on differences in flowering and/or morphological characteristics and nonmorphological characteristics, such as molecular sequences, that distinguish a plant from related species. An example is the specific epithet *scoparium* for the species *Schizachyrium scoparium*, which differs from all other species of *Schizachyrium* in specific morphological characters.

3. NOMENCLATURE AUTHORITY

The scientific name, for reasons of completeness and accuracy, is followed by the abbreviation or whole name of the person or persons who first applied that name to the plant. For example, (Michx.) Nash are the authorities for *Schizachyrium scoparium*. The French botanist Andre Michaux (1746–1802) first described and applied the specific epithet to that species. The name used was *Andropogon scoparius* Michx. American agrostologist George Nash (1864–1921) later transferred the species to the genus *Schizachyrium*. Nash became the author of a new combination/classification involving that specific epithet. Hence, the name *Schizachyrium scoparium* (Michx.) Nash. A list of nomenclature authorities follows the glossary.

4. COMMON NAME

Common names have been given to many species of plants. Common names are usually simple and often descriptive of the plant, honor some person, or give a geographical location. However, one species may have more than one common name, even within the same language. In this book we use little bluestem as the common name of *Schizachyrium scoparium*, but it is also called prairie beardgrass in some parts of English-speaking North America, while in Mexico it is referred to as popotillo colorado or popotillo cañuelo. Another weakness of common names is that one common name may be applied to several species. For example, junegrass in Kentucky is *Poa pratensis*; junegrass in Nebraska is *Koeleria macrantha*; and junegrass in Utah is *Bromus tectorum*.

As noted, there is only one correct scientific name for each plant; nevertheless, the name for a given plant will change if that plant is reclassified or if it is discovered that another valid name for it was published earlier. Although date of publication is absolute, the assignment of rank and position in the classification process is a matter of taxonomic opinion, which is often annoying to the layman. Names other than the correct one are synonyms and can be found listed below the accepted scientific and common names on the page of each species illustration.

A summary of the classification system for little bluestem is:

FAMILY: POACEAE
    TRIBE: ANDROPOGONEAE
        GENUS: *Schizachyrium*
            SPECIFIC EPITHET: *scoparium*
                SPECIES: *Schizachyrium scoparium* (Michx.) Nash

Wildland plants may be divided into grasses, grasslike plants, forbs, and woody plants. These can be easily distinguished by certain characteristics. Figure 1 presents a comparison of plant groups.

Grasses have either hollow or solid stems with nodes. Leaves are two-ranked, sheathing, and have parallel veins. Flowers are small, inconspicuous, and occur in spikelets. Fruits are usually caryopses.

Grasslike plants resemble grasses but generally have solid or pithy stems without elongated internodes. Leaf veins are parallel, but the leaves are two-ranked or three-ranked. Stems are often triangular, and the flowers are small and inconspicuous. Fruits are either achenes or capsules.

Forbs are herbaceous plants other than grasses and grasslike plants. They usually have solid stems and generally have broad leaves with netted venation. Flowers are often large and showy colored, although they may be small and inconspicuous. Forbs have a variety of fruit types.

Woody plants have secondary growth of their aerial stems that live throughout the year, although they may be dormant part of the time. Leaves are often broad and net-veined. Flowers are often showy, but they may be inconspicuous. Both trees and shrubs fit into this category. Fruit types of woody plant species vary.

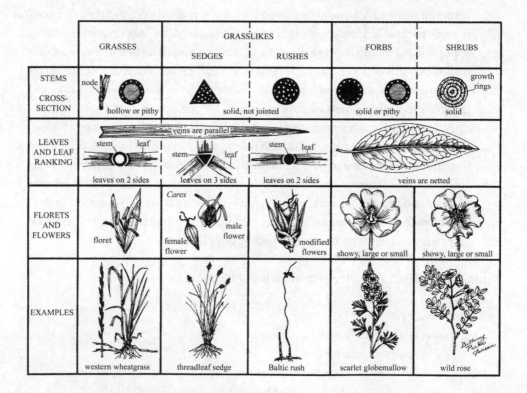

FIGURE 1. Comparison of plant groups

Figures 2 through 5 are a series of drawings illustrating various morphological features of the grass plant. See the glossary for definition of terms.

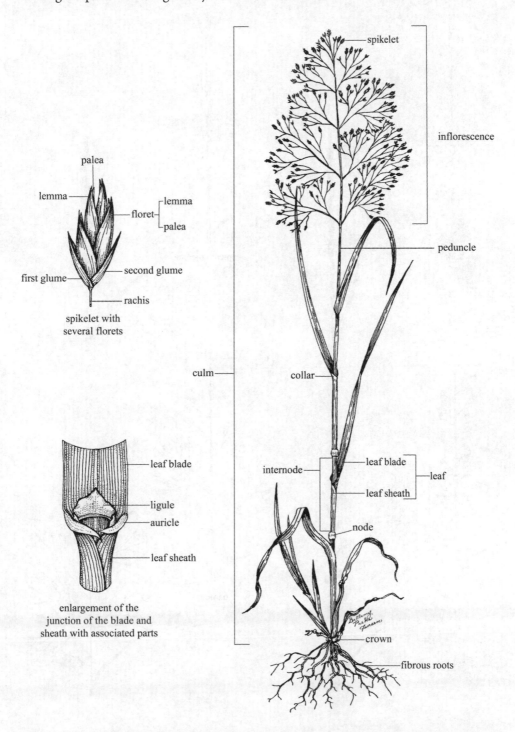

FIGURE 2. Grass plant and spikelet

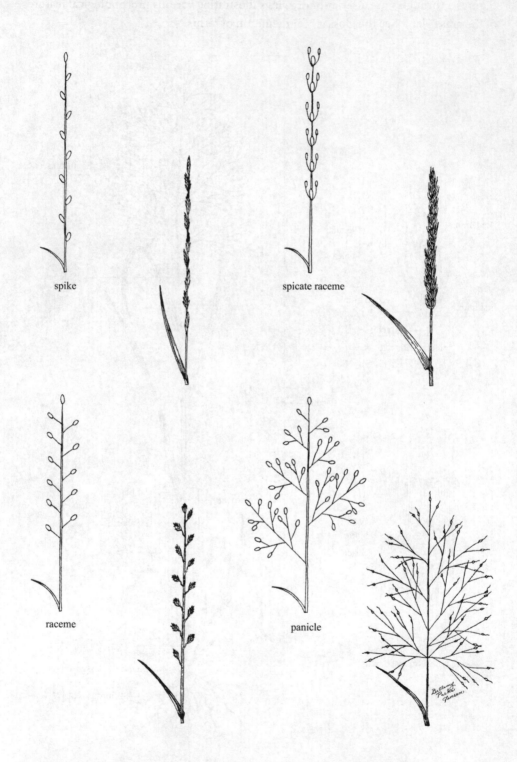

spike

spicate raceme

raceme

panicle

FIGURE 3A. Diagrammatic and actual representations of grass inflorescences

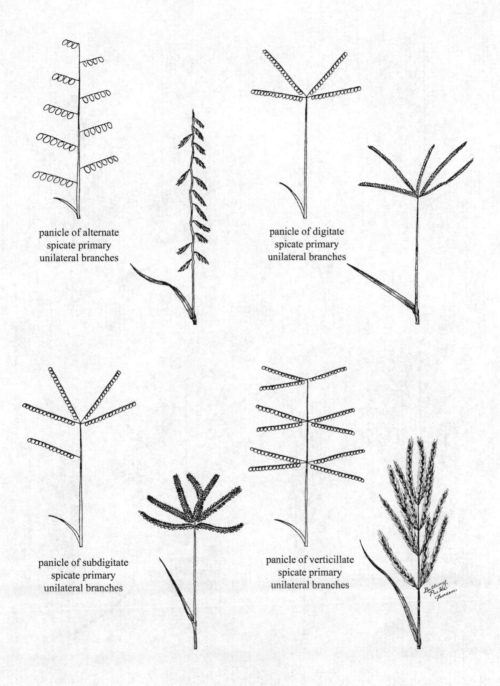

panicle of alternate
spicate primary
unilateral branches

panicle of digitate
spicate primary
unilateral branches

panicle of subdigitate
spicate primary
unilateral branches

panicle of verticillate
spicate primary
unilateral branches

FIGURE 3B. Diagrammatic and actual representations of grass inflorescences

11

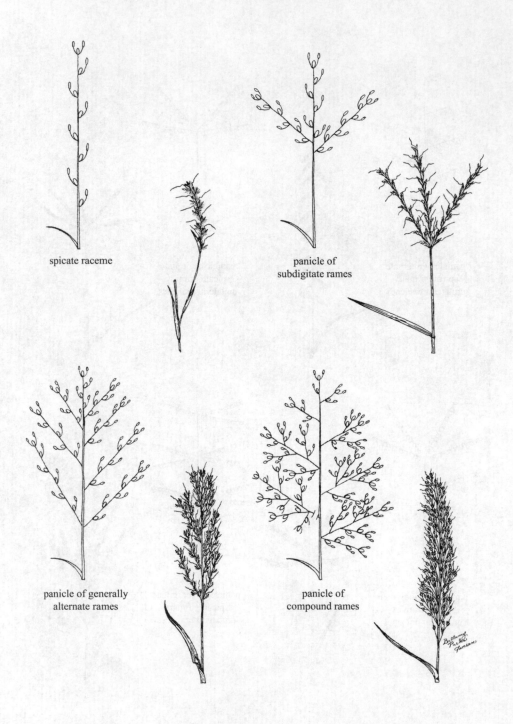

spicate raceme

panicle of
subdigitate rames

panicle of generally
alternate rames

panicle of
compound rames

FIGURE 3C. Diagrammatic and actual representations of grass inflorescences of the
ANDROPOGONEAE tribe

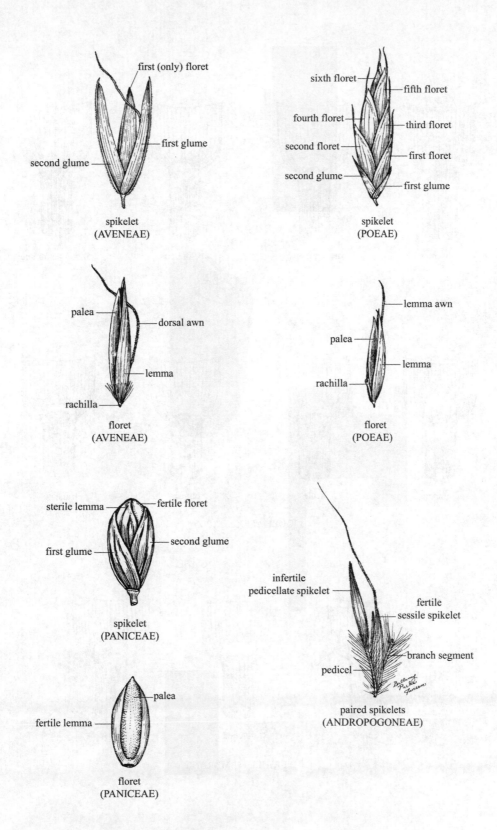

FIGURE 4. Grass spikelets and representative florets

13

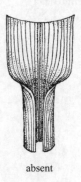
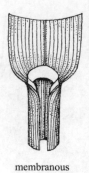

absent   membranous   line of hairs   ciliate membrane

LIGULE TYPE

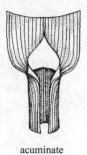
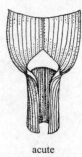

acuminate   acute   obtuse   truncate

LIGULE APEX SHAPE

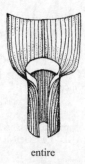

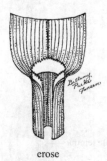

entire   notched   erose

LIGULE MEMBRANE

14   **FIGURE 5.** Grass ligule types, apex shapes, and margins

Figure 6 illustrates the morphological features of grasslike plants. See the glossary for definition of terms.

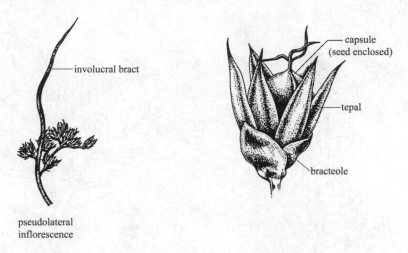

JUNCACEAE

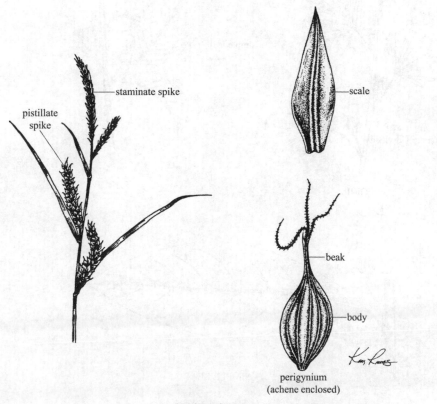

CYPERACEAE

FIGURE 6. Grasslike inflorescences and flowers

MORPHOLOGY OF FORBS AND WOODY PLANTS

Figures 7 through 19 are a series of drawings illustrating the various morphological features of forbs and woody plants. See the glossary for definition of terms.

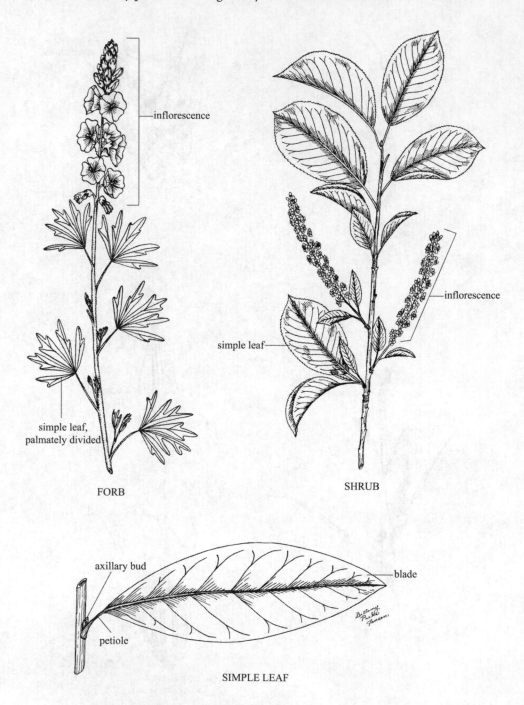

FORB

SHRUB

SIMPLE LEAF

FIGURE 7. Forb and woody plant leaf parts and comparison of forbs and shrubs

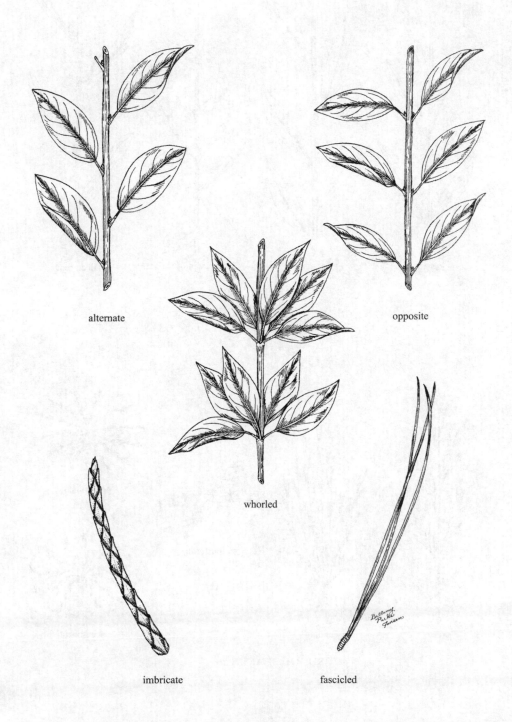

alternate

opposite

whorled

imbricate

fascicled

FIGURE 8. Forb and woody plant leaf arrangement

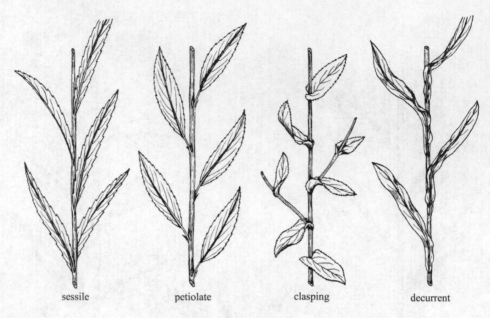

sessile   petiolate   clasping   decurrent

LEAF ATTACHMENT

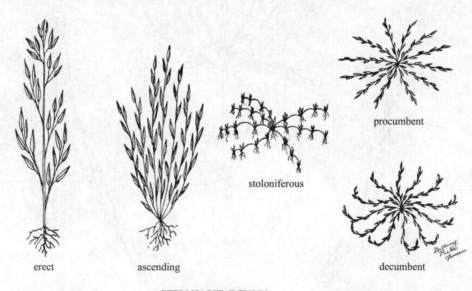

procumbent

stoloniferous

erect   ascending   decumbent

STEM HABITAT TYPES

FIGURE 9. Forb and woody plant leaf attachments and stem habit types

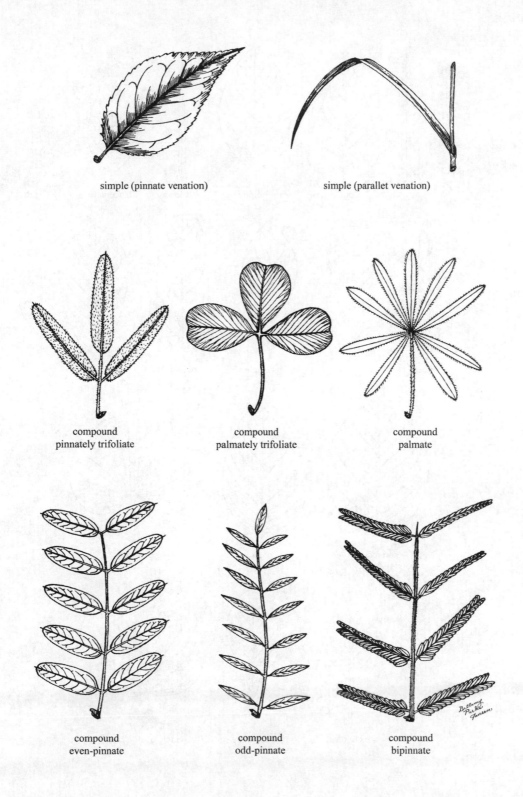

simple (pinnate venation)   simple (parallel venation)

compound
pinnately trifoliate

compound
palmately trifoliate

compound
palmate

compound
even-pinnate

compound
odd-pinnate

compound
bipinnate

FIGURE 10. Forb and woody plant types of simple and compound leaves
(the bud at the base of the petiole defines the start of the leaf)

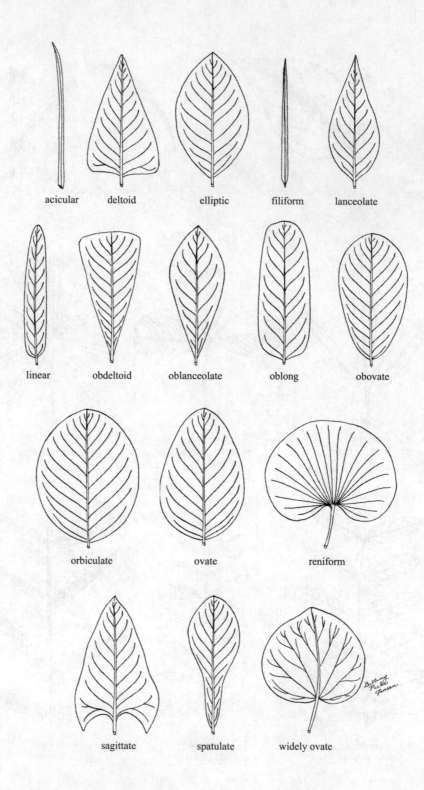

acicular deltoid elliptic filiform lanceolate

linear obdeltoid oblanceolate oblong obovate

orbiculate ovate reniform

sagittate spatulate widely ovate

FIGURE 11. Forb and woody plant leaf shapes

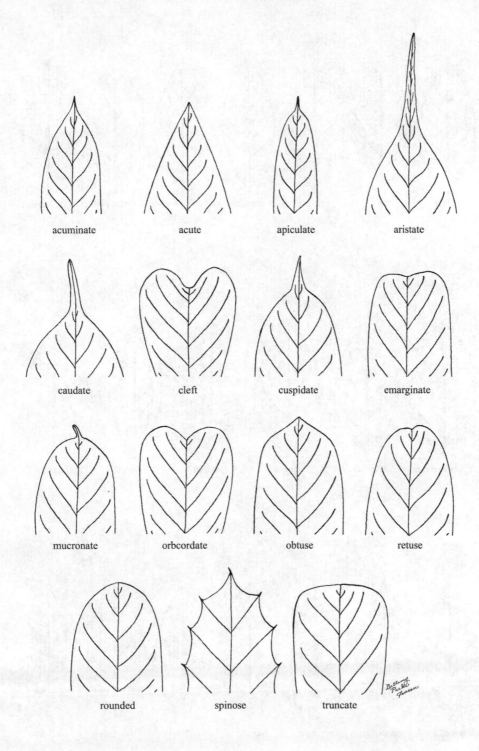

acuminate acute apiculate aristate

caudate cleft cuspidate emarginate

mucronate orbcordate obtuse retuse

rounded spinose truncate

FIGURE 12. Forb and woody plant leaf apices

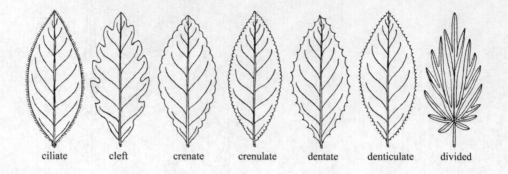

ciliate   cleft   crenate   crenulate   dentate   denticulate   divided

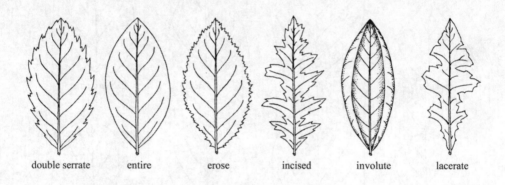

double serrate   entire   erose   incised   involute   lacerate

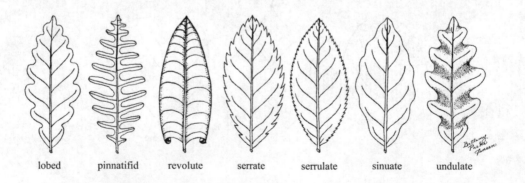

lobed   pinnatifid   revolute   serrate   serrulate   sinuate   undulate

FIGURE 13. Forb and woody plant leaf margins

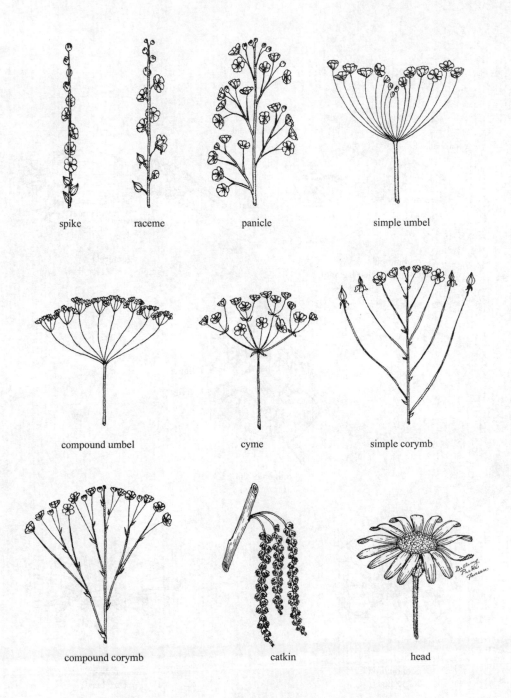

spike   raceme   panicle   simple umbel

compound umbel   cyme   simple corymb

compound corymb   catkin   head

FIGURE 14. Forb and woody plant inflorescence types

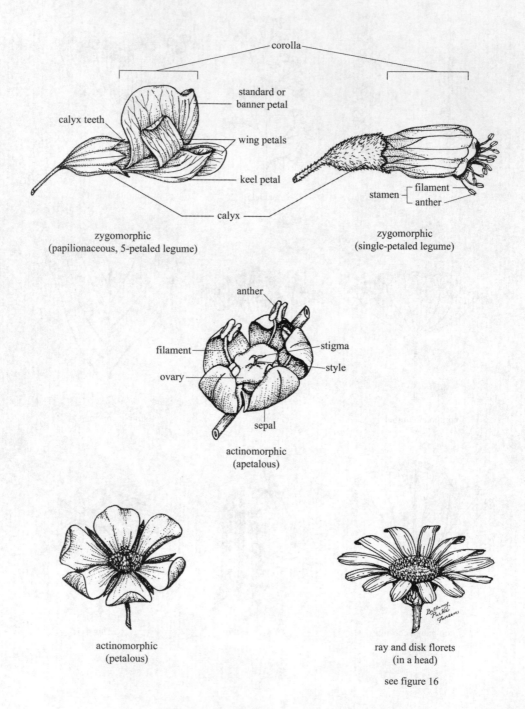

corolla

standard or
banner petal

wing petals

keel petal

calyx teeth

calyx

stamen — filament
— anther

zygomorphic
(papilionaceous, 5-petaled legume)

zygomorphic
(single-petaled legume)

anther

filament

ovary

stigma

style

sepal

actinomorphic
(apetalous)

actinomorphic
(petalous)

ray and disk florets
(in a head)

see figure 16

FIGURE 15. Selected types of forb and woody plant flowers

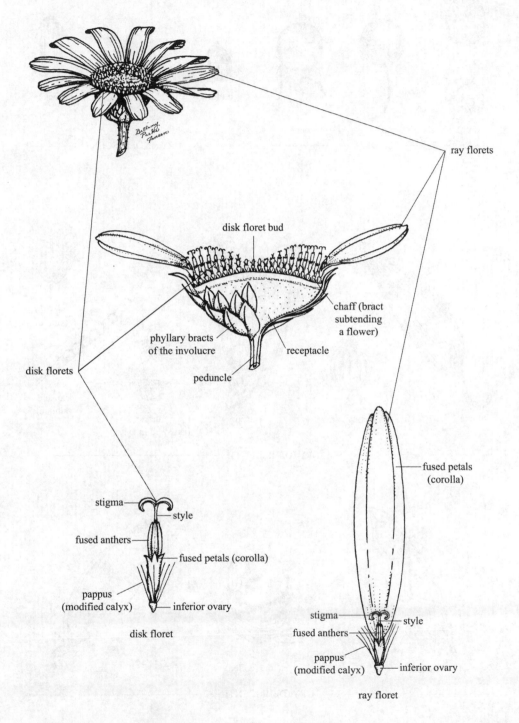

ray florets

disk floret bud

chaff (bract
subtending
a flower)

phyllary bracts
of the involucre

receptacle

disk florets

peduncle

stigma

style

fused anthers

fused petals (corolla)

pappus
(modified calyx)

inferior ovary

disk floret

fused petals
(corolla)

stigma

style

fused anthers

pappus
(modified calyx)

inferior ovary

ray floret

FIGURE 16. Forb and woody plant head inflorescence,
longitudinal section, and ray and disk florets

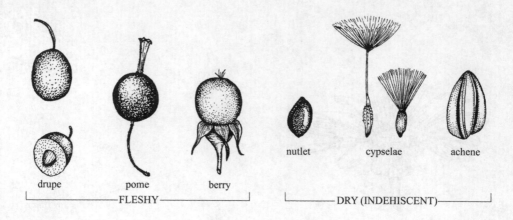

drupe   pome   berry

nutlet   cypselae   achene

└─────── FLESHY ───────┘   └─────── DRY (INDEHISCENT) ───────┘

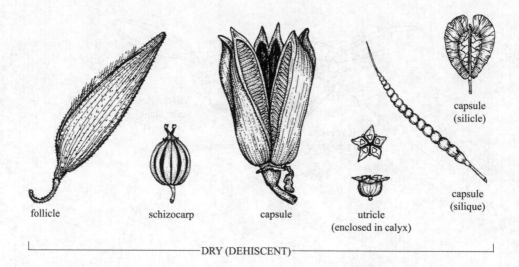

follicle   schizocarp   capsule   utricle (enclosed in calyx)   capsule (silicle)   capsule (silique)

└─────── DRY (DEHISCENT) ───────┘

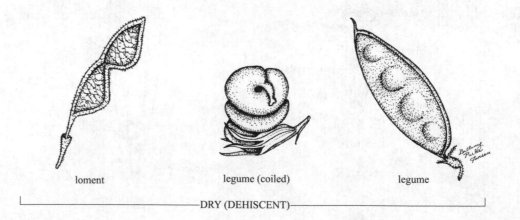

loment   legume (coiled)   legume

└─────── DRY (DEHISCENT) ───────┘

FIGURE 17. Forb and woody plant fruits

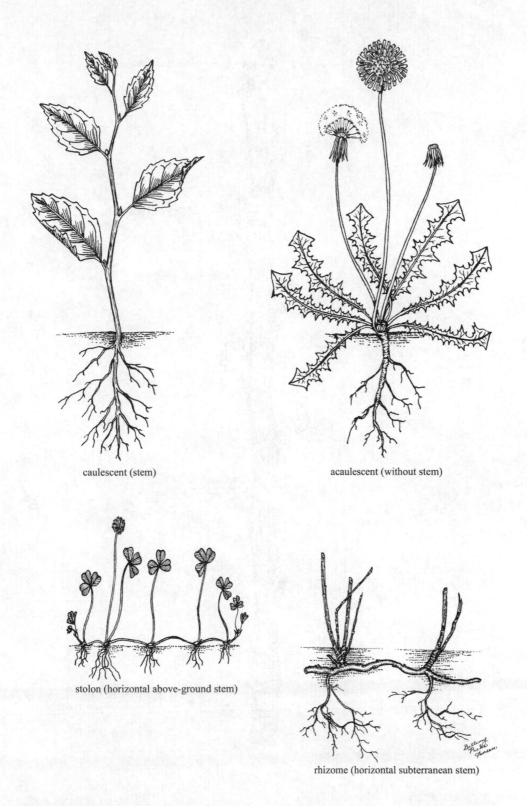

caulescent (stem)

acaulescent (without stem)

stolon (horizontal above-ground stem)

rhizome (horizontal subterranean stem)

FIGURE 18. Caulescent, acaulescent, and modified stems

27

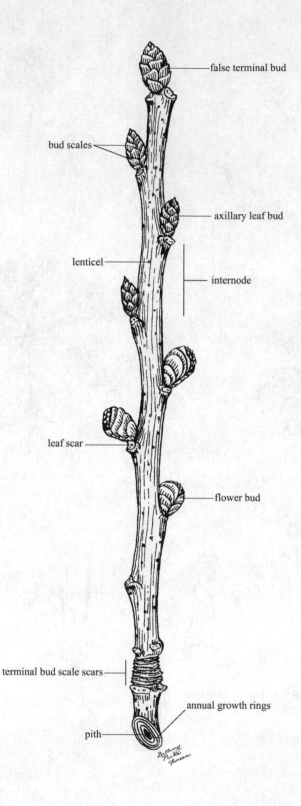

false terminal bud

bud scales

axillary leaf bud

lenticel

internode

leaf scar

flower bud

terminal bud scale scars

annual growth rings

pith

FIGURE 19. Twig parts

GRASSES

Big bluestem
Andropogon gerardi Vitman

SYN = *Andropogon furcatus* Muhl., *Andropogon gerardi* Vitman var. *chrysocomas* (Nash) Fernald, *Andropogon gerardii* Vitman, *Andropogon gerardii* Vitman var. *paucipilus* (Nash) Fernald, *Andropogon hallii* Hack.

panicles of 2–7 subdigitate rames

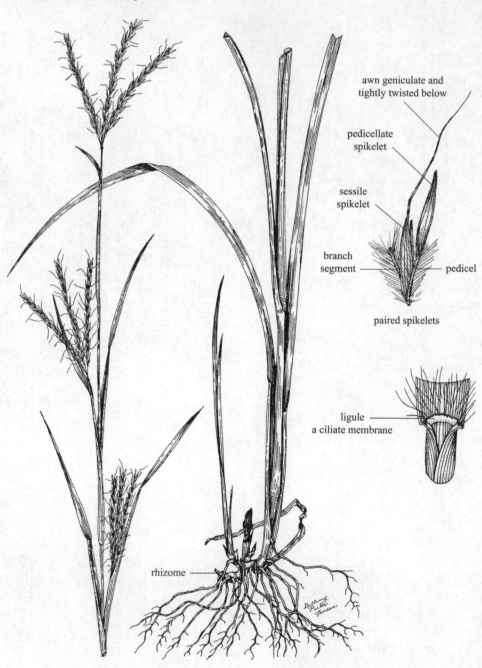

awn geniculate and tightly twisted below

pedicellate spikelet

sessile spikelet

branch segment

pedicel

paired spikelets

ligule
a ciliate membrane

rhizome

TRIBE:	ANDROPOGONEAE
SPECIES:	*Andropogon gerardi* Vitman
COMMON NAME:	Big bluestem (popotillo gigante, turkeyfoot)
LIFE SPAN:	Perennial
ORIGIN:	Native
SEASON:	Warm

INFLORESCENCE CHARACTERISTICS

type: panicles of 2–7 (**commonly 3, hence turkeyfoot**) digitate or subdigitate rames (4–15 cm long), long-exserted, terminal and axillary, fewer than 10 per culm, often purplish, sometimes yellowish

spikelets: paired; pedicellate and sessile spikelets nearly equal in length; lower spikelet sessile and perfect (7–10 mm long), florets 2; upper floret perfect; pedicellate spikelet neuter or staminate (4–10 mm long); rame nodes densely hairy

glumes: glumes of sessile spikelet subequal (7–10 mm long), **first slightly grooved or dished**; glumes of pedicellate spikelet not grooved (4–10 mm long)

awns: lemma of sessile spikelet awned; **awn** (1–2 cm long) **geniculate** and tightly twisted below; pedicellate spikelets awnless

VEGETATIVE CHARACTERISTICS

growth habit: rhizomatous, sometimes appearing cespitose

culms: erect or ascending (0.5–3 m tall), robust, sparingly branched toward apices, glabrous, glaucous, often grooved on one side

sheaths: compressed, purplish at bases, glabrous, glaucous, lower sheaths sometimes villous, margins hyaline

ligules: ciliate membranes (0.4–5 mm long)

blades: flat to involute (5–45 cm long, 2–10 mm wide), lower blades often villous; margins scabrous

GROWTH CHARACTERISTICS: grows rapidly from midspring to early autumn, many leaves produced in late spring and early summer, growing points stay near ground level until late summer, reproduces primarily from rhizomes

FORAGE VALUES: excellent and highly palatable to all classes of livestock when grazed or consumed in hay, commonly selected by livestock in preference to other grasses in summer, becomes coarse late in the season

HABITATS: deep soils of upland and lowland prairies, open woodlands, and wet overflow sites; most abundant in lowland prairies; adapted to all soil textures; frequently used in seed mixtures for prairie reconstruction and for forage production

Broomsedge bluestem
Andropogon virginicus L.

SYN = *Andropogon perangustatus* Nash

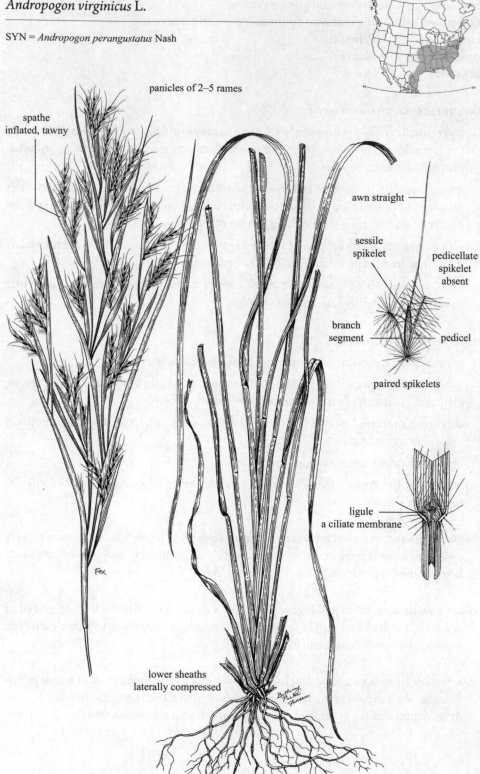

panicles of 2–5 rames

spathe
inflated, tawny

awn straight

sessile
spikelet

pedicellate
spikelet
absent

branch
segment

pedicel

paired spikelets

ligule
a ciliate membrane

lower sheaths
laterally compressed

TRIBE:	ANDROPOGONEAE
SPECIES:	*Andropogon virginicus* L.
COMMON NAME:	Broomsedge bluestem (popotillo, broomsedge)
LIFE SPAN:	Perennial
ORIGIN:	Native
SEASON:	Warm

INFLORESCENCE CHARACTERISTICS

type: panicles of 2–5 (commonly 2) rames (2–3 cm long), paired or digitate, numerous (6–25), broomlike appearance; **bases of panicle branches enclosed in an inflated, tawny spathe** (3–6 cm long, 2–6 mm wide, usually 3–4 mm wide)

spikelets: paired; lower spikelet sessile and perfect (3–4 mm long), glabrous, florets 2; upper floret perfect; pedicellate spikelet absent or vestigial, usually represented only by the villous pedicel

glumes: acuminate (3–4 mm long), green to yellowish

awns: upper lemma of sessile spikelet with delicate awn (1–2 cm long), **awns straight**

VEGETATIVE CHARACTERISTICS

growth habit: cespitose

culms: erect (0.5–1.5 m tall), branched above, slender, sulcate on one side, glabrous or with a few short hairs; basal nodes flat

sheaths: imbricate, lower sheaths laterally compressed, strongly keeled, usually wider than the blades; glabrous, scabrous, or pilose; margins hairy

ligules: ciliate membranes (0.3–1 mm long), truncate

blades: flat or folded (8–55 cm long, mostly 2–6 mm wide), tan to stramineous at maturity, midveins prominent, glabrous to pilose adaxially and near collar

GROWTH CHARACTERISTICS: starts growth when daytime temperatures average 16–17°C, produces seeds mostly from August or September until frost, reproduces from seeds and tillers, grows in infertile soils, not shade tolerant; may rapidly increase with improper grazing; indicator of early stages of plant succession

FORAGE VALUES: poor for livestock and wildlife, except in early growth stages during spring and early summer, nearly unpalatable when mature; may provide important wildlife habitat

HABITATS: open ground, old fields, open woodlands, lowlands, and sterile hills; sandy to rocky moist soils; most common in abandoned fields and on improperly grazed rangelands; will invade improved pastures

Silver bluestem
Bothriochloa laguroides (DC.) Herter

SYN = *Andropogon longipaniculata* (Gould) Allred & Gould, *Andropogon saccharoides* Sw., *Bothriochloa longipaniculata* (Gould) Allred & Gould, *Bothriochloa saccharoides* (Sw.) Rydb.

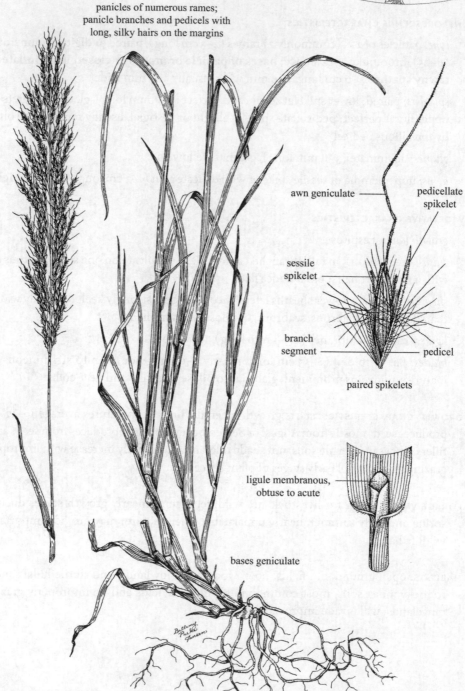

panicles of numerous rames;
panicle branches and pedicels with
long, silky hairs on the margins

awn geniculate

pedicellate spikelet

sessile spikelet

branch segment

pedicel

paired spikelets

ligule membranous, obtuse to acute

bases geniculate

TRIBE:	ANDROPOGONEAE
SPECIES:	*Bothriochloa laguroides* (DC.) Herter
COMMON NAME:	Silver bluestem (popotillo plateado, silver beardgrass)
LIFE SPAN:	Perennial
ORIGIN:	Native
SEASON:	Warm

INFLORESCENCE CHARACTERISTICS

type: panicles of 6 to many rames or compound rames, long-exserted, elongate (usually 7–15 cm long), terminal and axillary, usually fewer than 5 per culm, **silvery-white** in color; rames erect to ascending (less than 5 cm long), aromatic, margins fringed with long silky hairs; rame joints and pedicels long-villous; pedicels sulcate or dumbbell-shaped in cross section

spikelets: paired; lower spikelet sessile and perfect (3–4 mm long), florets 2; pedicellate spikelet neuter (1.5–3 mm long), narrow

glumes: unequal (2.5–4.5 mm long), firm but papery, without glandular pit; first with 2 keels; second with 1 keel; veins 3

awns: upper lemmas of sessile spikelets with **delicate geniculate awns** (8–25 mm long), twisted below

other: can be confused with *Digitaria californica*; compare inflorescence arrangements, spikelets, and awns

VEGETATIVE CHARACTERISTICS

growth habit: cespitose

culms: erect from geniculate bases (0.6–1.3 m tall), branched below, sulcate; hairs at the nodes short-appressed

sheaths: keeled near collar, glabrous, glaucous; collars usually with a few long hairs on margin; hairs may extend up the leaf margins

ligules: membranous (1–3 mm long), obtuse to acute, erose to entire

blades: flat or folded (2–25 cm long, 3–9 mm wide), linear, acuminate, glaucous; mid-veins prominent; margins white; often brownish toward the margin

GROWTH CHARACTERISTICS: starts growth in spring when daytime temperatures reach 21–24°C, inflorescences emerge 3–4 weeks later; produces abundant seeds, reproduces from seeds and tillers, seedlings must be protected from grazing to enhance establishment

FORAGE VALUES: fair to good for all classes of livestock and fair for wildlife, only lightly grazed following maturity

HABITATS: prairies, pastures, rocky slopes, waste grounds, and roadsides; adapted to a broad range of soil textures, does not grow well on moist sites

Tanglehead
Heteropogon contortus (L.) P. Beauv. *ex* Roem. & Schult.

SYN = *Andropogon contortus* L.

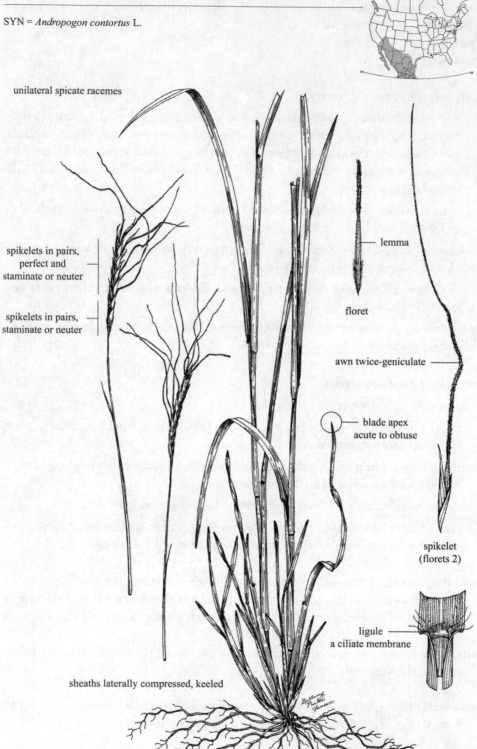

unilateral spicate racemes

spikelets in pairs,
perfect and
staminate or neuter

spikelets in pairs,
staminate or neuter

lemma

floret

awn twice-geniculate

blade apex
acute to obtuse

spikelet
(florets 2)

ligule
a ciliate membrane

sheaths laterally compressed, keeled

TRIBE:	ANDROPOGONEAE
SPECIES:	*Heteropogon contortus* (L.) P. Beauv. *ex* Roem. & Schult.
COMMON NAME:	Tanglehead (barba negra, barba, zacate colorado)
LIFE SPAN:	Perennial
ORIGIN:	Native
SEASON:	Warm

INFLORESCENCE CHARACTERISTICS

type: **unilateral spicate racemes** (3–8 cm long, excluding awns), terminal and axillary, few; **unilateral arrangement lends to a braided appearance**

spikelets: paired, imbricate; sessile spikelets perfect (5–8 mm long); calluses well developed (1.5–3 mm long); pedicellate spikelets staminate or neuter (7–10 mm long); few to several pairs at the base of the inflorescence staminate or neuter

glumes: sessile spikelet glumes rounded, dark brownish, papillose-hispid; pedicellate spikelet glumes thin, green, sparingly hispid to glabrous; veins many

awns: upper lemma of sessile spikelet awned; **awns weakly twice-geniculate** (5–12 cm long), hispid (hairs 0.5–1 mm long), **dark brown to black at maturity**, tangled with other awns; lemma of pedicellate spikelet awnless

VEGETATIVE CHARACTERISTICS

growth habit: cespitose

culms: erect to ascending (20–80 cm tall), flat, simple or branched at the base and upper nodes

sheaths: laterally compressed, keeled; margins glandular; glandular depressions on keel; collars with short hairs

ligules: ciliate membranes (about 1 mm long), acute to truncate

blades: flat (6–20 cm long, usually 3–7 mm wide); apices acute to obtuse; midveins prominent adaxially; margins white-glandular, usually ciliate; apices and bases reddish at maturity; somewhat aromatic

GROWTH CHARACTERISTICS: starts growth in early spring, produces inflorescences June through November, low seed production; relatively easy to establish from seeds, reproduces from seeds and tillers

LIVESTOCK LOSSES: awns may be troublesome to eyes and mouths of grazing animals, especially to sheep

FORAGE VALUES: fair to good for cattle and horses before maturity; little value to sheep due to coarseness

HABITATS: open, dry, rocky hills and canyons, usually in sandy soils; most abundant on heavily grazed rangelands

Little bluestem
Schizachyrium scoparium (Michx.) Nash

SYN = *Andropogon divergens* (Hack.) Andersson *ex* Hitchc., *Andropogon littoralis* Nash, *Andropogon scoparius* Michx.

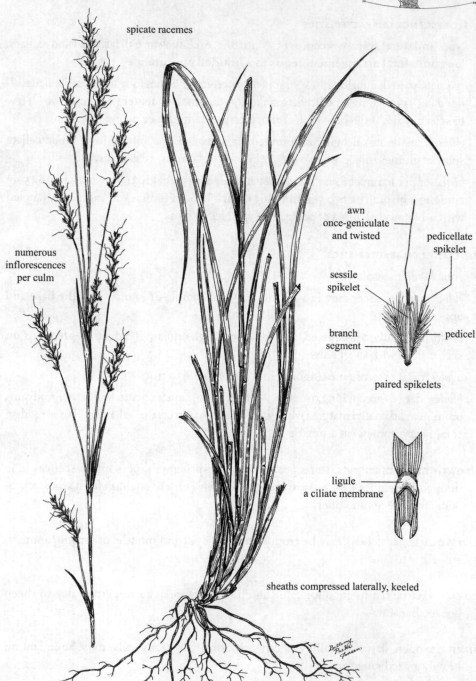

spicate racemes

awn
once-geniculate
and twisted

pedicellate
spikelet

sessile
spikelet

numerous
inflorescences
per culm

branch
segment

pedicel

paired spikelets

ligule
a ciliate membrane

sheaths compressed laterally, keeled

TRIBE:	ANDROPOGONEAE
SPECIES:	*Schizachyrium scoparium* (Michx.) Nash
COMMON NAME:	Little bluestem (popotillo colorado, popotillo cañuelo, prairie beardgrass)
LIFE SPAN:	Perennial
ORIGIN:	Native
SEASON:	Warm

INFLORESCENCE CHARACTERISTICS

type: spicate racemes (3–8 cm long), several per culm, terminal and axillary, jointed, **breaking apart into spikelet pairs as the rachis disarticulates**; peduncle included in sheaths; pedicels semiterete

spikelets: paired; sessile spikelets perfect (4–10 mm long), florets 2; upper floret perfect; pedicellate spikelets staminate or neuter; rachises and pedicels pilose

glumes: sessile spikelet glumes thickened (4–10 mm long), subequal, rounded on the back, glabrous to scabrous, firm

awns: upper lemma of sessile spikelet awned; **awns once-geniculate and twisted** (3–15 mm long); lemma of pedicellate spikelet awnless or with a short awn

other: spicate racemes have a zigzag pattern at maturity

VEGETATIVE CHARACTERISTICS

growth habit: cespitose or rhizomatous; rhizomes short or well developed

culms: erect (0.4–1.2 m tall), sometimes from decumbent bases, slender to robust, flat, not grooved, branching above; bases leafy, green to purplish, glaucous

sheaths: **keeled, compressed laterally**, glabrous to rarely pubescent

ligules: ciliate membrane (1–3 mm long), usually truncate

blades: flat (8–60 cm long, 2–8 mm wide), linear, acute, glabrous to hispid, scabrous adaxially and on margins; midveins strongly thickened below

GROWTH CHARACTERISTICS: starts growth in late spring, inflorescences appear in midsummer, matures in early autumn, seeds mature October to November; reproduces from tillers, rhizomes, and seeds

FORAGE VALUES: good while immature for all classes of livestock; after inflorescences mature, forage is fair for cattle and horses but too coarse for sheep, goats, and wildlife; can be an important component of upland hay

HABITATS: prairies, open woodlands, and dry hills in all soil textures; most conspicuous with season-long, moderate grazing which allows spot grazing leaving some plants ungrazed; frequently used in seed mixtures for revegetation

Indiangrass
Sorghastrum nutans (L.) Nash

SYN = *Sorghastrum avenaceum* (Michx.) Nash

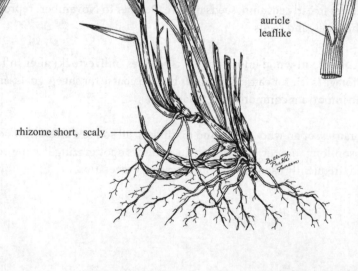

panicles of compound rames, narrow, dense

awn once-geniculate

sessile spikelet (florets 2)

pedicellate spikelet absent

branch segment

pedicel

paired spikelets

node pubescent

ligule membranous

auricle leaflike

rhizome short, scaly

TRIBE:	ANDROPOGONEAE
SPECIES:	*Sorghastrum nutans* (L.) Nash
COMMON NAME:	Indiangrass (zacate indio, yellow indiangrass)
LIFE SPAN:	Perennial
ORIGIN:	Native
SEASON:	Warm

INFLORESCENCE CHARACTERISTICS

type: panicles of compound rames (15–35 cm long, 6–8 cm wide), loosely contracted, **yellowish or tawny**; apices of rames, rame nodes, and pedicels grayish-hirsute

spikelets: paired; sessile spikelet perfect (5–8 mm long); pedicellate spikelet absent, represented only by the hairy pedicel

glumes: subequal (5–8 mm long), coriaceous, tawny or yellowish, first hirsute with edges inflexed over the second, abaxial surfaces flat or convex

awns: upper lemma of perfect spikelets awned; awns once-geniculate (1.5–2.8 cm long), tightly twisted below the bend, loosely twisted above

VEGETATIVE CHARACTERISTICS

growth habit: cespitose or rhizomatous; rhizomes short, scaly

culms: erect (1–2.4 m tall), robust to slender; nodes pubescent

sheaths: round or sometimes compressed, not keeled, glabrous to rarely pilose; **auricles erect; extending into firm, pointed projections (2–7 mm long) flanking and adnate to the ligules**

ligules: membranous (3–6 mm long)

blades: flat or somewhat keeled (to 60 cm long, 5–10 mm wide), constricted at the bases, apices acute to acuminate, midveins conspicuous abaxially

GROWTH CHARACTERISTICS: starts growth in midspring from rhizomes, matures from September to November, reproduces from rhizomes and seeds

FORAGE VALUES: excellent, palatable to cattle and horses throughout the summer, but does not cure well and is generally considered only moderately palatable after maturity; fair forage for winter grazing; produces good hay if cut before maturity

HABITATS: prairies, bottomlands, open woodlands, and meadows (moderately salt tolerant) in all soil textures; withstands occasional flooding; sometimes grown and managed with fertilization and irrigation in pure stands; a common component of seed mixtures for revegetation

Eastern gamagrass
Tripsacum dactyloides (L.) L.

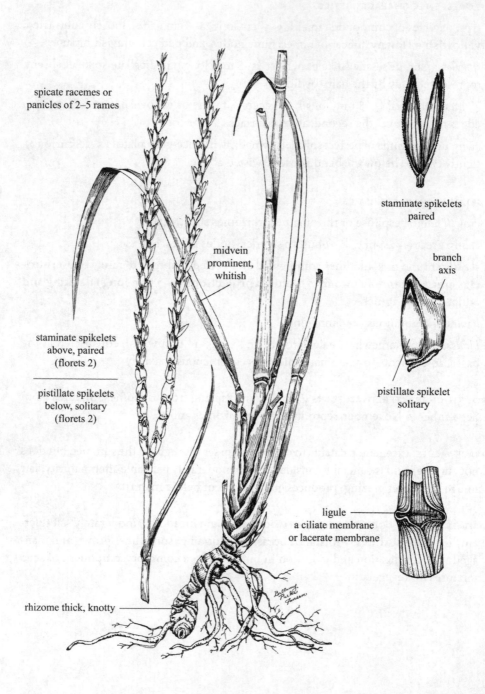

spicate racemes or
panicles of 2–5 rames

staminate spikelets
paired

midvein
prominent,
whitish

branch
axis

staminate spikelets
above, paired
(florets 2)

pistillate spikelets
below, solitary
(florets 2)

pistillate spikelet
solitary

ligule
a ciliate membrane
or lacerate membrane

rhizome thick, knotty

TRIBE:	ANDROPOGONEAE
SPECIES:	*Tripsacum dactyloides* (L.) L.
COMMON NAME:	Eastern gamagrass (zacate maicero, zacate maizero, maicillo)
LIFE SPAN:	Perennial
ORIGIN:	Native
SEASON:	Warm

INFLORESCENCE CHARACTERISTICS

type: spicate racemes or panicles (12–40 cm long) of 2–5 rames, terminal (occasionally axillary)

spikelets: unisexual; staminate spikelets above, paired, florets 2 (7–12 mm long), coriaceous, sessile or one slightly pedicellate in **2 rows on the branch**, often breaking apart at maturity; pistillate spikelets below (7–10 mm long), solitary, indurate, **beadlike**, sunken in depressions of the branch or rachis, breaking into single spikelet segments at maturity

glumes: staminate spikelet glumes equal (7–12 mm long), somewhat pyriform, coriaceous, keeled; pistillate spikelet glumes equal (5–9 mm long), indurate, shiny, often embedded in the rachis

awns: none

other: plants monoecious; pistillate portion of inflorescences one-fourth or less of the entire length

VEGETATIVE CHARACTERISTICS

growth habit: rhizomatous; rhizomes thick, knotty; forming extensive colonies

culms: erect to decumbent (1.5–3 m tall), stout, solid, **slightly compressed**, glabrous

sheaths: round to prominently keeled, glabrous, shiny, usually shorter than the internodes

ligules: ciliate membranes or lacerate membranes (to 2.5 mm long), truncate

blades: flat (30–75 cm long, 1–4.5 cm wide); midveins prominent, whitish; margins scabrous

GROWTH CHARACTERISTICS: most growth is in the spring and early summer, stays green until late autumn; produces seeds from July to September, although few seeds are produced; most reproduction is from rhizomes

FORAGE VALUES: excellent for all classes of livestock and wildlife throughout the growing season, makes good hay, foliage breaks down rapidly and it is not dependable for winter grazing; usually one of the first species to be eliminated by improper grazing

HABITATS: well-drained soils of upland and lowland prairies, swales, stream banks, and grasslands; most abundant in fertile soils; does not tolerate standing water for long periods; occasionally seeded for pastures

Prairie threeawn
Aristida oligantha Michx.

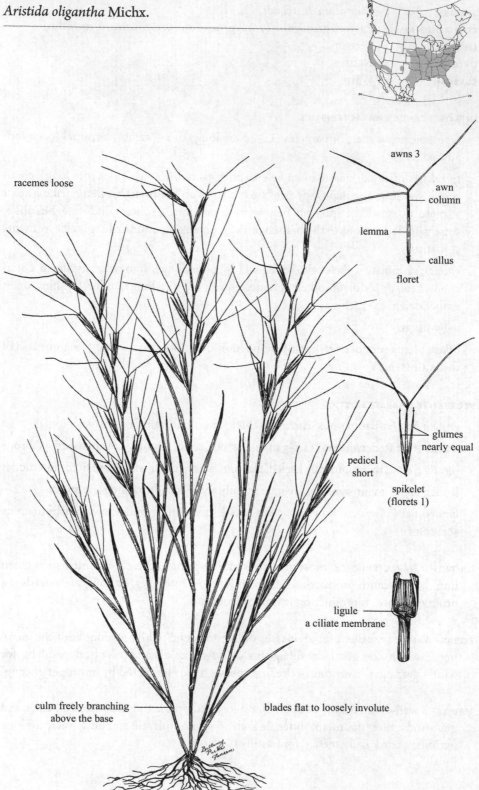

awns 3

awn
column

lemma

callus

floret

racemes loose

glumes
nearly equal

pedicel
short

spikelet
(florets 1)

ligule
a ciliate membrane

culm freely branching
above the base

blades flat to loosely involute

TRIBE:	ARISTIDEAE
SPECIES:	*Aristida oligantha* Michx.
COMMON NAME:	Prairie threeawn (tres barbas anual, oldfield threeawn)
LIFE SPAN:	Annual
ORIGIN:	Native
SEASON:	Warm

INFLORESCENCE CHARACTERISTICS

type: racemes (5–20 cm long, 1–4 cm wide), loose, purplish

spikelets: widely spaced, **florets 1**; lemma firm (6–28 mm long, excluding the awns); callus well developed, pilose; pedicel short, scabrous or pubescent

glumes: **nearly equal** (2–3 cm long); first veins 3–7; second usually veins 1, slightly longer than the first (compare to *Aristida purpurea*)

awns: **lemma awn columns branch into 3 awns**, awns divergent (3–7 cm long), nearly equal or central awn longest; first glume awnless or short-awned; second glume mucronate

VEGETATIVE CHARACTERISTICS

growth habit: cespitose

culms: ascending to geniculate (to 80 cm tall), freely branching above the bases, wiry, glabrous to scabrous; bases often purplish

sheaths: round, shorter than internodes, glabrous or slightly scabrous or sparsely pilose on the sides of the collars

ligules: ciliate membranes (0.1–0.5 mm long)

blades: flat to loosely involute (10–25 cm long, 1–3 mm wide), involute at the apices, reduced upward, glabrous to scabrous

GROWTH CHARACTERISTICS: seeds germinate in spring, completes life cycle in 2–3 months, seeds may be rapidly spread by wind and animals; an indicator of disturbance or improper grazing

LIVESTOCK LOSSES: long awns are tough, brittle, and can cause injury to eyes, nostrils, and mouths of livestock; awns also may decrease fleece value; losses may occur while animals are grazing or eating contaminated hay

FORAGE VALUES: fair to poor for a brief period in early growth stages, otherwise worthless; its presence in hay greatly reduces its value

HABITATS: disturbed areas in dry soils, most common on sandy and sandy calcareous soils of deteriorated rangelands, waste places, and abandoned fields; considered to be a disturbance species

Purple threeawn
Aristida purpurea Nutt.

SYN = *Aristida brownii* Warnock, *Aristida fendleriana* Steud., *Aristida longiseta*
Steud., *Aristida roemeriana* Scheele, *Aristida wrightii* Nash

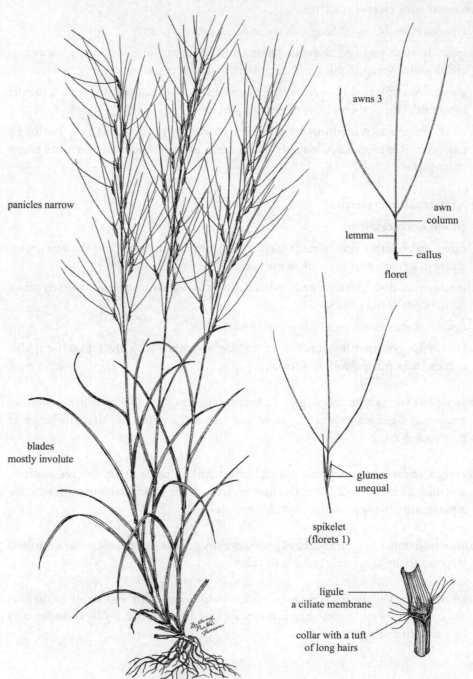

awns 3

awn
column

lemma

callus

floret

panicles narrow

glumes
unequal

spikelet
(florets 1)

blades
mostly involute

ligule
a ciliate membrane

collar with a tuft
of long hairs

TRIBE:	ARISTIDEAE
SPECIES:	*Aristida purpurea* Nutt.
COMMON NAME:	Purple threeawn (tres barbas púrpura, red threeawn, dogtowngrass)
LIFE SPAN:	Perennial
ORIGIN:	Native
SEASON:	Warm

INFLORESCENCE CHARACTERISTICS

type: panicles or rarely racemes (2–30 cm long, to 14 cm wide), open at maturity, often **purplish to reddish**, narrow; branches erect or somewhat flexuous

spikelets: **florets 1**; lemmas firm (typically 10–12 mm long, excluding the awns), veins 3; calluses short-bearded

glumes: **unequal**, broad; first about one-half as long (4–15 mm long) as the second (8–26 mm long) (compare to *Aristida oligantha*)

awns: first glume awned or awnless; lemma awn columns divide into **3 nearly equal awns**; awns divergent (2–10 cm long); awn columns not obvious or poorly defined

VEGETATIVE CHARACTERISTICS

growth habit: cespitose, leaves mostly basal

culms: erect to ascending (10–90 cm tall), glabrous, not branching above the base

sheaths: glabrous to weakly scabrous; collars often pubescent with a tuft of long, soft hairs on both sides

ligules: ciliate membranes (0.1–0.6 mm long)

blades: mostly involute (2–30 cm long, 1–2 mm wide), curved; apices acute; scabrous adaxially, sometimes hirsute, villous in tufts at the collar

GROWTH CHARACTERISTICS: starts growth in late spring, strong competitor; produces abundant seeds, reproduces from seeds and tillers

LIVESTOCK LOSSES: awns may cause irritation and abscesses in the mouths and nostrils and may damage eyes of grazing animals; awns also may decrease fleece value

FORAGE VALUES: poor (rarely fair) for livestock and wildlife, grazed only in early growth stages before awn development; afterward relatively unpalatable, nearly worthless in winter, but sometimes grazed because it often remains partially green

HABITATS: prairies, wastelands, and disturbed sites; in soils of all textures, especially dry sandy soils; most abundant on abused rangelands; an indicator of rangeland deterioration

Redtop
Agrostis stolonifera L.

SYN = *Agrostis alba* var. *palustris* (Huds.) Pers., *Agrostis alba* var. *stolonifera* (L.) Sm., *Agrostis maritima* Lam.

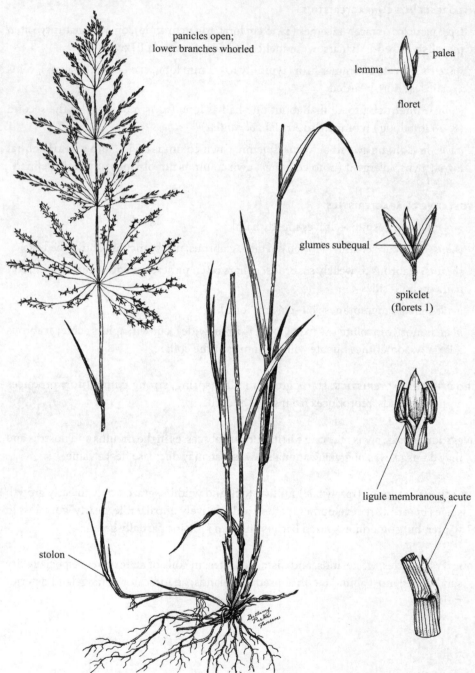

panicles open; lower branches whorled

lemma — palea

floret

glumes subequal

spikelet (florets 1)

ligule membranous, acute

stolon

TRIBE:	AVENEAE
SPECIES:	*Agrostis stolonifera* L.
COMMON NAME:	Redtop (zacate punta café, zacate de piedras castillitos, creeping bentgrass, redtop bent)
LIFE SPAN:	Perennial
ORIGIN:	Introduced (from Europe)
SEASON:	Cool

INFLORESCENCE CHARACTERISTICS

type: panicles (5–25 cm long, 5–20 cm wide), open, narrowly pyramidal to ovate, dense; branches spreading or ascending to appressed, densely flowered, purplish to reddish; **lower branches whorled**, without dense spikelets at bases; panicle nodes reddish

spikelets: **florets 1**; lemma tip blunt (1.8–3.1 mm long); palea one-half to two-thirds as long as the lemma; pedicellate; **readily shattering**

glumes: subequal (1.8–3.1 mm long), acute, veins 1, glabrous, keels sometimes scabrous, longer than the lemmas; **empty glumes form a distinctive V-shape**

awns: none (lemmas rarely awned)

VEGETATIVE CHARACTERISTICS

growth habit: stoloniferous, forming a sod

culms: erect to ascending (0.2–1.5 m tall), sometimes decumbent at the base

sheaths: round, glabrous, frequently purplish to reddish

ligules: membranous (2–7 mm long), acute, erose to entire

blades: flat (4–25 cm long, 1–10 mm wide), narrow; apices acute; midveins prominent abaxially; margins smooth to scabrous

GROWTH CHARACTERISTICS: starts growth in early spring, flowers in early summer, seeds mature by August; reproduces from stolons and seeds

FORAGE VALUES: good to very good for cattle and horses, fair to good for sheep, fair for elk and deer; commonly a component of meadow hay; hay quality is acceptable if cut no later than the early flowering stage, after which quality declines rapidly

HABITATS: pastures, low areas, and moist meadows; can withstand flooding for extended periods, widely naturalized, will grow on acidic soils, moderately salt tolerant, adapted to a wide range of soil and climatic conditions; commonly seeded in pastures, moist meadows, and irrigated pastures

Slender oats
Avena barbata Pott *ex* Link

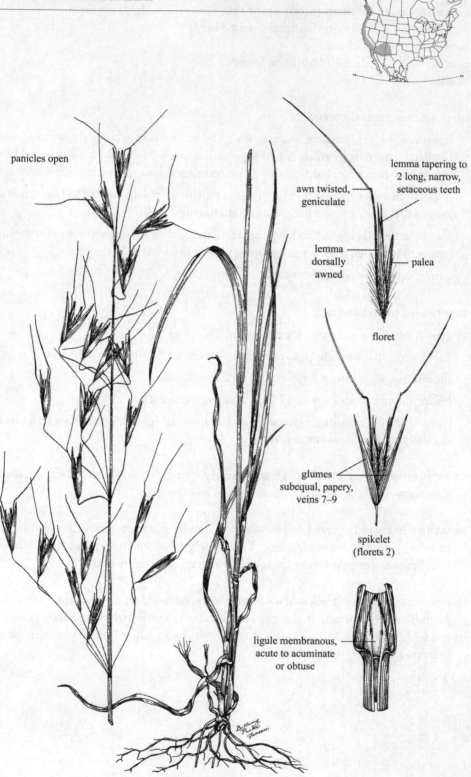

panicles open

awn twisted, geniculate

lemma tapering to 2 long, narrow, setaceous teeth

lemma dorsally awned

palea

floret

glumes subequal, papery, veins 7–9

spikelet (florets 2)

ligule membranous, acute to acuminate or obtuse

TRIBE:	AVENEAE
SPECIES:	*Avena barbata* Pott *ex* Link
COMMON NAME:	Slender oats (avena salvaje, avenilla)
LIFE SPAN:	Annual
ORIGIN:	Introduced (from Europe)
SEASON:	Cool

INFLORESCENCE CHARACTERISTICS

type: panicles (20–40 cm long), open, loose, erect or nodding

spikelets: florets 2, drooping with maturity; lemmas with stiff red hairs to the middle, tapering to a pair of long, narrow setaceous teeth (3–4 mm long); pedicels curved, capillary

glumes: subequal (1.5–3 cm long), **papery**, glabrous, acute, **veins 7–9**; veins prominent; narrower than those of *Avena fatua*

awns: lemmas dorsally awned; awns twisted (3–4.5 cm long), geniculate, stout, **reddish-brown to black**

VEGETATIVE CHARACTERISTICS

growth habit: cespitose, sometimes solitary culms

culms: initially prostrate, becoming erect (to 1.2 m tall), slender (compare to *Avena fatua*)

sheaths: round to somewhat keeled, upper sheaths glabrous or nearly so; lower sheaths usually pilose

ligules: membranous (4–6 mm long), acute to acuminate or obtuse, erose

blades: flat to slightly keeled (10–40 cm long, 5–10 mm wide), scabrous on both sides; margins often pilose

GROWTH CHARACTERISTICS: germinates in late autumn or early winter; most growth is in early spring when it grows more rapidly than associated species; flowers March to June; trampling by grazing animals after seeds have matured helps to plant seeds

FORAGE VALUES: good for all classes of livestock and wildlife during the winter and during the spring growth period, low quality and palatability after it matures in summer; produces hay with acceptable forage quality if cut before maturity, although cutting at that time will eliminate seed production; considered to be a weed on disturbed sites

HABITATS: foothills, fields, roadsides, disturbed areas, and waste places; adapted to a wide variety of soils; most abundant in dry, coarsely textured soils

Wild oats
Avena fatua L.

SYN = *Avena fatua* var. *sativa* (L.) Hausskn.

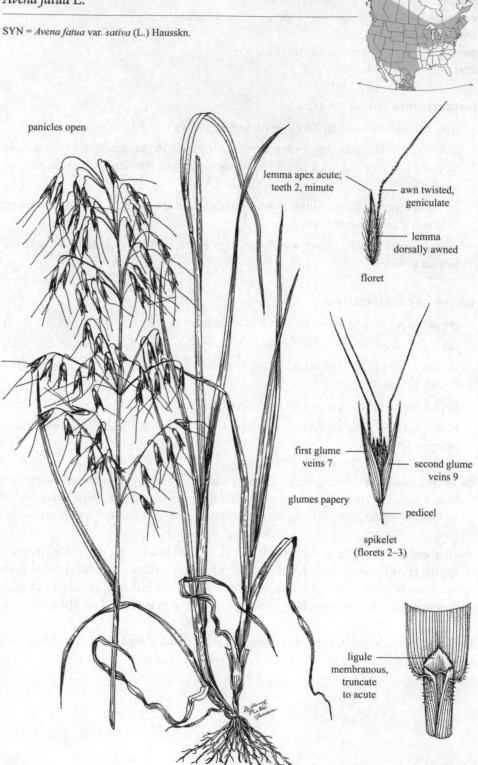

panicles open

lemma apex acute;
teeth 2, minute

awn twisted,
geniculate

lemma
dorsally awned

floret

first glume
veins 7

second glume
veins 9

glumes papery

pedicel

spikelet
(florets 2–3)

ligule
membranous,
truncate
to acute

TRIBE:	AVENEAE
SPECIES:	*Avena fatua* L.
COMMON NAME:	Wild oats (avena silvestre, avena guacha, avena loca)
LIFE SPAN:	Annual
ORIGIN:	Introduced (from Europe)
SEASON:	Cool

INFLORESCENCE CHARACTERISTICS

type: panicles (10–40 cm long, to 15 cm wide) or rarely racemes, loose, open, nodding, spikelets usually 8–35; branches unequal, horizontally spreading to ascending; spikelets pendulous

spikelets: florets 2–3 (rarely 4); lemmas firm (lowermost lemmas 1.5–2 cm long), apices acute, teeth 2; teeth minute (0.3–1.5 mm long), with long hairs; paleas thin, slightly shorter than lemma

glumes: subequal (1.5–3 cm long), longer than the florets, papery, acute, glabrous; first veins 7, **second veins 9**

awns: lemma awns dorsal; awns twisted (1.5–4 cm long), geniculate, stout, reddish-brown to black, **terminal section tan**

VEGETATIVE CHARACTERISTICS

growth habit: cespitose, sometimes solitary culms; initially prostrate, becoming erect

culms: erect (to 1.5 m tall), stout (compare to *Avena barbata*), glabrous

sheaths: round to somewhat keeled, open, glabrous to pubescent; margins broad and thick; collars pilose on front margins

ligules: membranous (4–6 mm long), truncate to acute, erose

blades: flat when mature (10–45 cm long, 5–15 mm wide), scabrous on both sides; margins glabrous to pilose (especially near the bases)

GROWTH CHARACTERISTICS: germinates in early winter, most growth occurs in early spring, flowers March to May, seeds are set in June, reproduces from seeds; highest production is on moist and rich soils

FORAGE VALUES: good to excellent for grazing by all classes of livestock and wildlife until after the florets are shed and the herbage dies, quality of hay is good if cut before maturity, production is relatively low; commonly makes up the majority of early forage for grazing on annual rangelands

HABITATS: valleys and open slopes of foothill rangelands; cultivated soils; disturbed soils in fields, waste places, and roadsides

Spikeoats
Avenula hookeri (Scribn.) Holub

SYN = *Helictotrichon hookeri* (Scribn.) Henrard

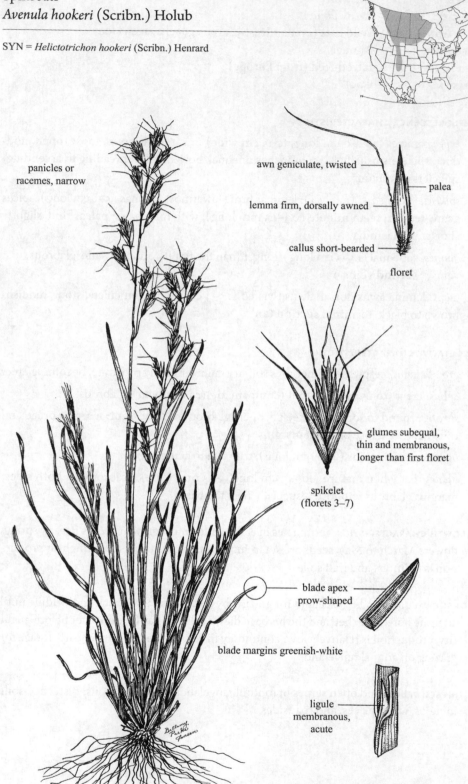

panicles or
racemes, narrow

awn geniculate, twisted

palea

lemma firm, dorsally awned

callus short-bearded

floret

glumes subequal,
thin and membranous,
longer than first floret

spikelet
(florets 3–7)

blade apex
prow-shaped

blade margins greenish-white

ligule
membranous,
acute

TRIBE:	AVENEAE
SPECIES:	*Avenula hookeri* (Scribn.) Holub
COMMON NAME:	Spikeoats (spike oatgrass)
LIFE SPAN:	Perennial
ORIGIN:	Native
SEASON:	Cool

INFLORESCENCE CHARACTERISTICS

type: panicles (5–12 cm long) or racemes, narrow, **elongate**, long-exserted at maturity; branches erect or ascending, upper branches with a single spikelet, lower branches with 2 spikelets

spikelets: florets 3–7; lemmas firm (1–1.2 cm long), veins 5, brown, toothed at the apices, round on the back; calluses short-bearded; paleas well developed, shorter than the lemmas; rachillas villous

glumes: subequal (8–15 mm long), apices acute, thin and membranous; first veins 3; second veins 3–5; longer than the lowermost florets

awns: lemma awns dorsal; awns (1–1.8 cm long) attached slightly above the middle of the abaxial side of the lemma, **geniculate (often twice)**, twisted, compressed below the bend

VEGETATIVE CHARACTERISTICS

growth habit: cespitose; leaves mostly basal

culms: erect (10–75 cm tall), glabrous, smooth or scaberulous

sheaths: round, closed less than one-half their length, smooth to scabridulous

ligules: membranous (4–7 mm long), acute, erose to lacerate, sometimes entire, whitish

blades: flat or involute (4–22 cm long, 1–4 mm wide), folded in the bud, **twisted with age**; apices prow-shaped; **midveins prominent and whitish abaxially**; scabrous abaxially; slightly pubescent adaxially; margins somewhat thickened, **greenish-white**

GROWTH CHARACTERISTICS: starts growth in early spring, flowers June to July, seeds mature July to August, reproduces from seeds and tillers, may regrow in autumn if adequate soil moisture is available

FORAGE VALUES: good for cattle and horses, fair for sheep and big game; quality rapidly declines with maturity

HABITATS: open prairies, meadows, foothills, hillsides, open woodlands, and mountain-tops at or near the tree line, in all soil textures; most abundant in moist to moderately dry soils

Bluejoint
Calamagrostis canadensis (Michx.) P. Beauv.

SYN = *Calamagrostis anomala* Suksd.

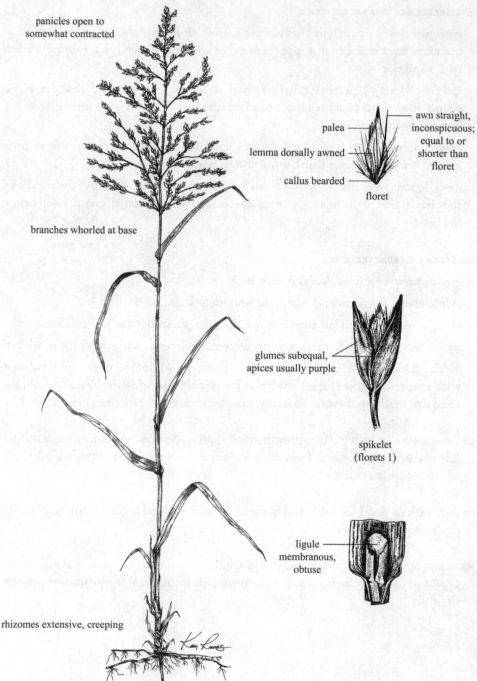

panicles open to somewhat contracted

branches whorled at base

rhizomes extensive, creeping

palea

lemma dorsally awned

callus bearded

floret

awn straight, inconspicuous; equal to or shorter than floret

glumes subequal, apices usually purple

spikelet (florets 1)

ligule membranous, obtuse

TRIBE:	AVENEAE
SPECIES:	*Calamagrostis canadensis* (Michx.) P. Beauv.
COMMON NAME:	Bluejoint (meadow pinegrass, Canadian reedgrass, marsh pinegrass)
LIFE SPAN:	Perennial
ORIGIN:	Native
SEASON:	Cool

INFLORESCENCE CHARACTERISTICS

type: panicles (7–20 cm long, 2–10 cm wide), open to somewhat contracted, often nodding, **stramineous or purplish or greenish**; branches ascending, mostly visible, whorled

spikelets: florets 1 (2–5 mm long); lemma (1.5–4 mm long) nearly as long as the glumes, teeth 2–4; **callus bearded**; hairs (1–3.5 mm long) often as long as the lemma (compare to *Calamagrostis rubescens*)

glumes: subequal (2–5 mm long), second slightly shorter than the first; acute or acuminate, rounded or keeled; midveins scabrous; **apices usually purple**

awns: lemmas dorsally awned from the middle or below; awns slender, weak, inconspicuous, usually straight (0.5–3 mm long), equal to or shorter than the floret, nearly as long as the glumes

VEGETATIVE CHARACTERISTICS

growth habit: extensive creeping rhizomes, cespitose to solitary

culms: erect (0.7–1.3 m tall), often branching above the base, stout, glabrous

sheaths: round, smooth, open, glabrous to scabrous; collar yellowish, constricted, characteristic burlaplike texture following senescence

ligules: membranous (3–7 mm long), obtuse, entire to erose-ciliate or lacerate

blades: flat to slightly involute (8–40 cm long and 2–8 mm wide), loosely spreading, often drooping, glabrous to scabrous, widely spaced ridges and furrows

other: a highly variable species with several varieties

GROWTH CHARACTERISTICS: flowers June to August; reproduces by seeds, rhizomes, and tillers

FORAGE VALUES: fair to good for cattle and horses in the spring, fair for wildlife; palatability decreases in the summer; makes good hay if cut early; production can be relatively high; it is sensitive to continuous heavy use

HABITATS: marshes, sloughs, meadows, ravines, and other wet places

Pine reedgrass
Calamagrostis rubescens Buckley

SYN = *Calamagrostis fasciculata* Kearney

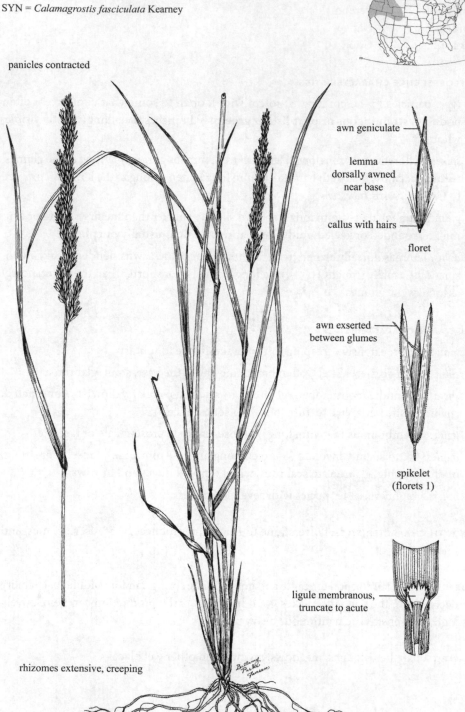

panicles contracted

awn geniculate

lemma
dorsally awned
near base

callus with hairs

floret

awn exserted
between glumes

spikelet
(florets 1)

ligule membranous,
truncate to acute

rhizomes extensive, creeping

TRIBE:	AVENEAE
SPECIES:	*Calamagrostis rubescens* Buckley
COMMON NAME:	Pine reedgrass (pinegrass)
LIFE SPAN:	Perennial
ORIGIN:	Native
SEASON:	Cool

INFLORESCENCE CHARACTERISTICS

type: panicles (7–18 cm long), contracted, dense to occasionally loose, interrupted, pale greenish to purplish

spikelets: florets 1; lemma pale (2.5–4 mm long), nearly as long as the glumes, smooth to scaberulous; callus with tuft of hairs (about 1 mm long) (compare to *Calamagrostis canadensis*); palea shorter than the lemma; **darkened at the bases when mature**

glumes: equal to subequal (3–6 mm long), narrow, acuminate, glabrous to minutely scaberulous; first veins 1; second veins 3, **midveins prominent**, lateral veins obscure

awns: lemmas awned dorsally from near the base; **awns delicate (2.5–3.5 mm long), exserted between the glumes**, nearly as long to slightly longer than glumes, geniculate

VEGETATIVE CHARACTERISTICS

growth habit: rhizomatous; rhizomes extensive and creeping; leaves mostly basal

culms: erect (0.4–1 m tall), mostly solitary, slender, smooth; nodes may be dark; basal internodes may be reddish

sheaths: round, smooth, distinctly veined, often purplish at bases, generally glabrous, collars pubescent (although sometimes obscurely so)

ligules: membranous (3–5 mm long), truncate to acute, erose

blades: flat (8–35 cm long, 2–6 mm wide) or involute at the apices, ascending with curved or drooping apices; scabrous to glabrous or hairy

GROWTH CHARACTERISTICS: starts growth in early spring, flowers in July or August, reproduces from seeds and rhizomes; remains green late into the autumn; quickly declines with excessive grazing or haying

FORAGE VALUES: poor to fair for sheep, fair for cattle and horses, less desirable for big game, although elk and deer graze it in the spring when it is immature and most palatable; herbage becomes unpalatable with maturity

HABITATS: open to dense pine woodlands, prairies, meadows, and stream banks in dry to moderately moist soils; usually does not persist in open sunlight

Tufted hairgrass
Deschampsia cespitosa (L.) P. Beauv.

SYN = *Aira cespitosa* L., *Deschampsia bottnica* (Wahlenb.) Trin., *Deschampsia caespitosa* (L.) P. Beauv., *Deschampsia glauca* Hartm., *Deschampsia mackenzieana* Raup, *Deschampsia obensis* Roshev., *Deschampsia paramushirensis* Honda, *Deschampsia pumila* (Griseb.) Ostenf., *Deschampsia sukatschewii* (Popl.) Roshev.

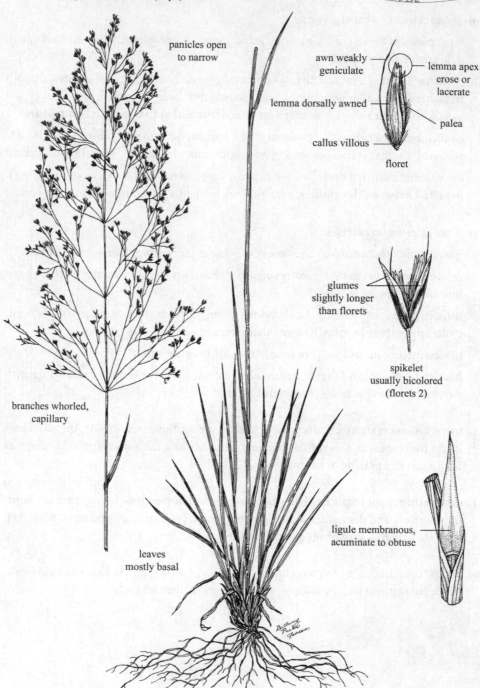

panicles open to narrow

awn weakly geniculate

lemma apex erose or lacerate

lemma dorsally awned

palea

callus villous

floret

glumes slightly longer than florets

spikelet usually bicolored (florets 2)

branches whorled, capillary

leaves mostly basal

ligule membranous, acuminate to obtuse

TRIBE:	AVENEAE
SPECIES:	*Deschampsia cespitosa* (L.) P. Beauv.
COMMON NAME:	Tufted hairgrass (zacate de montaña, saltandpeppergrass)
LIFE SPAN:	Perennial
ORIGIN:	Native
SEASON:	Cool

INFLORESCENCE CHARACTERISTICS

type: panicles (10–35 cm long), open to narrow, loose and often nodding, shiny black and light brown when mature; branches whorled, capillary

spikelets: florets 2 (rarely 3), **usually bicolored** pale brown or dark purplish-black, (hence the name saltandpeppergrass); lemmas purplish-black at base (2.5–5.4 mm long), **membranous with erose or lacerate apices**; apices usually light brown; calluses villous (hairs to 2 mm long); paleas about equaling the lemmas

glumes: subequal, slightly longer than florets, lanceolate, acute, glabrous or scaberulous; shiny; first 2.5–7 mm long, veins 1; second 2–7.5 mm long, veins 1 or 3

awns: dorsally awned from near the base to midlemma; awns weakly geniculate (2–6 mm long), occasionally straight, length varies from short to twice as long as spikelet

VEGETATIVE CHARACTERISTICS

growth habit: strongly cespitose, sometimes forms a dense tuft; leaves mostly basal

culms: erect (0.2–1.4 m tall), smooth

sheaths: keeled, open, margins decurrent, glabrous to scabrous, veins prominent

ligules: membranous (5–12 mm long), acuminate to obtuse, lacerate to entire

blades: flat or folded (5–30 cm long, 1–4 mm wide), firm, constricted at the collars; margins scabrous; ridged and scabrous adaxially, glabrous abaxially

GROWTH CHARACTERISTICS: starts growth early in the spring, flowers from July to September, seeds mature August to September, reproduces from seeds and tillers

FORAGE VALUES: good to excellent for all classes of livestock and fair to good for wildlife, produces good hay, not tolerant of continuous heavy grazing

HABITATS: wet meadows, prairies, stream banks, ditches, and open bogs; common in the spruce-fir zone; adapted to a wide range of soil textures; occasionally occurs in moderately saline and alkaline soils; commonly seeded for hay or grazing

Prairie junegrass
Koeleria macrantha (Ledeb.) Schult.

SYN = *Koeleria albescens* DC., *Koeleria cristata* Pers., *Koeleria gracilis* Pers., *Koeleria nitida* Nutt., *Koeleria pyramidata* (Lam.) P. Beauv., *Koeleria yukonensis* Hultén

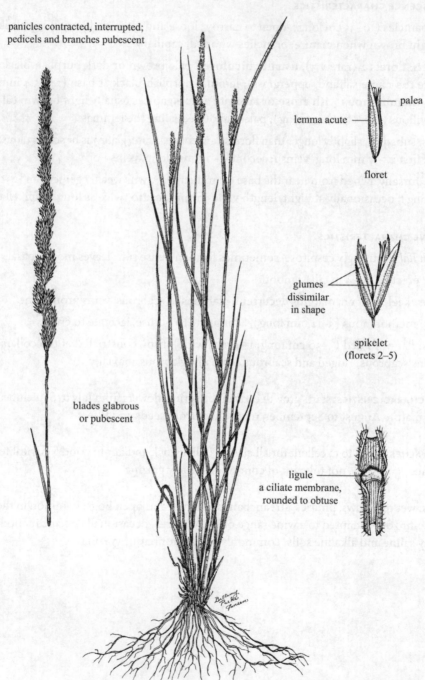

panicles contracted, interrupted;
pedicels and branches pubescent

palea

lemma acute

floret

glumes
dissimilar
in shape

spikelet
(florets 2–5)

blades glabrous
or pubescent

ligule
a ciliate membrane,
rounded to obtuse

TRIBE:	AVENEAE
SPECIES:	*Koeleria macrantha* (Ledeb.) Schult.
COMMON NAME:	Prairie junegrass (zacate de Junio, zacate de cresta, junegrass)
LIFE SPAN:	Perennial
ORIGIN:	Native
SEASON:	Cool

INFLORESCENCE CHARACTERISTICS

type: panicles (3–12 cm long, 1–3 cm wide) dense, contracted, interrupted, tapered toward the apices; branches ascending to spreading at anthesis, pubescent; pedicels pubescent; one of the most variable grasses

spikelets: florets 2–5, compressed laterally; lemmas narrow (3–6 mm long), lanceolate, acute (often appearing mucronate), midveins scabrous; rachillas pubescent

glumes: **subequal (3–6 mm long), dissimilar in shape;** first veins 1, narrow; second veins 3, broadened about the middle, shiny and scarious, shorter than first floret; scabrous at least on the keel

awns: usually awnless, lemmas rarely awned

VEGETATIVE CHARACTERISTICS

growth habit: cespitose; leaves mostly basal

culms: erect (20–60 cm tall), with a few fine hairs below the inflorescences

sheaths: round, distinctly veined, hispid or retrorsely pubescent, upper sheaths may be glabrous; collars pilose on margins

ligules: ciliate membrane or membranous (about 1 mm long), rounded to obtuse, erose to entire, whitish

blades: flat or involute (5–25 cm long, 1–3 mm wide), folded in bud; apices blunt or prow-shaped; keeled, coarsely veined adaxially, glabrous or pubescent abaxially; may have long hairs near collar

GROWTH CHARACTERISTICS: starts growth in early spring, flowers in June and July, produces seed through September, may regrow in the autumn if soil moisture is adequate; reproduces from seeds and tillers

FORAGE VALUES: excellent for all classes of livestock, although its forage production is low; good for wildlife in spring and in the autumn after curing, less palatable during seed production; hay quality is fair to good

HABITATS: mesic to semiarid prairies, open woodlands, foothills, savannas, and subalpine areas in all soil textures

Reed canarygrass
Phalaris arundinacea L.

SYN = *Phalaroides arundinacea* (L.) Rauschert

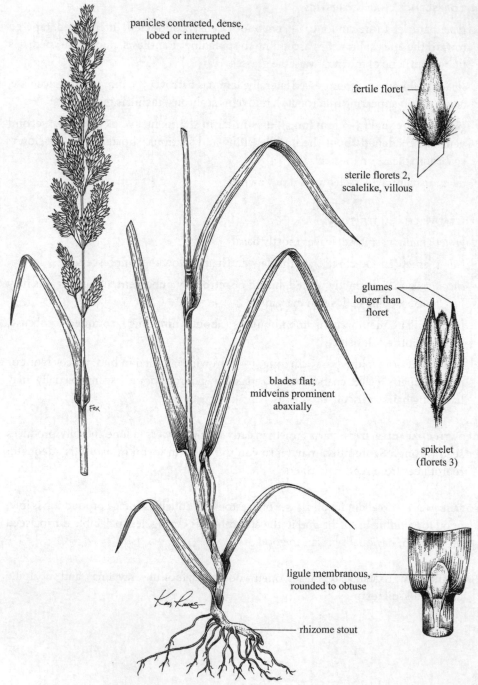

panicles contracted, dense, lobed or interrupted

fertile floret

sterile florets 2, scalelike, villous

glumes longer than floret

blades flat; midveins prominent abaxially

spikelet (florets 3)

ligule membranous, rounded to obtuse

rhizome stout

Fox

64

TRIBE:	AVENEAE
SPECIES:	*Phalaris arundinacea* L.
COMMON NAME:	Reed canarygrass
LIFE SPAN:	Perennial
ORIGIN:	Native
SEASON:	Cool

INFLORESCENCE CHARACTERISTICS

type: panicles (6–18 cm long, 1–3 cm wide), contracted, dense, cylindrical, lobed or interrupted; spikelets occurring in clusters on short ascending scabrous branches, reddish or purplish in anthesis, becoming stramineous

spikelets: florets 3; fertile floret 1 (2.7–4.2 mm long); sterile florets 2, below, unlike fertile upper floret; **fertile floret glabrous** to appressed-pubescent; lemma ovate (2.5–4.1 mm long), **firm and shiny**, villous on margins; sterile lemmas scalelike (1–2.5 mm long), villous, appressed to fertile floret bases

glumes: subequal (3.4–7 mm long), longer than the fertile floret, veins 3, ovate to lanceolate, laterally compressed, keeled, greenish-white to purplish becoming stramineous; midveins scabrous, not winged

awns: **none**

VEGETATIVE CHARACTERISTICS

growth habit: strongly rhizomatous; rhizomes stout, forming large colonies even in shallow water

culms: erect (0.6–2.1 m tall) to occasionally geniculate below, glabrous

sheaths: round, open, lower sheaths longer than the internodes, with obvious air chambers, cross-septate, glabrous, sometimes purplish

ligules: membranous (4–9 mm long), rounded to obtuse, entire to lacerate, decurrent

blades: flat (7–40 cm long, 4–18 mm wide), **glabrous** to scabrous, midveins prominent abaxially

GROWTH CHARACTERISTICS: reproduces by rhizomes, tillers, and seeds; forms dense colonies, often seeded or transplanted in wetlands; an ecotype introduced from Europe is invasive and aggressive, it may dominate wetlands; a form of this species with white-striped leaves is used as an ornamental

FORAGE VALUES: good to fair for livestock, provides good hay if cut prior to maturity, fair for wildlife, provides important nesting cover for waterfowl

HABITATS: wet meadows, marshes, ditches, edges of ponds and streams, and roadside wetlands

Alpine timothy
Phleum alpinum L.

SYN = *Phleum commutatum* Gaudin

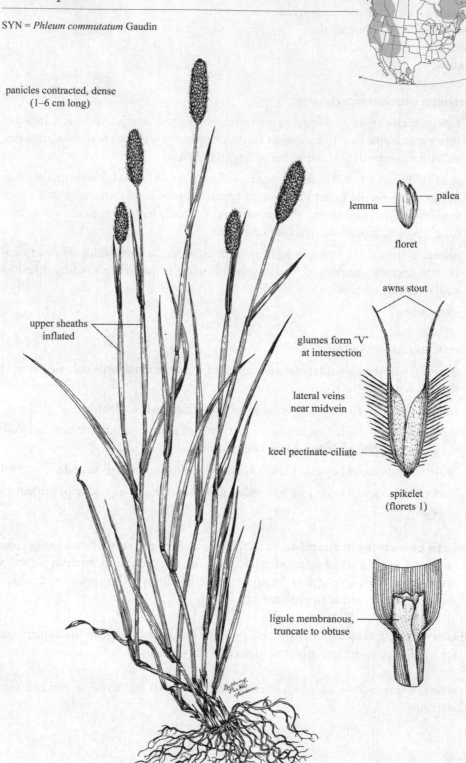

panicles contracted, dense
(1–6 cm long)

upper sheaths
inflated

lemma — palea

floret

awns stout

glumes form "V"
at intersection

lateral veins
near midvein

keel pectinate-ciliate

spikelet
(florets 1)

ligule membranous,
truncate to obtuse

TRIBE:	AVENEAE
SPECIES:	*Phleum alpinum* L.
COMMON NAME:	Alpine timothy (timothy alpino, mountain timothy, wild timothy)
LIFE SPAN:	Perennial
ORIGIN:	Native
SEASON:	Cool

INFLORESCENCE CHARACTERISTICS

type: panicles (1–6 cm long, 6–12 mm wide), contracted, dense, ovoid to cylindrical, branches tightly appressed, **usually from 1.5–3 times as long as wide, often purplish in color** (compare to *Phleum pratense*), bristly

spikelets: florets 1, elliptic, small, somewhat compressed; lemma (1.7–2.5 mm long) shorter than glumes, lance-ovate, glabrous to puberulent, veins 5–7; apices minutely erose; palea slightly shorter than the lemma; usually disarticulating below the glumes

glumes: subequal (3–5 mm long), keeled, veins 3, pectinate-ciliate on keel, hyaline, **gradually tapering to form a "V" where the two glumes intersect** (compare to *Phleum pratense*); lateral veins near the midveins; first glumes sometimes with ciliate margins

awns: glumes awned from apices; awns stout (1–3 mm long)

VEGETATIVE CHARACTERISTICS

growth habit: usually cespitose, sometimes with short rhizomes, and may form a sod

culms: erect or occasionally decumbent bases (15–60 cm tall), glabrous, internodes visible, lower culm nodes not swollen or bulbous

sheaths: round; upper sheaths inflated, glabrous, prominently veined, open; collars yellowish

ligules: membranous (2–4 mm long), truncate to obtuse, erose

blades: flat (2–16 cm long, 3–8 mm wide), linear, tapering; scabrous abaxially, margins scabrous

GROWTH CHARACTERISTICS: starts growth in early spring, flowers June to August, remains green throughout the summer, reproduces from seeds and tillers; requires occasional rest from heavy grazing to sustain productivity

FORAGE VALUES: good to excellent for all classes of livestock, as well as for elk and deer; most valuable in summer

HABITATS: mountain meadows, bogs, stream banks, floodplains, and moist woodlands in deep, poorly drained soils to deep stony loam soils; generally above 1250 m elevation

Timothy
Phleum pratense L.

SYN = *Phleum nodosum* L.

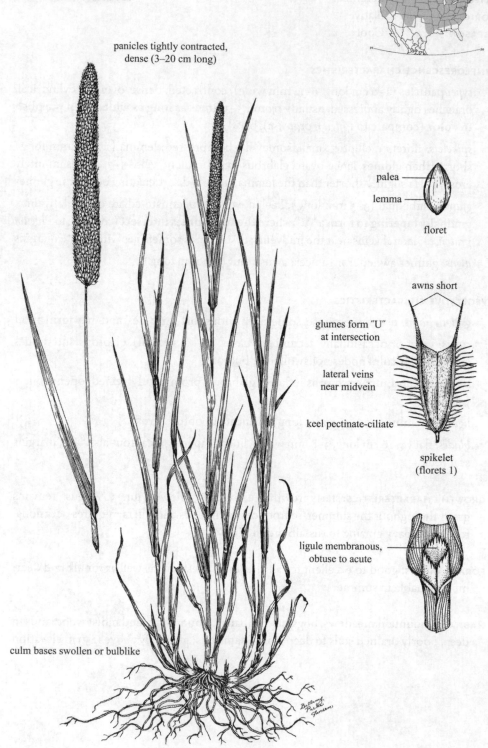

panicles tightly contracted,
dense (3–20 cm long)

palea

lemma

floret

awns short

glumes form "U"
at intersection

lateral veins
near midvein

keel pectinate-ciliate

spikelet
(florets 1)

ligule membranous,
obtuse to acute

culm bases swollen or bulblike

TRIBE:	AVENEAE
SPECIES:	*Phleum pratense* L.
COMMON NAME:	Timothy (zacate cola de gato, common timothy)
LIFE SPAN:	Perennial
ORIGIN:	Introduced (from Europe)
SEASON:	Cool

INFLORESCENCE CHARACTERISTICS

type: panicles (3–20 cm long, 5–9 mm wide), spicate, tightly contracted, dense, cylindrical, branches tightly appressed, **several times longer than wide**, bristly, **green or stramineous** (compare to *Phleum alpinum*)

spikelets: florets 1, elliptic, laterally compressed, small; lemma shorter (1.3–2.5 mm long) than glumes, delicate; usually disarticulating below the glumes

glumes: subequal to equal (2–3.5 mm long, excluding the awns), laterally compressed and keeled, pectinate-ciliate on keels, veins 3, lateral veins near midvein; **abruptly taper forming "U" where the two glumes intersect** (compare to *Phleum alpinum*)

awns: veins of glumes extended into awns; awns short (0.5–1.5 mm long), thick; central vein of lemmas sometimes extended into a mucro

VEGETATIVE CHARACTERISTICS

growth habit: cespitose, erect from a geniculate base, culms rarely solitary

culms: erect (0.5–1.2 m tall), sometimes geniculate below, glabrous; **bases swollen or bulblike**

sheaths: round, open, summits glabrous, distinctly veined, often purplish at the base; collars indistinct

ligules: membranous (3–6 mm long), obtuse to acute, entire to erose

blades: flat or loosely involute (5–30 cm long, 3–10 mm wide); apices acute; glabrous or scabrous; margins retrorsely scabrous

GROWTH CHARACTERISTICS: starts growth in early spring, flowers May to August, reproduces from seeds and tillers; cold tolerant, poor drought tolerance, responds to nitrogen fertilizer

FORAGE VALUES: good to excellent for all classes of livestock, as well as for deer, pronghorn, bighorn sheep, and elk; produces leafy, palatable hay for cattle and horses; not tolerant of heavy grazing

HABITATS: seeded in pastures and meadows; commonly escapes from cultivation into roadsides, fields, open woodlands, and waste places from the plains to subalpine elevations; most abundant on moist, fertile sites with fine-textured, nonsaline soils

Spike trisetum
Trisetum spicatum (L.) K. Richt.

SYN = *Aira spicata* L., *Trisetum molle* Kunth, *Trisetum montanum* Vasey, *Trisetum spicatum* var. *villosissimum* (Lange) Louis-Marie, *Trisetum subspicatum* (L.) P. Beauv., *Trisetum triflorum* (Bigelow) Á. Löve & D. Löve

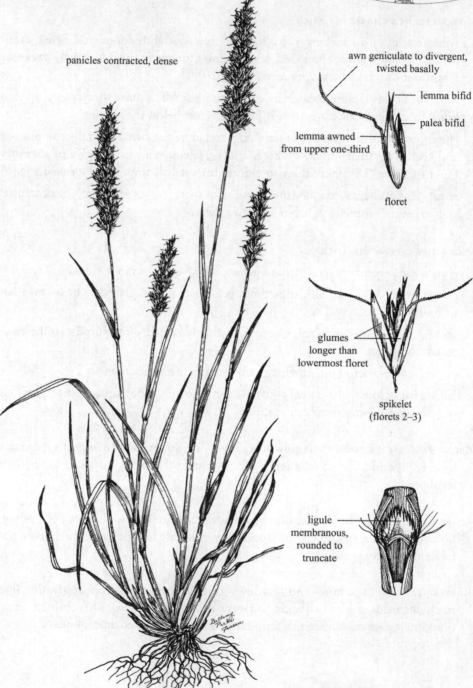

panicles contracted, dense

awn geniculate to divergent, twisted basally

lemma bifid

palea bifid

lemma awned from upper one-third

floret

glumes longer than lowermost floret

spikelet (florets 2–3)

ligule membranous, rounded to truncate

TRIBE:	AVENEAE
SPECIES:	*Trisetum spicatum* (L.) K. Richt.
COMMON NAME:	Spike trisetum (zacate tres cerdas)
LIFE SPAN:	Perennial
ORIGIN:	Native
SEASON:	Cool

INFLORESCENCE CHARACTERISTICS

type: panicles (2–25 cm long), **contracted, narrow, dense, interrupted** (compare to *Avenula hookeri*), branches ascending, purplish-green, bristly

spikelets: florets 2–3; lemmas keeled (3.5–5.5 mm long), veins 5, bifid, scaberulous, callus hairs short; paleas bifid; rachillas with short hairs

glumes: unequal, lanceolate, acute, glabrous to scabrous, longer than the lowermost floret; first 3.5–6 mm long, veins 1; second 3.5–7 mm long, broader than the first, veins 3

awns: lemmas awned from the upper one-third; **awns conspicuous (3–8 mm long), geniculate to divergent, twisted basally**

VEGETATIVE CHARACTERISTICS

growth habit: cespitose; leaves mostly basal

culms: erect (0.2–1 m tall); internodes glabrous to downy-pubescent; nodes pubescent

sheaths: keeled at summit, glabrous to pubescent (except on margins), with scattered long hairs near collars

ligules: membranous (1–3.5 mm long), rounded to truncate, erose or sometimes finely ciliate

blades: flat to folded (3–30 cm long, 2–5 mm wide), glabrous to pubescent, tapering to blunt apices, distinctly veined

GROWTH CHARACTERISTICS: starts growth in early spring, remains green until August; reproduces from seeds and tillers; low production of viable seeds; seldom occurs in dense stands; does not withstand heavy use

FORAGE VALUES: good for all classes of livestock and wildlife throughout the growing season, cures well, furnishes forage late into autumn if not covered by snow; one of the most important grasses in the mountains

HABITATS: moist to moderately moist alpine and subalpine meadows, open woodlands, bottomlands, and gentle slopes; abundant on well drained, medium-textured soils

Mountain brome
Bromus carinatus Hook. & Arn.

SYN = *Bromus marginatus* Nees *ex* Steud., *Ceratachloa carinata* (Hook. & Arn.)
Tutin

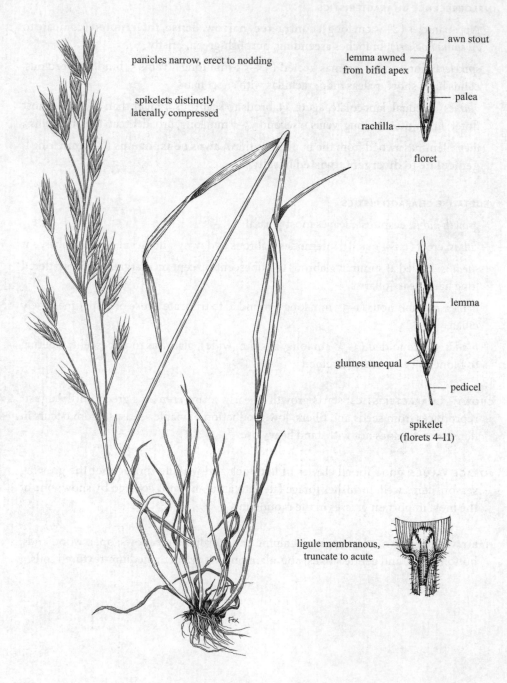

panicles narrow, erect to nodding

spikelets distinctly
laterally compressed

awn stout

lemma awned
from bifid apex

palea

rachilla

floret

lemma

glumes unequal

pedicel

spikelet
(florets 4–11)

ligule membranous,
truncate to acute

TRIBE:	BROMEAE
SPECIES:	*Bromus carinatus* Hook. & Arn.
COMMON NAME:	Mountain brome (bromo triste, bromo, California brome)
LIFE SPAN:	Perennial
ORIGIN:	Native
SEASON:	Cool

INFLORESCENCE CHARACTERISTICS

type: panicles (10–25 cm long), narrow, erect to nodding; branches appressed to spreading, drooping at maturity

spikelets: 2–3.5 cm long; **distinctly laterally compressed** (compare to *Bromus inermis*); florets 4–11; lemmas keeled (11–17 mm long), glabrous to slightly hirsute; apices bifid at maturity

glumes: unequal, strongly keeled, uniformly pubescent, sometimes glabrous to scabrous and apically pubescent; first 7–11 mm long, veins 3–5; second 9–13 mm long, shorter than lowermost lemmas, veins 5–7

awns: **lemmas awned** from bifid apices; awns stout (4–18 mm long), straight, sometimes slightly divergent

VEGETATIVE CHARACTERISTICS

growth habit: cespitose

culms: erect to ascending (0.3–1 m tall), coarse, glabrous to pubescent

sheaths: round, retrorsely pubescent to pilose or nearly glabrous, pilose near the throats; margins connate to near the apices

ligules: membranous (1.5–4 mm long), truncate to acute, erose or lacerate

blades: flat (15–30 cm long, 6–10 mm wide), scabrous to pilose or glabrous

GROWTH CHARACTERISTICS: starts growth in early spring, seeds mature by August, reproduces from seeds and tillers; seeded for rangeland revegetation and erosion control; does not withstand continuous heavy grazing, reducing grazing pressure during the reproductive stages will help to maintain this plant; plants have a relatively short life span

FORAGE VALUES: excellent for cattle, horses, and elk; good for sheep and deer; becoming coarse and fibrous at maturity; seeds furnish food for many kinds of birds and small mammals

HABITATS: mountain slopes, ridgetops, valleys, meadows, open woodlands, and waste places; grows in a broad range of soils; most abundant in moderately moist, well-developed, deep, medium-textured soils and in partial shade; may be found in poorly drained soils, moderately salt tolerant

Ripgut brome
Bromus diandrus Roth

SYN = *Anisantha rigida* (Roth) Hyl., *Bromus rigidus* Roth, *Genea rigida* (Roth) Dumort.

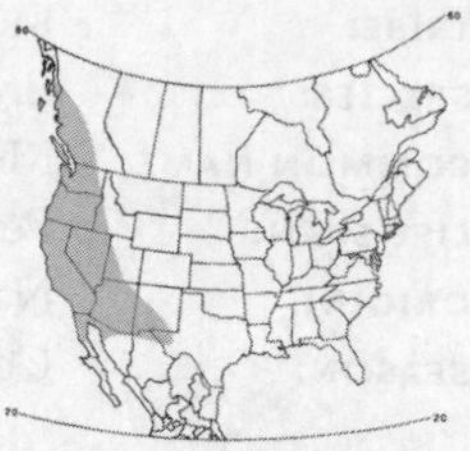

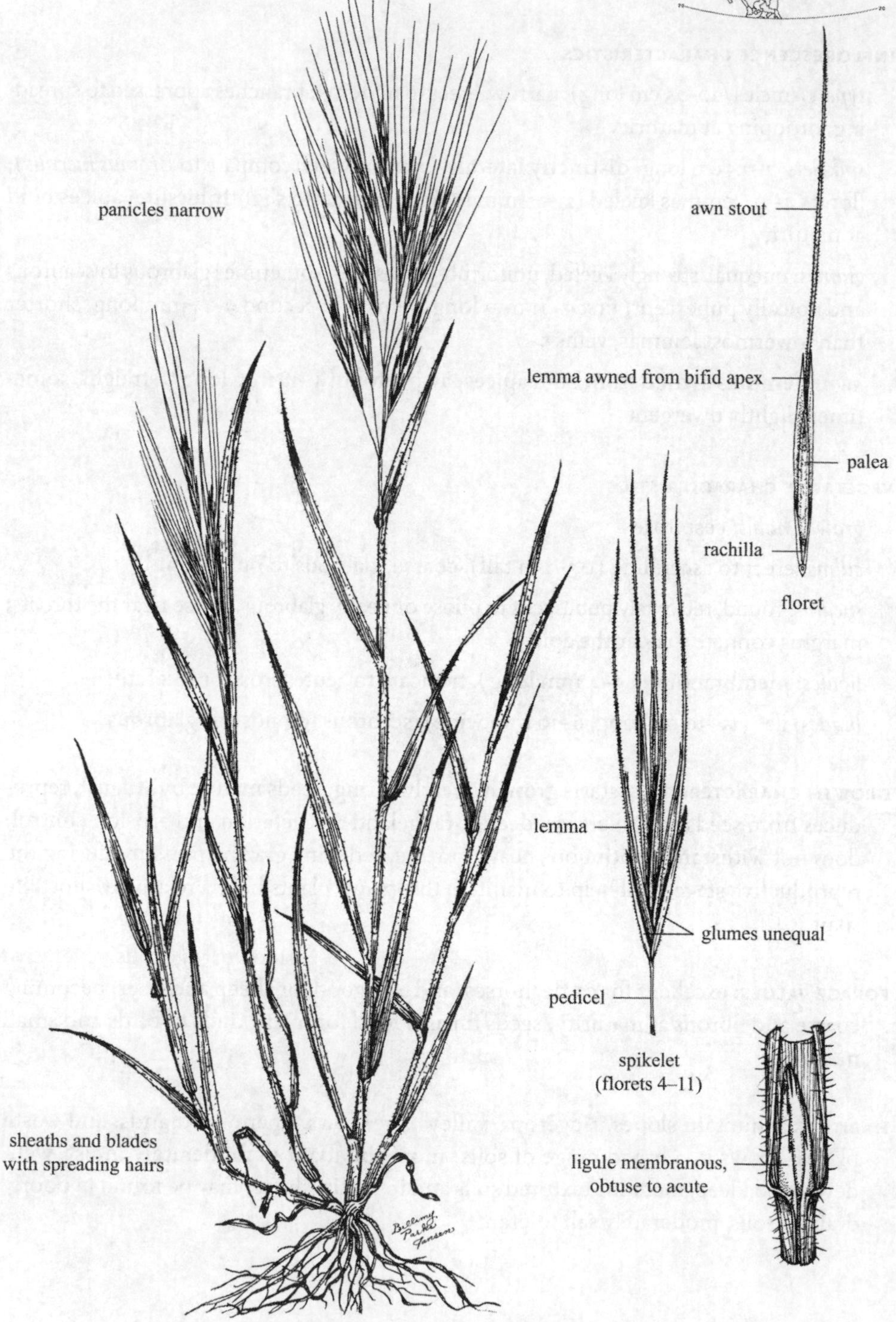

panicles narrow

awn stout

lemma awned from bifid apex

palea

rachilla

floret

lemma

glumes unequal

pedicel

spikelet
(florets 4–11)

sheaths and blades
with spreading hairs

ligule membranous,
obtuse to acute

TRIBE:	BROMEAE
SPECIES:	*Bromus diandrus* Roth
COMMON NAME:	Ripgut brome (bromo frágil, great brome)
LIFE SPAN:	Annual
ORIGIN:	Introduced (from Europe)
SEASON:	Cool

INFLORESCENCE CHARACTERISTICS

type: panicles (7–25 cm long, 2–11 cm wide), narrow, stout; branches elongate, **erect above**, spreading below, each with 1 or 2 spikelets

spikelets: 3–7 cm long, excluding awns, moderately laterally compressed, florets 4–11; lemmas linear-lanceolate (2–3 cm long), veins 7, glabrous or scabrous; margins broad and hyaline; apices bifid; apical teeth 3–6 mm long

glumes: unequal; first 1.5–2.5 cm long, veins 1–3; second 2–3.5 cm long, veins 3–5; smooth to scabrous, margins hyaline, scabrous

awns: **lemmas awned; awns stout (3–6 cm long)**

VEGETATIVE CHARACTERISTICS

growth habit: cespitose or solitary

culms: erect to decumbent at bases (20–90 cm tall), thick but weak, easily broken after maturity, puberulent below the panicles

sheaths: round, softly pilose with **spreading hairs**; margins connate

ligules: membranous (2–4 mm long), obtuse to acute, erose to lacerate

blades: flat (10–25 cm long, 3–9 mm wide), sometimes involute on drying, soft, erect to spreading, pubescent with spreading hairs on both surfaces

GROWTH CHARACTERISTICS: seeds germinate in late autumn, grows rapidly in the spring, matures 2–3 months later, reproduces from seeds

LIVESTOCK LOSSES: awns may be injurious to livestock, sometimes working into the nostrils and eyes of grazing animals

FORAGE VALUES: excellent in seedling stage and during periods of vigorous vegetative growth but becomes poor for sheep and wildlife and fair for cattle at flowering, worthless at maturity

HABITATS: open ground, dunes, waste places, roadsides, field borders, and disturbed sites; adapted to nearly all soil types, most abundant on dry sites

Soft brome
Bromus hordeaceus L.

SYN = *Bromus mollis* L.

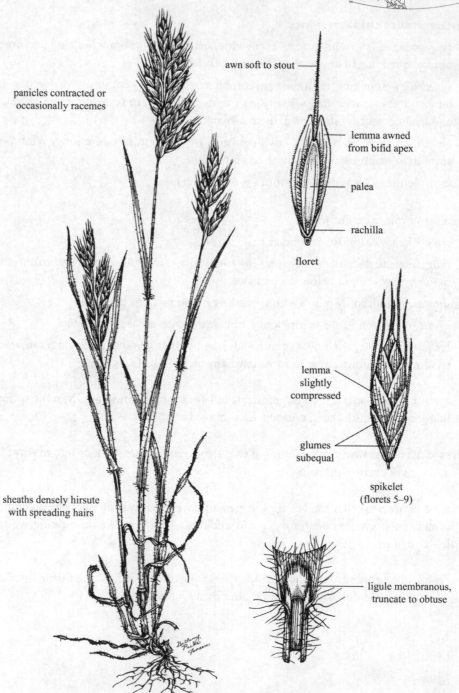

panicles contracted or
occasionally racemes

awn soft to stout

lemma awned
from bifid apex

palea

rachilla

floret

lemma
slightly
compressed

glumes
subequal

spikelet
(florets 5–9)

sheaths densely hirsute
with spreading hairs

ligule membranous,
truncate to obtuse

TRIBE:	BROMEAE
SPECIES:	*Bromus hordeaceus* L.
COMMON NAME:	Soft brome (bromo, soft chess, bald brome)
LIFE SPAN:	Annual
ORIGIN:	Introduced (from Europe)
SEASON:	Cool

INFLORESCENCE CHARACTERISTICS

type: panicles or occasionally racemes (3–10 cm long), erect, contracted, usually ovoid; branches ascending to appressed; pedicels shorter than spikelets

spikelets: 1.3–2.5 cm long, 4–7 mm wide, nearly terete, pubescent to glabrous, florets 5–9; lemmas slightly compressed (7–10 mm long, 1.5–2.5 mm wide), obtuse, pilose or scabrous; veins 7, prominent; margins hyaline; apices bifid; teeth short (1 mm); paleas subequal to the lemma

glumes: subequal, broad, obtuse, coarsely pilose to scabrous, rarely glabrous; first 4–7.5 mm long, veins 3–5; second 5–8.5 mm long, veins 5–7; **surfaces appear crinkly**

awns: lemmas awned; awns straight or divergent at maturity (5–9 mm long), soft to stout

VEGETATIVE CHARACTERISTICS

growth habit: cespitose or solitary

culms: erect to ascending (to 60 cm tall), bases geniculate, weak; glabrous to retrorsely pubescent, especially at the nodes

sheaths: round, **densely hirsute with spreading hairs**; margins connate

ligules: membranous (1–2 mm long), truncate to obtuse, erose

blades: flat (3–15 cm long, 2–6 mm wide), abaxial surfaces glabrous or hairy, adaxial surfaces pubescent

GROWTH CHARACTERISTICS: germinates in late autumn when moisture is adequate, flowers in early spring, seeds mature in May and June; growth period of about 12 weeks, reproduces from seeds

FORAGE VALUES: excellent for livestock and good for wildlife while immature, good to fair when mature; soft awns allow it to be grazed without injury even after seed maturity; seeds remain on the plants providing good winter grazing

HABITATS: open grounds, fields, waste places, and disturbed sites; most abundant on clay loam and sandy soils but grows well on nearly any well-drained soils

Smooth brome
Bromus inermis Leyss.

SYN = *Bromopsis inermis* (Leyss.) Holub, *Bromopsis pumpelliana* (Scribn.) Holub, *Bromus pumpellianus* Scribn.

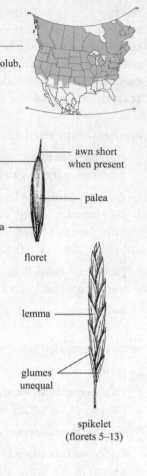

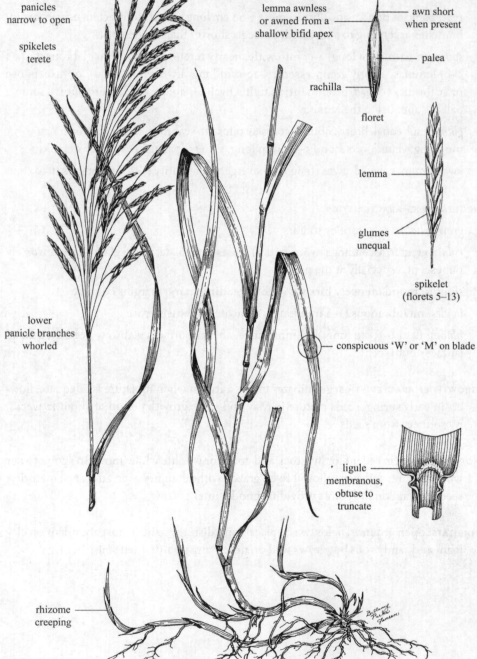

panicles
narrow to open

spikelets
terete

lower
panicle branches
whorled

rhizome
creeping

lemma awnless
or awned from a
shallow bifid apex

awn short
when present

palea

rachilla

floret

lemma

glumes
unequal

spikelet
(florets 5–13)

conspicuous 'W' or 'M' on blade

ligule
membranous,
obtuse to
truncate

TRIBE:	BROMEAE
SPECIES:	*Bromus inermis* Leyss.
COMMON NAME:	Smooth brome (zacate bromo, bromo suave)
LIFE SPAN:	Perennial
ORIGIN:	Introduced (from Europe)
SEASON:	Cool

INFLORESCENCE CHARACTERISTICS

type: panicles (7–24 cm long), erect, narrow to open, usually contracted at maturity; branches ascending; lower branches whorled

spikelets: 1.5–3.5 cm long, 2.5–7 mm wide, terete to oval (compare to *Bromus carinatus*), pointed, florets 5–13; lemmas rounded on back (9–14 mm long), **glabrous** to puberulent, greenish to purplish; apices shallowly bifid

glumes: unequal, papery, lanceolate, **glabrous** to puberulent; first 4–8 mm long, veins usually 1; second 6–10 mm long, veins usually 3

awns: **lemmas awnless** or with a short awn (1–3 mm), from between the teeth of bifid apices

VEGETATIVE CHARACTERISTICS

growth habit: **strongly rhizomatous**; rhizomes creeping, forming a sod

culms: erect to rarely decumbent at base (0.4–1.2 m tall), **glabrous**

sheaths: round, **glabrous** to rarely pilose, prominently veined; margins connate

ligules: membranous (1–3 mm long), obtuse to truncate, minutely erose-ciliate

blades: flat (15–40 cm long, 4–15 mm wide), **glabrous** to rarely pilose, margins scabrous; **conspicuous "W" or "M" constriction**; auricles rarely present (to 0.5 mm long)

GROWTH CHARACTERISTICS: starts growth in early spring; flowers May to July; reproduces from seeds, tillers, and rhizomes; responds to nitrogen fertilizer; may regrow and reflower in the autumn if moisture is sufficient

FORAGE VALUES: excellent for livestock and wildlife, quality and palatability rapidly decline after inflorescence development, produces excellent hay, regrowth may furnish valuable autumn grazing

HABITATS: cultivated as a dryland or irrigated hay and pasture grass; roadsides, waterways, field borders, and waste places on all soil types; often invades native prairies where it is difficult to control

Cheatgrass
Bromus tectorum L.

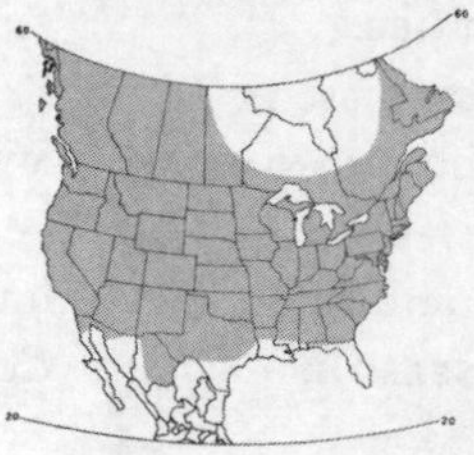

SYN = *Anisantha tectorum* (L.) Nevski

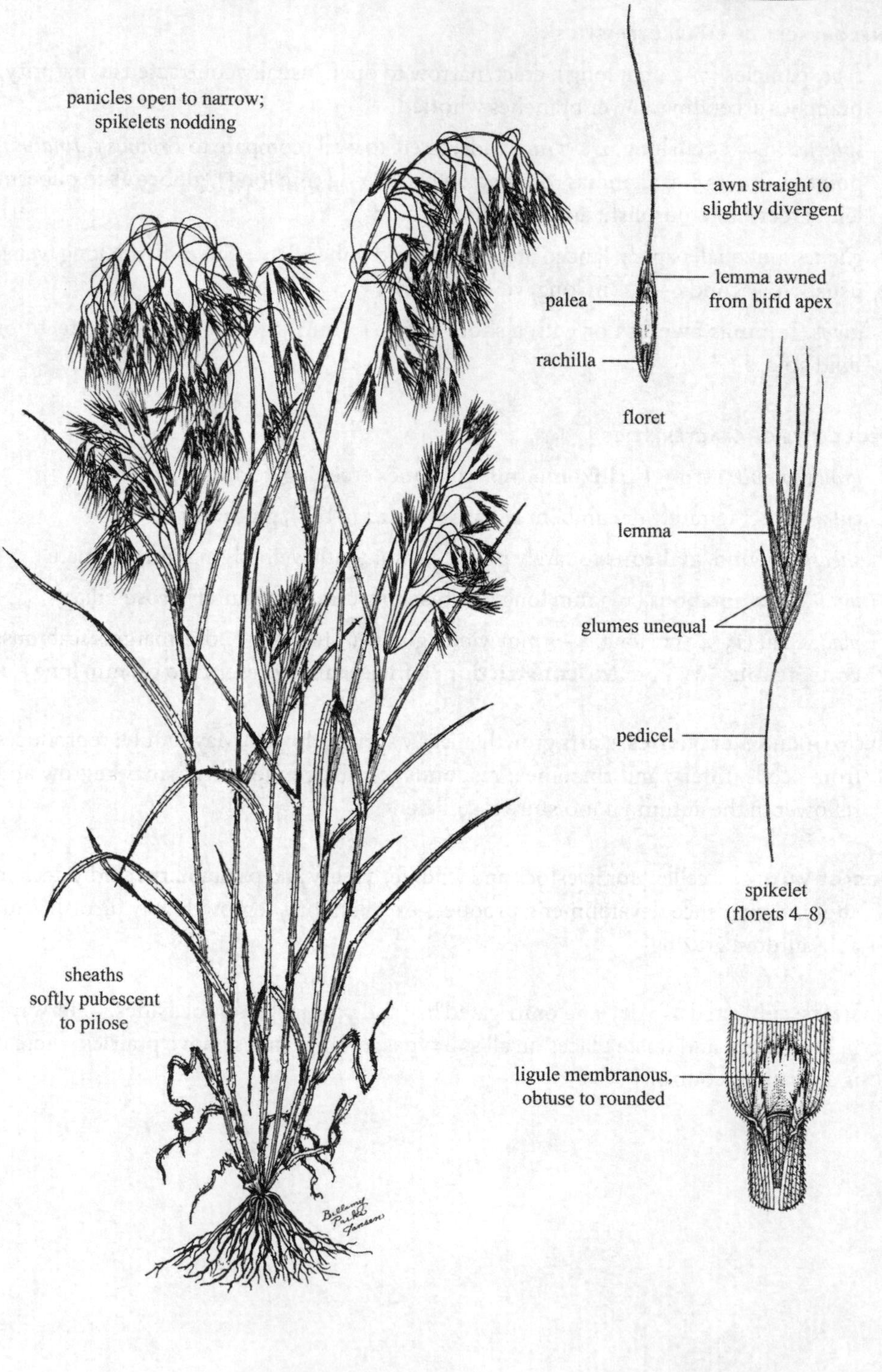

panicles open to narrow;
spikelets nodding

awn straight to
slightly divergent

lemma awned
from bifid apex

palea

rachilla

floret

lemma

glumes unequal

pedicel

spikelet
(florets 4–8)

sheaths
softly pubescent
to pilose

ligule membranous,
obtuse to rounded

TRIBE:	BROMEAE
SPECIES:	*Bromus tectorum* L.
COMMON NAME:	Cheatgrass (bromo velloso, downy brome, broncograss, junegrass)
LIFE SPAN:	Annual
ORIGIN:	Introduced (from Europe)
SEASON:	Cool

INFLORESCENCE CHARACTERISTICS

type: panicles (5–20 cm long), open to narrow, dense, much branched, **nodding**, often purplish; **branches and pedicels slender, flexuous**

spikelets: 1.2–2 cm long, 3–6 mm wide excluding awns, florets 4–8; lemmas lanceolate (9–15 mm long), rounded on back, glabrous to hirsute; margins thin, membranous; apices bifid; teeth slender (1–3 mm long)

glumes: unequal, first 4–7 mm long, veins 1; second 8–11 mm long, veins 3–5; glabrous to hirsute; margins broad, hyaline

awns: **lemmas awned** from between the teeth of bifid apices; awn straight to slightly divergent (1–1.8 cm long)

VEGETATIVE CHARACTERISTICS

growth habit: cespitose or solitary

culms: erect to ascending or decumbent at bases (to 70 cm tall), weak

sheaths: round, keeled toward collar, **softly pubescent** to pilose; upper sheaths sometimes glabrous; margins connate

ligules: membranous (1.5–3.5 mm long), obtuse to rounded, erose to lacerate

blades: flat (5–15 cm long, 2–7 mm wide), softly pubescent

GROWTH CHARACTERISTICS: seeds germinate in the late autumn or early spring, rapid spring growth, seeds mature about 2 months later, reproduces from seeds; an aggressive weed

LIVESTOCK LOSSES: awns may injure eyes and mouths of grazing animals and contaminate fleece

FORAGE VALUES: fair to good for livestock for the short period before the inflorescence emerges, then is practically worthless; deer and pronghorn graze it in the spring while it is actively growing; furnishes food for some upland birds and small mammals

HABITATS: heavily grazed rangelands, roadsides, waste places, and disturbed sites; adapted to a broad range of soil textures, most abundant on dry sites; dry plants can be a severe fire hazard

Sideoats grama
Bouteloua curtipendula (Michx.) Torr.

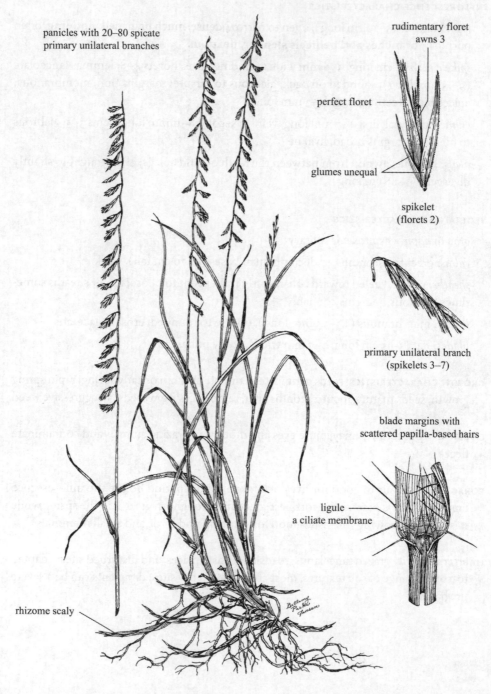

panicles with 20–80 spicate
primary unilateral branches

rudimentary floret
awns 3

perfect floret

glumes unequal

spikelet
(florets 2)

primary unilateral branch
(spikelets 3–7)

blade margins with
scattered papilla-based hairs

ligule
a ciliate membrane

rhizome scaly

TRIBE:	CYNODONTEAE
SPECIES:	*Bouteloua curtipendula* (Michx.) Torr.
COMMON NAME:	Sideoats grama (banderilla, banderita)
LIFE SPAN:	Perennial
ORIGIN:	Native
SEASON:	Warm

INFLORESCENCE CHARACTERISTICS

type: panicles (10–30 cm long) of **20–80 spicate primary unilateral branches; branches** (1–4 cm long) **distant, pendant** (compare to *Bouteloua repens*), spikelets in 2 rows; individual branches usually turned to one side of inflorescence, spicate branches fall as a unit, **branch bases remain on culm after disarticulation**; spikelets 3–7 per branch, crowded

spikelets: florets 2; fertile lemma (3–6 mm long) below, veins 3; sterile floret (rudiment) above, variable, usually a short lemma, awns 3

glumes: unequal; first short (2.5–6 mm long), thin; second longer (5–8 mm long), thick, tapering, glabrous or scabrous, purplish

awns: fertile floret lemmas teeth 3, awn tips 3 (1–2 mm long); rudimentary lemmas with awns 3 (3–6 mm long), unequal

VEGETATIVE CHARACTERISTICS

growth habit: cespitose, rhizomatous, or stoloniferous; rhizomes scaly; leaves mostly basal

culms: erect (0.2–1 m tall), smooth; nodes purplish

sheaths: round, glabrous below to somewhat pilose above, prominently veined; collars pilose on margin

ligules: ciliate membranes (0.3–0.7 mm long), truncate

blades: flat to sub-involute (2–40 cm long, 2–7 mm wide), linear; usually scabrous adaxially, smooth abaxially; **margins with scattered papilla-based hairs**

GROWTH CHARACTERISTICS: starts growth in early spring and flowers July to September; reproduces from seeds, tillers, stolons, and rhizomes; commonly included in seeding mixtures for revegetation

FORAGE VALUES: good for all classes of livestock and wildlife throughout summer and autumn, remains moderately palatable into winter; makes good hay

HABITATS: dry plains, prairies, woodland openings, and rocky hills; most abundant in fine-textured soils, seldom grows in coarse-textured soils, better adapted to calcareous and moderately alkaline soils than to neutral or acidic soils

Buffalograss
Bouteloua dactyloides (Nutt.) Columbus

SYN = *Buchloë dactyloides* (Nutt.) Engelm.

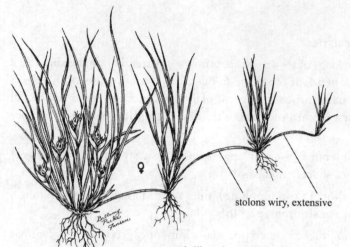

pistillate spikelet
(florets 1)

stolons wiry, extensive

pistillate plants with burlike clusters (spikelets 3–5)

plants dioecious, occasionally monoecious

staminate plants with panicles of 1–4 spicate
primary unilateral branches (spikelets 6–12)

lemma veins 3 — palea

staminate floret

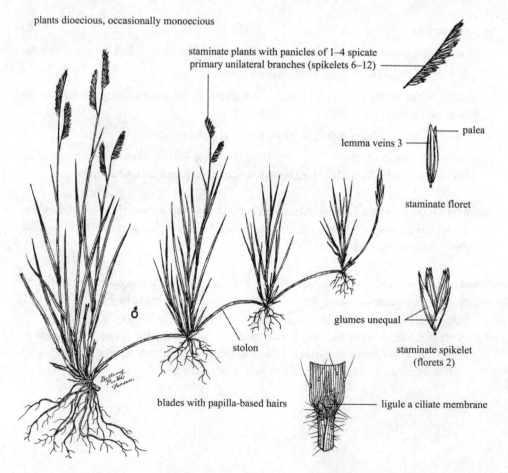

glumes unequal

staminate spikelet
(florets 2)

stolon

blades with papilla-based hairs

ligule a ciliate membrane

TRIBE:	CYNODONTEAE
SPECIES:	*Bouteloua dactyloides* (Nutt.) Columbus
COMMON NAME:	Buffalograss (pasto chino, búfalo)
LIFE SPAN:	Perennial
ORIGIN:	Native
SEASON:	Warm

INFLORESCENCE CHARACTERISTICS

type: **staminate plants** with elevated panicles of 1–4 **spicate primary unilateral branches**; branches (6–15 mm long) erect to spreading, spikelets 6–12 in 2 rows on each branch, pectinate; **pistillate plants with burlike clusters** (up to 7.5 mm long) near the midculm or within the leaves, spikelets 3–5 per cluster; burs fall as a unit

spikelets: staminate spikelets 4–5.5 mm long, florets 2; lemmas longer than the glumes, glabrous; pistillate spikelets usually with a single floret, yellowish; lemma veins 3, totally enclosed in glumes

glumes: staminate spikelet glumes unequal (first about 1 mm long, second 1.2–3 mm long), veins 1–2, acute; pistillate glumes unequal (to 7.5 mm long), first highly reduced, second indurate, enclosing lemma, bearing 3 mucronate lobes

awns: second glumes of pistillate spikelets mucronate

other: plants dioecious, occasionally monoecious

VEGETATIVE CHARACTERISTICS

growth habit: stoloniferous; stolons wiry, extensive, forming mats

culms: culms of staminate plants erect (5–25 cm tall), nodes glabrous; culms of pistillate plants much shorter; mostly unbranched

sheaths: round, open, glabrous except for a few marginal hairs near the collar

ligules: ciliate membranes (0.5–1 mm long), truncate to obtuse; often flanked by long hairs

blades: flat (1–17 cm long, 1–2.5 mm wide), curling when dry, with papilla-based hairs on both surfaces

GROWTH CHARACTERISTICS: starts growth in midspring when adequate moisture is available, flowers in summer, reproduces from stolons and seeds; can withstand heavy grazing and dry conditions; cannot tolerate shading

FORAGE VALUES: good for all classes of livestock and fair for wildlife, cures well and provides winter grazing when not covered by snow; not tall enough to be hayed

HABITATS: dry plains and prairies on medium- to fine-textured soils, rare on sandy soils, most abundant on heavily used rangelands

Black grama
Bouteloua eriopoda (Torr.) Torr.

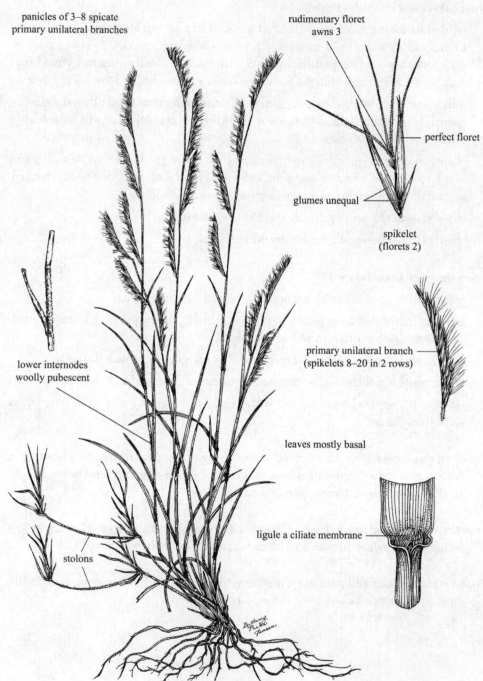

panicles of 3–8 spicate
primary unilateral branches

rudimentary floret
awns 3

perfect floret

glumes unequal

spikelet
(florets 2)

primary unilateral branch
(spikelets 8–20 in 2 rows)

lower internodes
woolly pubescent

leaves mostly basal

ligule a ciliate membrane

stolons

TRIBE:	CYNODONTEAE
SPECIES:	*Bouteloua eriopoda* (Torr.) Torr.
COMMON NAME:	Black grama (navajita negra)
LIFE SPAN:	Perennial
ORIGIN:	Native
SEASON:	Warm

INFLORESCENCE CHARACTERISTICS

type: panicles of 3–8 spicate primary unilateral branches; branches ascending (2–6 cm long), slender, delicate, white-lanate, branch not projecting beyond the spikelet-bearing portion (compare to *Bouteloua hirsuta*); spikelets 8–20 in 2 rows on each branch, not crowded, somewhat pectinate

spikelets: florets 2, perfect floret with a rudiment above; fertile lemma 4.5–6.5 mm long, veins 3, **usually bearded at the base**; rudiment may be reduced to 3 awns

glumes: unequal; first shorter (3–6 mm long) than the second (6–10 mm long), acute to acuminate, glabrous to scabrous, purplish

awns: fertile floret lemmas awns 3 (1.5–3 mm long), central awn longest; rudiment awns 3 (3.5–8 mm long)

VEGETATIVE CHARACTERISTICS

growth habit: stoloniferous, often short-rhizomatous; leaves mostly basal

culms: ascending (20–60 cm tall) from a decumbent base, spreading, wiry, slender; internodes arched, **lower culm internodes woolly pubescent**, appearing as alternating glabrous and pubescent areas; bases swollen, knotty, woolly

sheaths: round, shorter than the internodes, glabrous

ligules: ciliate membranes (0.3–0.7 mm long), truncate

blades: flat or folded below, twisted and involute above (2–7 cm long, 0.5–2 mm wide), flexuous; apices acute; lower margins with papilla-based hairs

GROWTH CHARACTERISTICS: starts growth when sufficient moisture is available in late spring, reproduces mainly from stolons and tillers, low seed productivity and viability; frequently grows in nearly pure stands; not tolerant of heavy grazing

FORAGE VALUES: excellent for all classes of livestock and wildlife throughout the year, especially valuable for winter grazing, produces excellent hay during years with adequate moisture

HABITATS: rocky or sandy mesas, dry hills, and dry, open prairie from 1000 to 2000 m altitude; seldom grows in fine-textured soils

Blue grama
Bouteloua gracilis (Kunth) Lag. *ex* Griffiths

SYN = *Bouteloua oligostachya* (Nutt.) Torr. *ex* A. Gray, *Chondrosum gracile* Kunth

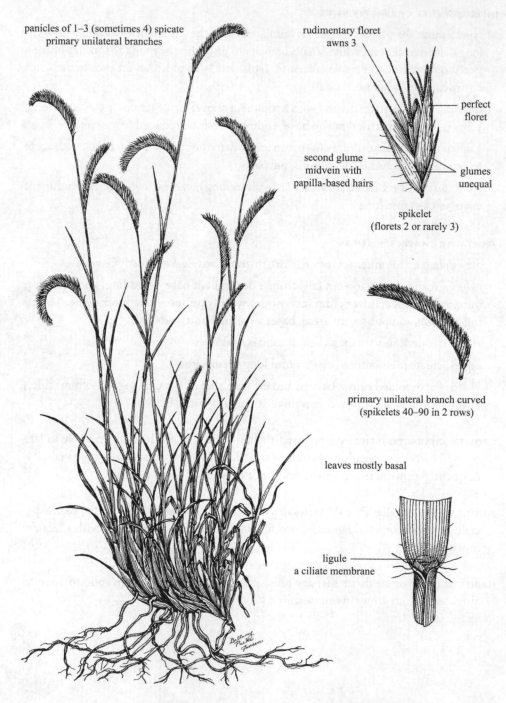

panicles of 1–3 (sometimes 4) spicate primary unilateral branches

rudimentary floret awns 3

perfect floret

second glume midvein with papilla-based hairs

glumes unequal

spikelet (florets 2 or rarely 3)

primary unilateral branch curved (spikelets 40–90 in 2 rows)

leaves mostly basal

ligule a ciliate membrane

TRIBE:	CYNODONTEAE
SPECIES:	*Bouteloua gracilis* (Kunth) Lag. *ex* Griffiths
COMMON NAME:	Blue grama (navajita azúl, navajita común)
LIFE SPAN:	Perennial
ORIGIN:	Native
SEASON:	Warm

INFLORESCENCE CHARACTERISTICS

type: panicles of 1–3 (sometimes 4) spicate primary unilateral branches; **branches (1.5–5 cm long) curved and spreading at maturity, not extending beyond the spikelet-bearing portion** (compare to *Bouteloua hirsuta*); spikelets 40–90 in 2 rows on each branch, crowded, pectinate

spikelets: florets 2 (rarely 3); fertile florets 1 (3.5–6 mm long) pubescent on back, veins 3; rudiments 1 (rarely 2) above (0.8–3 mm long), highly variable, may be reduced to 1–3 awns

glumes: unequal, veins 1; first 1.5–3.5 mm long, glabrous or with hairs on the midveins; second 3.5–6 mm long, papilla-based hairs on the midveins, may be purplish

awns: fertile floret lemma awns 3 (1–3 mm long), central awn longest; rudimentary lemma awns 1–3 (2.5–5 mm long) or may be awnless

VEGETATIVE CHARACTERISTICS

growth habit: cespitose, occasionally with short rhizomes; forming mats; leaves mostly basal

culms: erect (20–65 cm tall), slender, often geniculate below, glabrous; nodes glabrous or puberulent

sheaths: round, glabrous to pilose, long-pilose at collars

ligules: ciliate membranes (0.1–0.5 mm long), truncate; often with marginal tufts of long hairs

blades: flat at base, loosely involute above at maturity (2–25 cm long, 1–3 mm wide), tapering, puberulent to scabrous above, smooth to slightly scabrous below, sometimes sparingly pilose on both surfaces

GROWTH CHARACTERISTICS: starts growth in May or June, flowers June to October, reproduces primarily from tillers; cannot tolerate shading by taller plants, withstands relatively heavy grazing

FORAGE VALUES: good for all classes of livestock and wildlife, quality is highest when it is green, but it retains much of its value when dry and furnishes autumn and winter grazing

HABITATS: open plains, prairies, mesas, foothills, and open woodlands; in all soil textures, but most abundant in sandy or gravelly soils; not found in wet, poorly drained soils

Hairy grama
Bouteloua hirsuta Lag.

SYN = *Chondrosum hirsutum* (Lag.) Sweet

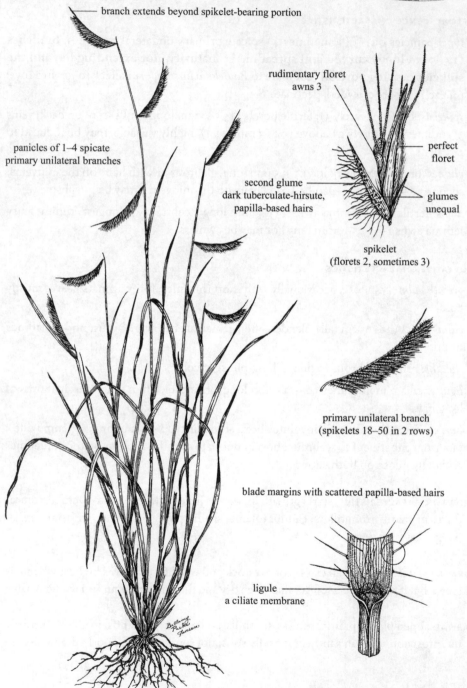

branch extends beyond spikelet-bearing portion

rudimentary floret awns 3

perfect floret

panicles of 1–4 spicate primary unilateral branches

second glume dark tuberculate-hirsute, papilla-based hairs

glumes unequal

spikelet (florets 2, sometimes 3)

primary unilateral branch (spikelets 18–50 in 2 rows)

blade margins with scattered papilla-based hairs

ligule a ciliate membrane

TRIBE:	CYNODONTEAE
SPECIES:	*Bouteloua hirsuta* Lag.
COMMON NAME:	Hairy grama (navajita velluda)
LIFE SPAN:	Perennial
ORIGIN:	Native
SEASON:	Warm

INFLORESCENCE CHARACTERISTICS

type: panicles of 1–4 spicate primary unilateral branches; branches (2–4 cm long) ascending to spreading, occasionally curved, **extending beyond spikelet-bearing portion in an obvious point (5–10 mm long)** (compare to *Bouteloua gracilis*); spikelets 18–50 in 2 rows on each branch, crowded, pectinate

spikelets: florets 2 (sometimes 3); fertile floret below, lemma 4–6 mm long, deeply 3–cleft, veins 3, hairy; rudiments 1 (sometimes 2), above, lemmas of rudiments often highly reduced to only 1–3 awns

glumes: unequal; first 1.5–3.5 mm long; second 3–6 mm long, **dark tuberculate-hirsute**, papilla-based hairs

awns: second glumes mucronate; fertile floret lemma awns 3 (0.2–2.5 mm long), central awns longest; rudiment lemma awns 1–3 (2–4.5 mm long), may be awnless, dark brown

VEGETATIVE CHARACTERISTICS

growth habit: cespitose; leaves basal and cauline

culms: erect to spreading (15–70 cm tall), somewhat geniculate below; nodes 4–8

sheaths: round, glabrous to sparsely pilose; margins with papilla-based hairs; collars hairy

ligules: ciliate membranes (0.2–0.4 mm long), truncate

blades: flat to loosely involute (2–18 cm long, 1–3 mm wide), narrow; apices acute; surfaces smooth to scabrous; margins with scattered papilla-based hairs (often on both surfaces as well), thickened

GROWTH CHARACTERISTICS: starts growth in May or June or as soon as moisture is available, flowers in July and August, reproduces primarily from tillers and seeds

FORAGE VALUES: fair to good for livestock and fair for wildlife, palatability is highest late in the growing season, cures well and furnishes winter grazing when not covered by snow, not tall enough to be hayed

HABITATS: prairies, dunes, shallow uplands, and rocky ridges; most abundant on dry, loose sands and neutral to slightly calcareous soils

Slender grama
Bouteloua repens (Kunth) Scribn. & Merr.

SYN = *Bouteloua filiformis* (E. Fourn.) Griffiths

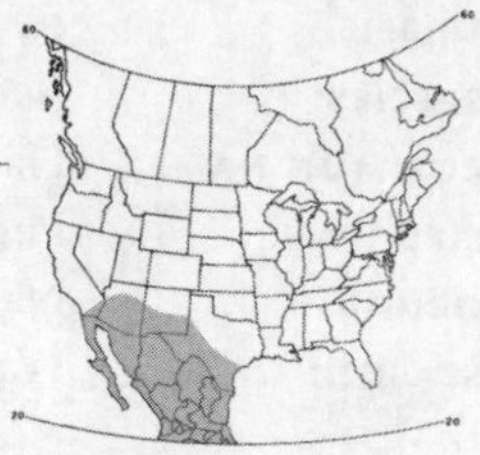

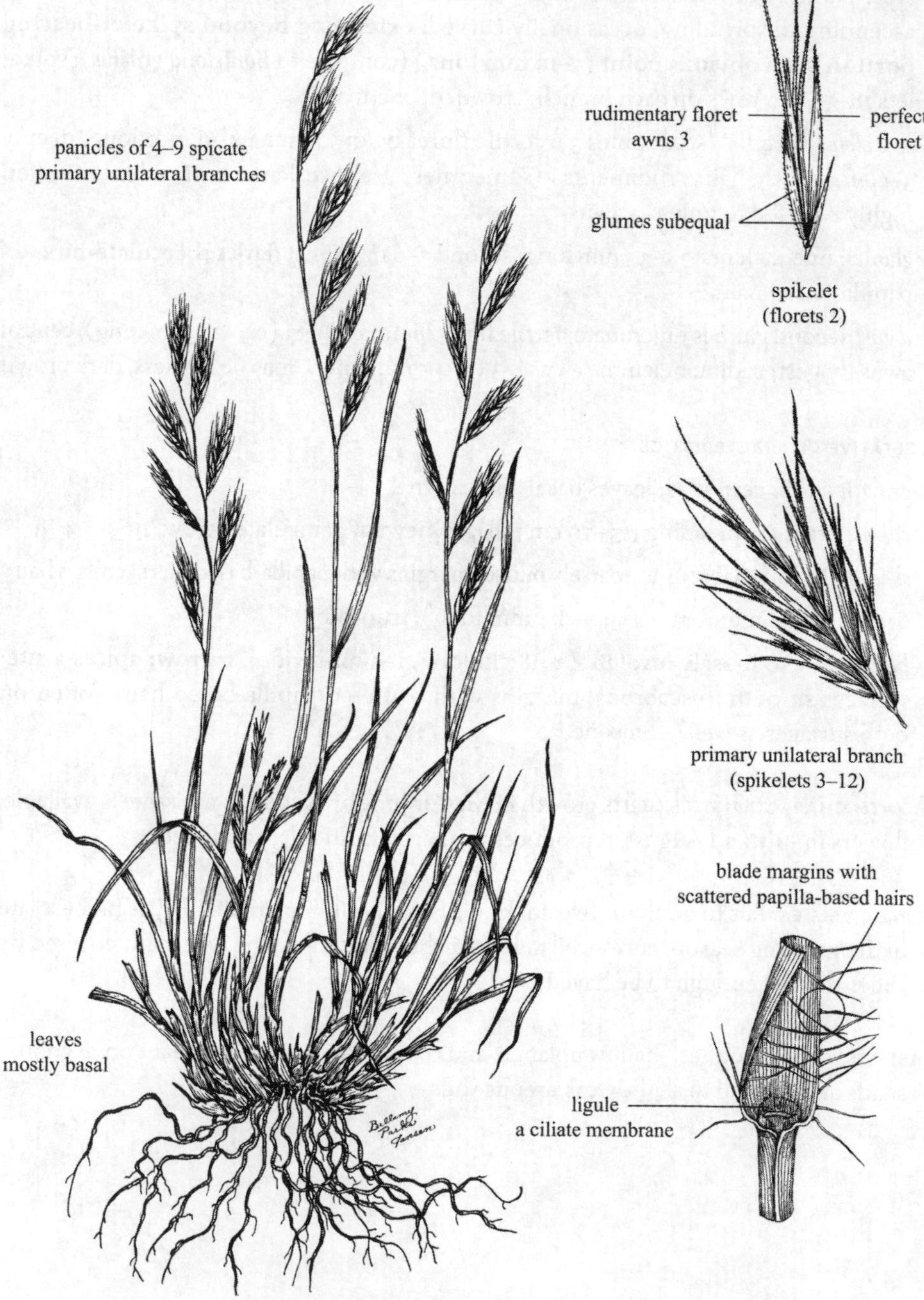

panicles of 4–9 spicate
primary unilateral branches

rudimentary floret
awns 3

perfect
floret

glumes subequal

spikelet
(florets 2)

primary unilateral branch
(spikelets 3–12)

blade margins with
scattered papilla-based hairs

leaves
mostly basal

ligule
a ciliate membrane

TRIBE:	CYNODONTEAE
SPECIES:	*Bouteloua repens* (Kunth) Scribn. & Merr.
COMMON NAME:	Slender grama (navajita esbelta, navajita rastrera)
LIFE SPAN:	Perennial
ORIGIN:	Native
SEASON:	Warm

INFLORESCENCE CHARACTERISTICS

type: panicles of 4–9 spicate primary unilateral branches; branches (1–2 cm long) flattened, **ascending to spreading** (compare to *Bouteloua curtipendula*); extending beyond the uppermost spikelets (2–6 mm); branches fall as a unit; spikelets in 2 rows, typically 3–12 per branch, not crowded, slightly pectinate

spikelets: florets 2; fertile floret below; lemma 4–7 mm long, veins 3, 3-cleft, glabrous; lemma of rudiment well developed, veins 3

glumes: subequal (4–8 mm long), broad, scabrous or strigose on the midveins, otherwise glabrous

awns: fertile floret lemma short-awned (4–6 mm long); lemma of rudiment with 3 stout awns (3–8 mm long), central awn the longest

VEGETATIVE CHARACTERISTICS

growth habit: cespitose, not dense; leaves mostly basal

culms: erect to ascending (15–50 cm tall) from a decumbent base, slender, weak, usually branching from the upper nodes

sheaths: round, strongly veined, glabrous to pubescent; collars sparsely pilose

ligules: ciliate membranes (0.5–3 mm long), truncate

blades: flat (5–18 cm long, 2–5 mm wide), becoming involute apically, thin; margins with scattered papilla-based hairs below the middle; surfaces glabrous to pubescent

GROWTH CHARACTERISTICS: starts growth in April when adequate moisture is available, flowers May through November, reproduces from seeds and tillers; withstands extended moderate grazing; may persist where other grasses have been eliminated; relatively short-lived; not highly drought tolerant

FORAGE VALUES: good for livestock and wildlife, cures well and is moderately palatable even when dry; especially valuable in autumn and winter

HABITATS: open or brushy pastures, dry slopes, roadsides, and along streams; adapted to a broad range of soil textures, most abundant on sandy or rocky soils

Hooded windmillgrass
Chloris cucullata Bisch.

panicles of 6–20 digitate or subdigitate spicate primary unilateral branches

primary unilateral branch with 25–90 spikelets in 2 rows

sterile floret inflated

fertile floret elliptical

florets 2

second glume mucronate

glumes unequal

spikelet (florets 2)

sheaths keeled, margins white

ligule a ciliate membrane

rarely with short rhizomes

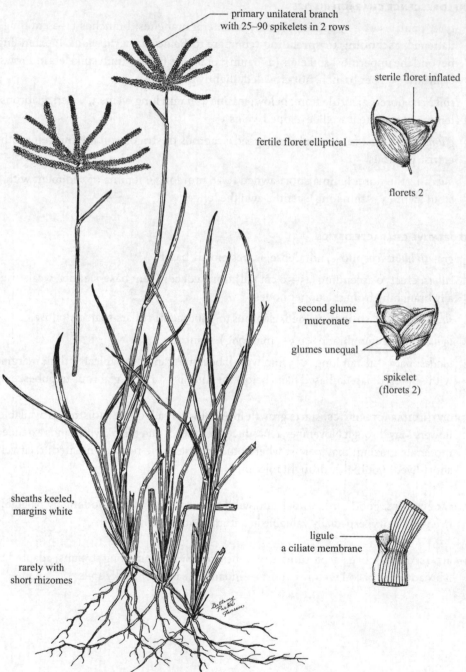

TRIBE:	CYNODONTEAE
SPECIES:	*Chloris cucullata* Bisch.
COMMON NAME:	Hooded windmillgrass (pata de gallo arenosa, verdillo papalote)
LIFE SPAN:	Perennial
ORIGIN:	Native
SEASON:	Warm

INFLORESCENCE CHARACTERISTICS

type: panicles of 6–20 digitate or subdigitate spicate primary unilateral branches, dark brown to tawny at maturity; branches (2–5 cm long) curled; spikelets 25–90 in 2 rows on each branch, crowded

spikelets: florets 2; perfect floret below, sterile floret above; fertile floret lemma elliptical (1.5–2.1 mm long); sterile lemma inflated (1–1.5 mm long), cup-shaped, veins 3, upper margins inrolled

glumes: unequal, first shorter (0.5–0.7 mm long) than the second (1–1.6 mm long), lanceolate to obovate, membranous, glabrous, midveins scabrous

awns: second glume mucronate, fertile floret lemmas awnless or awned (0.3–1.8 mm long) from below the apices, sterile lemmas may be awnless or with an awn (to 2 mm long)

VEGETATIVE CHARACTERISTICS

growth habit: cespitose, rarely with short rhizomes

culms: erect to ascending (15–70 cm tall), compressed, glabrous

sheaths: keeled, glabrous; margins white

ligules: ciliate membranes (0.8–1.1 mm long), truncate

blades: folded (2–25 cm long, 2–4 mm wide), keeled near the base, glabrous to scabrous; midveins white; apices often blunt

GROWTH CHARACTERISTICS: starts growth in early spring, stays green until autumn, may produce inflorescences several times each year; reproduces primarily from seeds

FORAGE VALUES: fair to good for livestock and wildlife, cures well; provides fair forage in winter, but animals may require protein supplementation

HABITATS: pastures, plains, roadsides, lawns, and disturbed areas; most abundant on acidic to neutral soils with a sandy to gravely texture; not adapted to calcareous or clay soils

Bermudagrass
Cynodon dactylon (L.) Pers.

SYN = *Capriola dactylon* (L.) Kuntze, *Cynodon aristiglumis* Caro & E.A. Sánchez, *Cynodon erectus* J. Presl, *Cynodon incompletus* Nees, *Cynosurus dactylon* (L.) Pers., *Panicum dactylon* L.

spikelets imbricate

primary unilateral branch
(spikelets numerous in 2 rows)

panicles of 2–7 digitate spicate
primary unilateral branches

lemma

palea

rachilla

floret

glumes subequal

spikelet
(florets 1)

ligule
a ciliate membrane

stolons flat and rhizomes creeping

96

TRIBE:	CYNODONTEAE
SPECIES:	*Cynodon dactylon* (L.) Pers.
COMMON NAME:	Bermudagrass (zacate bermuda común, zacate pata de gallo, bermuda, agrarista)
LIFE SPAN:	Perennial
ORIGIN:	Introduced (from Africa)
SEASON:	Warm

INFLORESCENCE CHARACTERISTICS

type: panicles of **2–7 digitate spicate primary unilateral branches**; branches (2–6 cm long) ascending to spreading, spikelet-bearing to the base; spikelets numerous in 2 rows on each branch, imbricate, glabrous or scabrous

spikelets: florets 1 (2–3 mm long), perfect; lemma 2–3 mm long, veins 3; laterally compressed; rachilla extending behind the palea, occasionally bearing a rudiment

glumes: subequal, first 1.3–1.8 mm long, veins 1; second 1.4–2.1 mm long, lanceolate, veins 1

awns: none

VEGETATIVE CHARACTERISTICS

growth habit: rhizomatous and stoloniferous; rhizomes creeping, extensive, forming mats; stolons flat

culms: creeping, weak; only the flowering culms erect (10–50 cm tall), flattened

sheaths: round, glabrous except for tufts of hair on either side of the collar and on either side of the ligules

ligules: ciliate membranes (0.2–0.5 mm long), truncate

blades: flat or folded (3–12 cm long, 1–4 mm wide), linear, often attached at a 45–90° angle, spreading, glabrous or occasionally pilose adaxially

GROWTH CHARACTERISTICS: may grow and flower throughout the year if temperatures and moisture permit; reproduces from seeds, rhizomes, and stolons; pastures are commonly started with vegetative sprigs rather than seeds

FORAGE VALUES: good for cattle, poor for wildlife; used for pasture in spring and early summer, quality declines in summer; infrequently cut for hay, hay quality is fair

HABITATS: open ground, fields, ditches, waste places, roadsides, along stream banks and lakes, marshy swales, planted in pastures, and is a common lawn grass; common in moist saline soils, grows in all soil textures; well adapted to clayey bottomlands that are occasionally flooded

Curly mesquite
Hilaria belangeri (Steud.) Nash

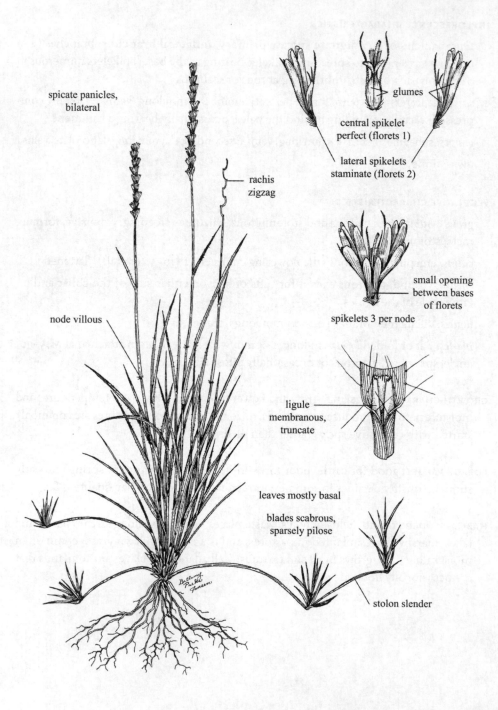

spicate panicles,
bilateral

rachis
zigzag

node villous

leaves mostly basal

blades scabrous,
sparsely pilose

glumes

central spikelet
perfect (florets 1)

lateral spikelets
staminate (florets 2)

small opening
between bases
of florets

spikelets 3 per node

ligule
membranous,
truncate

stolon slender

TRIBE:	CYNODONTEAE
SPECIES:	*Hilaria belangeri* (Steud.) Nash
COMMON NAME:	Curly mesquite (toboso menudo, espiga negra)
LIFE SPAN:	Perennial
ORIGIN:	Native
SEASON:	Warm

INFLORESCENCE CHARACTERISTICS

type: spicate panicles (2–4 cm long), bilateral, dense, exserted on narrow peduncles; **central axis flat and strongly angled at each node** (zigzag); inflorescence nodes usually 4–8

spikelets: spikelets in fascicles of 3 per node (4.5–6 mm long), central spikelet shortest with 1 perfect floret; lateral spikelets with 2 staminate florets each; small opening between bases of florets; lemmas of all florets similar, veins 3, narrowed above mid-lemma, lanceolate

glumes: lateral spikelet glumes unequal, scabrous, united below, usually shorter than lemmas; outer glumes broadened above, notched; central spikelet glumes subequal, glabrous or scabrous, lobed

awns: midveins of glumes extended into short awns; lateral glume awns shorter (< 1 mm long) than central glume awns (2.5–5 mm long)

VEGETATIVE CHARACTERISTICS

growth habit: usually stoloniferous, **stolons slender** (5–20 cm long), wiry; leaves mostly basal

culms: erect to ascending in tufts (10–30 cm tall), internodes wiry, glabrous to scabrous; nodes villous, especially lower ones

sheaths: round, shorter than the internodes, glabrous, striate

ligules: membranous (0.5–2.5 mm long), truncate, erose to occasionally lacerate

blades: flat or less commonly involute (5–20 cm long, 1–3 mm wide), ascending, scabrous; adaxial surfaces sparsely pilose; margins sparsely pilose basally

GROWTH CHARACTERISTICS: starts growth in late spring, inflorescences emerge 1 month or more later, highly drought tolerant; produces a relatively large amount of forage for the limited plant height; reproduces primarily from stolons, usually occurs in nearly pure stands, does not tolerate shade

FORAGE VALUES: fair for cattle, sheep, goats, deer, and pronghorn; cures well and furnishes forage in the autumn and winter

HABITATS: dry hillsides, plains, and grassy or brushy plains; grows in a wide range of soil textures; most abundant on medium- to fine-textured soils

Galleta
Hilaria jamesii (Torr.) Benth.

SYN = *Pleuraphis jamesii* Torr.

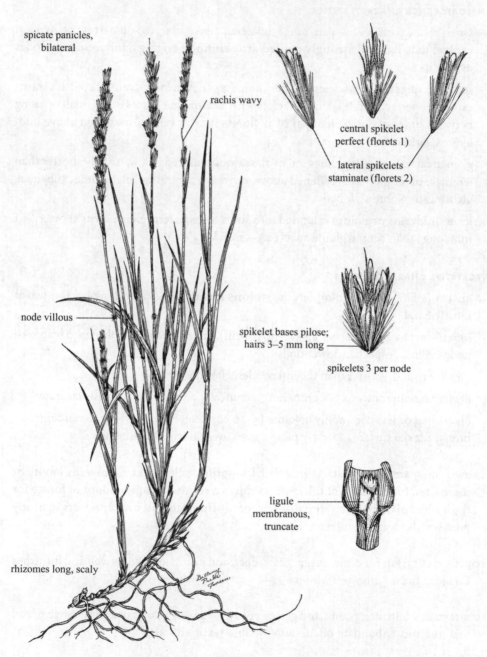

spicate panicles, bilateral

rachis wavy

central spikelet perfect (florets 1)

lateral spikelets staminate (florets 2)

node villous

spikelet bases pilose; hairs 3–5 mm long

spikelets 3 per node

ligule membranous, truncate

rhizomes long, scaly

TRIBE:	CYNODONTEAE
SPECIES:	*Hilaria jamesii* (Torr.) Benth.
COMMON NAME:	Galleta (zacate galleta)
LIFE SPAN:	Perennial
ORIGIN:	Native
SEASON:	Warm

INFLORESCENCE CHARACTERISTICS

type: spicate panicles (3–8 cm long), bilateral, dense; inflorescence nodes 10–25, more hairy than *Hilaria mutica*; rachises wavy

spikelets: spikelets in fascicles of 3 per node (6–9 mm long), **each node villous at the base** (hairs 3–5 mm long); central spikelet with 1 perfect floret set behind 2 lateral spikelets with 2 (rarely 3) staminate florets each; lemmas of perfect spikelets 5.8–7.6 mm long, bifid, veins 3; lemmas of lateral spikelets (4.4–7 mm long) ciliate at apices, inrolled; **often purplish** but bleaching with maturity

glumes: lateral spikelet glumes subequal, narrowing from the middle upward, **not fan-shaped** (compare with *Hilaria mutica*); central spikelet glumes subequal, short, cleft

awns: first glumes of lateral spikelets awned (3–6 mm long) from back; glumes and lemmas of central spikelet mucronate

VEGETATIVE CHARACTERISTICS

growth habit: rhizomatous; rhizomes long, scaly; rarely stoloniferous; leaves mostly basal

culms: erect (30–75 cm tall), coarse, wiry, glabrous; nodes villous; bases usually much-branched, decumbent

sheaths: round, longer than the internodes, prominently veined; collars glabrous to villous; collars pilose at the edges

ligules: membranous (1–5 mm long), truncate, erose to lacinate

blades: bases flat (3–20 cm long, 2–5 mm wide), upper two-thirds often involute, scabrous; curled when dry

GROWTH CHARACTERISTICS: grows mainly in summer after sufficient rain, reproduces from rhizomes and seeds, may occur in nearly pure or scattered stands, withstands heavy grazing

FORAGE VALUES: good for cattle, horses, and wildlife and fair for sheep while it is green; fair to worthless for all classes of livestock during the dormant periods; not grazed in autumn or winter unless other forage is unavailable

HABITATS: deserts, dry plains, canyons, and open valleys; most abundant on finely textured soils, but it will occur on coarsely textured soils

Tobosa
Hilaria mutica (Buckley) Benth.

SYN = *Pleuraphis mutica* Buckley

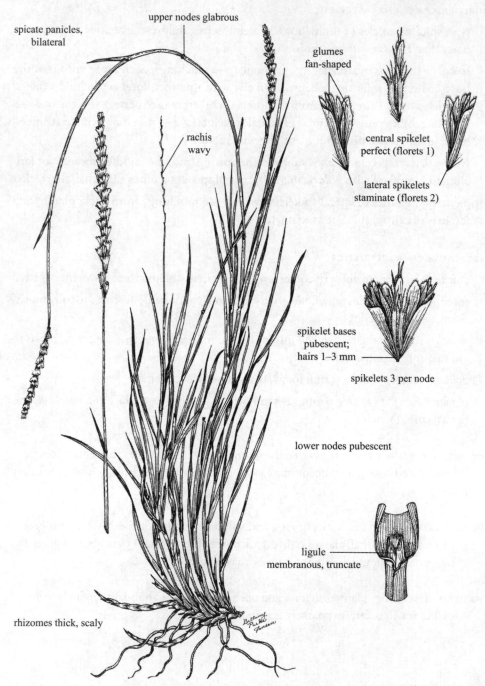

spicate panicles, bilateral

upper nodes glabrous

glumes fan-shaped

central spikelet perfect (florets 1)

lateral spikelets staminate (florets 2)

rachis wavy

spikelet bases pubescent; hairs 1–3 mm

spikelets 3 per node

lower nodes pubescent

ligule membranous, truncate

rhizomes thick, scaly

TRIBE:	CYNODONTEAE
SPECIES:	*Hilaria mutica* (Buckley) Benth.
COMMON NAME:	Tobosa (zacate toboso)
LIFE SPAN:	Perennial
ORIGIN:	Native
SEASON:	Warm

INFLORESCENCE CHARACTERISTICS

type: spicate panicles (4–6 cm long), bilateral, dense; inflorescence nodes 7–20; less hairy than *Hilaria jamesii*; rachises wavy

spikelets: spikelets in fascicles of 3 per node (6–9 mm long), short-pubescent to hirsute (hairs 1–3 mm long) at the bases; central spikelets with 1 perfect floret set behind 2 lateral spikelets with 2 staminate florets each; lemma veins 3, entire or irregularly erose and ciliate at apices, **pale to white**

glumes: lateral spikelet glumes subequal, **fan-shaped**, widest near the apices, ciliate; central spikelet glumes subequal, narrow, short, cleft

awns: inner glumes of lateral spikelets often with a rough or hairy awn (0.5–3 mm long); glumes of central spikelet mucronate

VEGETATIVE CHARACTERISTICS

growth habit: rhizomatous; rhizomes thick, stout, scaly

culms: erect to ascending (30–75 cm tall); bases decumbent, slender, wiry; lower nodes pubescent; upper nodes glabrous

sheaths: round, longer than the internodes, glabrous, veins prominent; collar pubescent on margins

ligules: membranous (1–2 mm long), truncate, erose to lacerate

blades: flat to involute (5–15 cm long, 2–5 mm wide), glabrous to scabrous, occasionally pubescent adaxially; papilla-based hairs behind the ligules

GROWTH CHARACTERISTICS: growth starts in late spring or summer after sufficient rain, grows vigorously from rhizomes, forms dense and nearly pure to scattered stands, responds quickly to extra moisture during the growing season, will increase with heavy grazing; low seed production

FORAGE VALUES: good to fair for cattle and horses, fair for sheep, poor for wildlife; becomes relatively unpalatable when mature; makes good hay if cut at about the time that the inflorescences appear

HABITATS: dry upland plains and plateaus, rocky slopes, alluvial flats, playas, desert valleys, and swales; most abundant on heavy clay soils, occasionally on sandy to gravelly soils

Tumblegrass
Schedonnardus paniculatus (Nutt.) Trel.

SYN = *Muhlenbergia paniculata* (Nutt.) Columbus

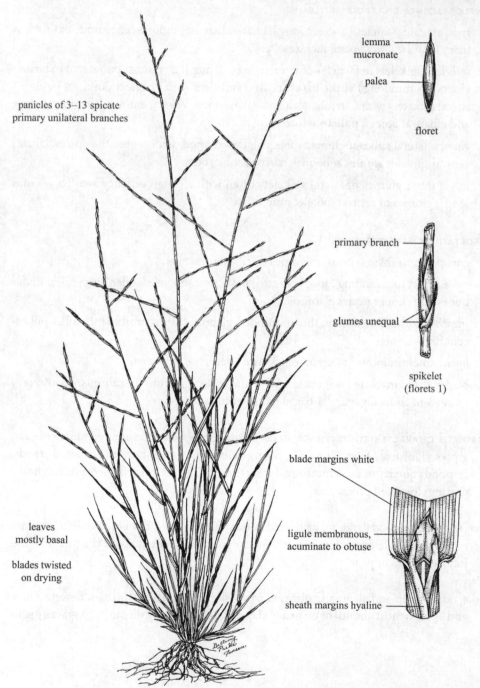

panicles of 3–13 spicate
primary unilateral branches

lemma
mucronate

palea

floret

primary branch

glumes unequal

spikelet
(florets 1)

blade margins white

ligule membranous,
acuminate to obtuse

sheath margins hyaline

leaves
mostly basal

blades twisted
on drying

TRIBE:	CYNODONTEAE
SPECIES:	*Schedonnardus paniculatus* (Nutt.) Trel.
COMMON NAME:	Tumblegrass (Texas crabgrass, wiregrass)
LIFE SPAN:	Perennial
ORIGIN:	Native
SEASON:	Warm

INFLORESCENCE CHARACTERISTICS

type: panicles (30–60 cm long) of 3–13 spicate primary unilateral branches; branches spreading (2–20 cm long), remote; central axes and branches curving at maturity; entire inflorescences break off at the base and are tumbled by the wind

spikelets: florets 1, widely spaced to only slightly imbricate, **imbedded in and appressed to branches**, slender, sessile; lemma 3–5 mm long, veins 3, glabrous to scabrous

glumes: unequal; first 1.5–2.2 mm long, veins 1, acuminate; glumes 2–4.5 mm long, veins 1, lanceolate, acuminate

awns: lemmas and second glumes mucronate, minute

VEGETATIVE CHARACTERISTICS

growth habit: cespitose; leaves mostly basal

culms: erect to ascending (8–70 cm tall), often decumbent at the bases, stiffly curving, branching from the bases

sheaths: laterally compressed, upper portion keeled, glabrous; margins connate, hyaline

ligules: membranous (1–3 mm long), acuminate to obtuse, erose

blades: flat or folded (2–15 cm long, 1–3 mm wide), stiff, **twisted on drying**, glabrous; midveins prominent abaxially; **margins scabrous, white, cartilaginous**

GROWTH CHARACTERISTICS: grows from early spring to late autumn when moisture is available; flowers throughout the growing period under favorable conditions, reproduces from seeds and tillers; regarded as an indicator of abusive grazing and disturbance

FORAGE VALUES: poor to worthless for livestock and wildlife, seldom grazed, develops little herbage

HABITATS: roadsides, disturbed areas, waste places, open prairies, and plains; adapted to a broad range of soil types; most frequent on dry clay or clay loam soils

Alkali cordgrass
Spartina gracilis Trin.

SYN = *Sporobolus hookerianus* P.M. Peterson & Saarela

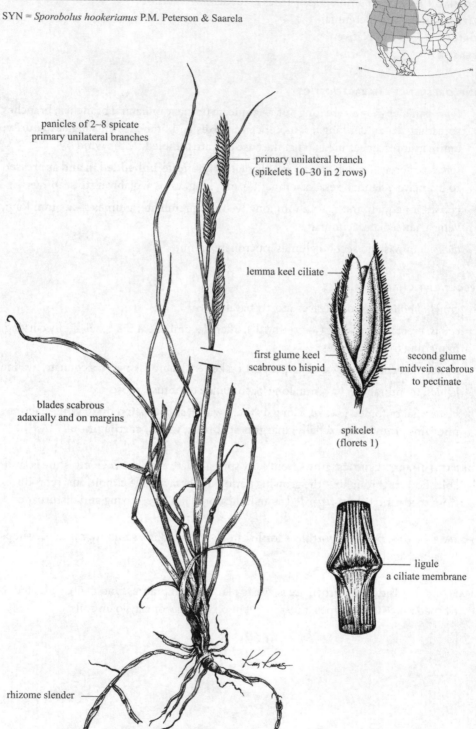

panicles of 2–8 spicate
primary unilateral branches

primary unilateral branch
(spikelets 10–30 in 2 rows)

lemma keel ciliate

first glume keel
scabrous to hispid

second glume
midvein scabrous
to pectinate

blades scabrous
adaxially and on margins

spikelet
(florets 1)

ligule
a ciliate membrane

rhizome slender

TRIBE:	CYNODONTEAE
SPECIES:	*Spartina gracilis* Trin.
COMMON NAME:	Alkali cordgrass (zacate espinilla)
LIFE SPAN:	Perennial
ORIGIN:	Native
SEASON:	Warm

INFLORESCENCE CHARACTERISTICS

type: panicles of 2–8 spicate primary unilateral branches; branches (2–8 cm long) appressed to ascending, lower branches not floriferous to the base; spikelets 10–30 per branch, **imbricate in 2 rows on abaxial side of the branch**

spikelets: 6–10 mm long, strongly compressed, sessile, florets 1; lemma lanceolate (5.5–8.5 mm long), veins 3 (lateral veins obscure), blunt; keels ciliate toward apices; palea subequal to lemma, papery

glumes: keeled, unequal; first awl-like (3.5–6.8 mm long), linear, shorter than lemmas, veins 1, scabrous to hispid on keels; second narrowly lanceolate (6–10 mm long), veins 3–5, lateral and upper midveins scabrous to pectinate

awns: **glumes awnless to mucronate** (to 1 mm long) (compare to *Spartina pectinata*)

VEGETATIVE CHARACTERISTICS

growth habit: strongly rhizomatous; rhizomes slender (3–5 mm in diameter), elongate, spreading, scaly, whitish

culms: erect (30–90 cm tall), solitary, terete to slightly compressed, glabrous

sheaths: round to slightly compressed, smooth to striate, glabrous; margins open; throats occasionally ciliate

ligules: ciliate membranes (0.5–1.8 mm long), obtuse to truncate

blades: flat (6–40 cm long, 2.5–6 mm wide), becoming involute when dry, strongly ribbed, scabrous adaxially and on margins, glabrous abaxially, attenuate

GROWTH CHARACTERISTICS: flowers June to September; reproduces from seeds, rhizomes, and tillers

FORAGE VALUES: fair to good for cattle in spring and summer, poor for wildlife; palatability rapidly declines as the plants mature and only the leaf tips are eaten, produces fair hay if cut while immature

HABITATS: moist alkaline or saline sites, shores of lakes or ponds, wet meadows, seepage areas, flats, marshes, sometimes on dry sandy soils

Prairie cordgrass
Spartina pectinata Link

SYN = *Spartina michauxiana* Hitchc., *Sporobolus michauxianus* (Hitchc.)
P.M. Peterson & Saarela

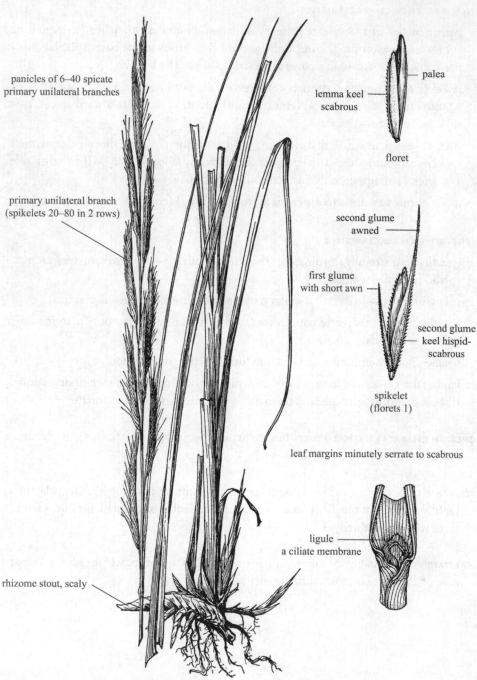

panicles of 6–40 spicate
primary unilateral branches

primary unilateral branch
(spikelets 20–80 in 2 rows)

palea

lemma keel
scabrous

floret

second glume
awned

first glume
with short awn

second glume
keel hispid-
scabrous

spikelet
(florets 1)

leaf margins minutely serrate to scabrous

ligule
a ciliate membrane

rhizome stout, scaly

TRIBE:	CYNODONTEAE
SPECIES:	*Spartina pectinata* Link
COMMON NAME:	Prairie cordgrass (espartillo, tall marshgrass, sloughgrass)
LIFE SPAN:	Perennial
ORIGIN:	Native
SEASON:	Warm

INFLORESCENCE CHARACTERISTICS

type: panicles of 6–40 spicate primary unilateral branches; branches (4–15 cm long) alternate, ascending or appressed to occasionally spreading, lower branches not floriferous to the base; spikelets 20–80 per branch, **imbricate in 2 rows** on abaxial side of the branches

spikelets: 10–25 mm long, sessile, florets 1; lemma (6–9 mm long) laterally compressed, veins 3, scabrous on keels, cleft at apices, shorter than the palea

glumes: unequal; first usually as long as floret (5–10 mm long), awl-like, veins 1; second exceeding the floret (10–25 mm long), veins 3, hispid-scabrous on keels

awns: **first glumes with short awns (0.9–4 mm long); second glumes awns 4–10 mm long** (compare to *Spartina gracilis*), stout

VEGETATIVE CHARACTERISTICS

growth habit: strongly rhizomatous; rhizomes stout (4–10 mm in diameter), elongate, widely spreading, scaly, sharply pointed, light brown to purplish, drying white

culms: erect (1–2.5 m tall), solitary or in small clusters, robust, terete

sheaths: round, may be keeled above, veins distinct, smooth to slightly striate, mostly glabrous, usually pubescent only on the throats, margins scabrous

ligules: ciliate membranes (2–4 mm long), truncate

blades: flat when green (20–120 cm long, 6–15 mm wide), becoming involute upon drying, attenuate, glabrous; margins minutely serrate to scabrous

GROWTH CHARACTERISTICS: growth starts in early spring, produces flowers in late summer, reproduces from seeds and rhizomes, may grow in nearly pure stands

FORAGE VALUES: herbage coarse and furnishes poor to fair forage for cattle and poor forage for wildlife; becomes unpalatable with maturity, produces fair hay if cut while immature

HABITATS: prairies, wet meadows, marshes, along swales, ditches, floodplains, and other moist areas; in both fresh and salt water

California oatgrass
Danthonia californica Bol.

SYN = *Danthonia americana* Scribn., *Merathrepta californica* (Bol.) Piper, *Pentameris californica* (Bol.) A. Nelson & J.F. Macbr.

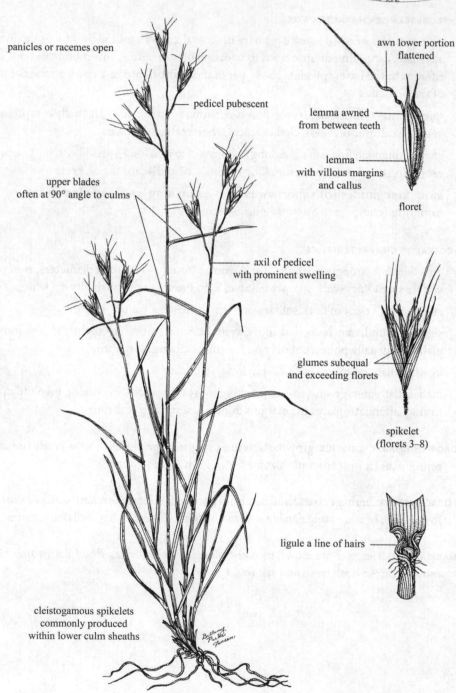

panicles or racemes open

pedicel pubescent

upper blades
often at 90° angle to culms

axil of pedicel
with prominent swelling

cleistogamous spikelets
commonly produced
within lower culm sheaths

awn lower portion
flattened

lemma awned
from between teeth

lemma
with villous margins
and callus

floret

glumes subequal
and exceeding florets

spikelet
(florets 3–8)

ligule a line of hairs

TRIBE:	DANTHONIEAE
SPECIES:	*Danthonia californica* Bol.
COMMON NAME:	California oatgrass
LIFE SPAN:	Perennial
ORIGIN:	Native
SEASON:	Cool

INFLORESCENCE CHARACTERISTICS

type: panicles or racemes (2–7 cm long), open, spikelets 3–7; pedicels flexuous (1–2 cm long), spreading to reflexed; pedicels on lowest branches longer than the spikelets, pubescent; **axils of pedicels with prominent swelling**; usually fewer spikelets than *Danthonia intermedia*

spikelets: 1–3 cm long, florets 3–8; lemmas bifid (9–13 mm long), **frequently purplish**, mostly glabrous except for villous margins and calluses; lemma teeth narrow (2–5 mm long), becoming aristate; cleistogamous spikelets commonly produced within lower culm sheaths

glumes: subequal (1.5–2 cm long), usually exceeding all florets, keeled, glabrous or scaberulous

awns: lemmas awned (8–12 mm long) from between teeth of bifid apices, lower portion flattened, geniculate; terminal segments 5–10 mm long

VEGETATIVE CHARACTERISTICS

growth habit: cespitose; bases retain old, brown sheaths

culms: erect (0.3–1 m tall), robust, stiff, glabrous; nodes abruptly constricted

sheaths: round; margins hyaline, glabrous to pilose; throats glabrous to pilose

ligules: line of hairs (0.3–1 mm long)

blades: flat to involute (5–25 cm long, 2–5 mm wide), scaberulous, margins glabrous to short-pilose; upper blades often diverging from the culm at nearly 90° angles

GROWTH CHARACTERISTICS: starts growth in the early spring, flowers May to July, seeds mature July to August; reproduces from seeds, tillers, and cleistogamous spikelets

FORAGE VALUES: good to excellent for cattle, horses, and wildlife; somewhat less palatable to sheep and goats; hay quality is good if cut before reaching maturity

HABITATS: mountain meadows, open woodlands, prairies, sagebrush hills, coastal prairies; occurs in both open and partially shaded areas; most abundant on dry sites

Timber oatgrass
Danthonia intermedia Vasey

SYN = *Danthonia canadensis* B.R. Baum & Findlay, *Merathrepta intermedia* (Vasey) Piper, *Pentameris intermedia* (Vasey) A. Nelson & J.F. Macbr.

panicles or racemes narrow

awn flattened

lemma awned from between teeth

lemma with pilose margins

callus pilose

floret

pedicels glabrous

glumes subequal and exceeding all florets

spikelet (florets 3–6)

ligule a line of hairs

cleistogamous spikelets infrequently produced within lower sheaths

TRIBE:	DANTHONIEAE
SPECIES:	*Danthonia intermedia* Vasey
COMMON NAME:	Timber oatgrass (timber danthonia)
LIFE SPAN:	Perennial
ORIGIN:	Native
SEASON:	Cool

INFLORESCENCE CHARACTERISTICS

type: panicles or racemes (2–8 cm long), narrow with numerous ascending branches; each branch bearing 1–2 spikelets; **pedicels glabrous**; pedicels on lowest branches shorter than the spikelets; usually more spikelets than *Danthonia californica*

spikelets: large (1.3–2 cm long), often purplish florets 3–6; lemmas (7–10 mm long) primarily glabrous, although pilose along margins and on calluses, bifid; teeth acuminate (1.5–2.5 mm long); paleas narrowed above, notched at apices; cleistogamous spikelets infrequently produced within lower leaf sheaths; florets less exserted from the glumes than in *Danthonia californica*

glumes: subequal (1.3–2 cm long), usually exceeding all florets, veins 3–5; veins irregular, lateral veins obscure; glabrous

awns: lemmas awned from between teeth (awn 7–10 mm long) of bifid apices, twisted and somewhat geniculate, flattened

VEGETATIVE CHARACTERISTICS

growth habit: cespitose; leaves mainly basal, a few brown sheaths persisting at the crowns

culms: erect (10–60 cm tall), glabrous

sheaths: round, glabrous (rarely pilose); throats pilose (hairs up to 2 mm long); collars usually pilose

ligules: line of hairs (0.3–1 mm long)

blades: flat to involute (5–15 cm long, 2–4 mm wide), mostly straight, ascending on upper part of culms, generally glabrous, occasionally pubescent abaxially

GROWTH CHARACTERISTICS: starts growth in early spring; seeds mature by September; reproduces from seeds, tillers, and cleistogamous spikelets

FORAGE VALUES: generally good to excellent for livestock, deer, and elk; utilized in the spring; occasionally has a low relative palatability and is not grazed or only lightly grazed; relatively tolerant to grazing

HABITATS: mountain meadows and bogs (above 1800 m), as understory in forests at lower elevations; most abundant in medium- to fine-textured soils

Parry oatgrass
Danthonia parryi Scribn.

SYN = *Merathrepta parryi* (Scribn.) A. Heller

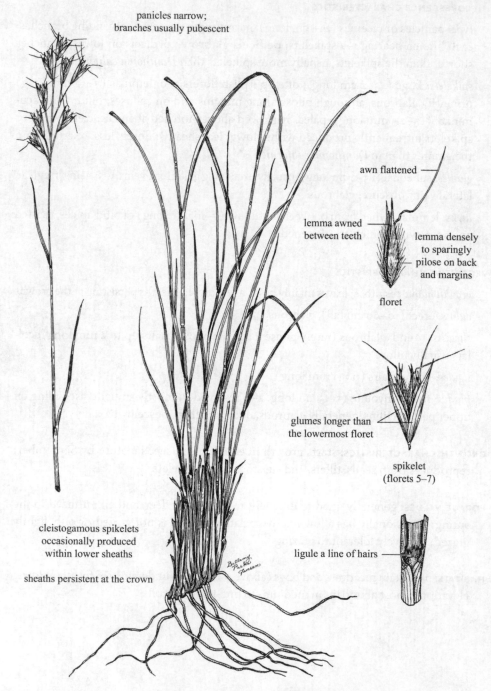

panicles narrow;
branches usually pubescent

awn flattened

lemma awned
between teeth

lemma densely
to sparingly
pilose on back
and margins

floret

glumes longer than
the lowermost floret

spikelet
(florets 5–7)

cleistogamous spikelets
occasionally produced
within lower sheaths

ligule a line of hairs

sheaths persistent at the crown

TRIBE:	DANTHONIEAE
SPECIES:	*Danthonia parryi* Scribn.
COMMON NAME:	Parry oatgrass (Parry danthonia)
LIFE SPAN:	Perennial
ORIGIN:	Native
SEASON:	Cool

INFLORESCENCE CHARACTERISTICS

type: panicles (3–10 cm long, 2–4 cm wide), narrow, spikelets 3–10 on ascending or appressed pedicels; pedicels on lowest branch equal to or shorter than the spikelets; **branches usually pubescent**

spikelets: 1.5–2.5 cm long, florets 5–7; lemmas 5–10 mm long, veins 11, **densely to sparingly pilose** (hairs to 1 cm long) on backs and margins, bifid; teeth acuminate (3–8 mm long); paleas narrowed above, nearly as long as the lemmas; cleistogamous spikelets occasionally produced within lower sheaths

glumes: equal to subequal (1.5–2.5 cm long), longer than the lowermost florets, veins 5, only the midveins distinct

awns: lemmas awned from between teeth; awns twisted (8–15 mm long), geniculate, flattened

VEGETATIVE CHARACTERISTICS

growth habit: cespitose; sheaths persistent and coarse at the crowns

culms: erect (20–80 cm tall), stout, stiff, glabrous, somewhat enlarged at the bases because of **numerous overlapping sheaths**

sheaths: round, persistent, glabrous or sparsely pubescent; margins glabrous or pubescent; throats pilose

ligules: line of hairs (0.3–0.7 mm long)

blades: flat or involute (10–25 cm long, 1–4 mm wide), erect to flexuous, not curled, glabrous, erect or not diverging more than 20° from the culms; margins scaberulous to scabrous

GROWTH CHARACTERISTICS: starts growth in spring, seeds mature in July and August; reproduces from seeds, tillers, and occasionally from cleistogamous spikelets

FORAGE VALUES: fair for cattle, horses, and wildlife; fair to poor for sheep; quality rapidly declines with maturity; seldom abundant enough to furnish large quantities of forage

HABITATS: prairies, grasslands, open woodlands, rocky hillsides, and valleys; most abundant at relatively high altitudes (2000–3000 m) in coarsely textured soils

Pine dropseed
Blepharoneuron tricholepis (Torr.) Nash

SYN = *Muhlenbergia tricholepis* (Torr.) Columbus, *Vilfa tricholepis* Torr.

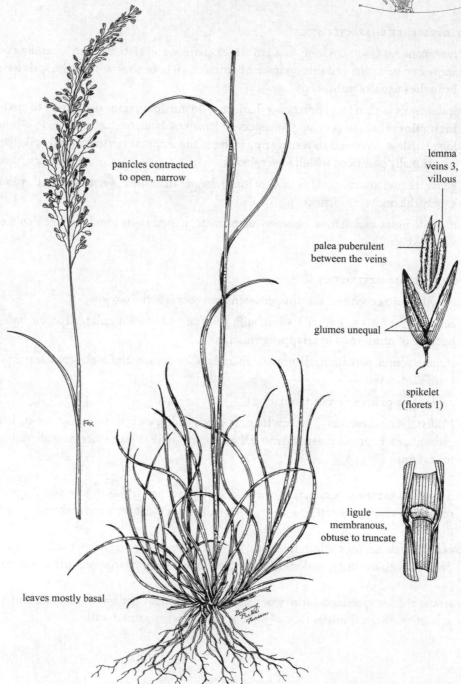

panicles contracted
to open, narrow

lemma
veins 3,
villous

palea puberulent
between the veins

glumes unequal

spikelet
(florets 1)

ligule
membranous,
obtuse to truncate

leaves mostly basal

Fox

TRIBE:	ERAGROSTEAE
SPECIES:	*Blepharoneuron tricholepis* (Torr.) Nash
COMMON NAME:	Pine dropseed (popotillo del pinar, hairy dropseed)
LIFE SPAN:	Perennial
ORIGIN:	Native
SEASON:	Warm

INFLORESCENCE CHARACTERISTICS

type: panicles (5–22 cm long, 2–7 cm wide), oblong to elliptical, contracted to open, narrow, long-exserted or sometimes enclosed in the uppermost sheaths; branches ascending, capillary, glabrous; branch bases free of spikelets

spikelets: florets 1, slightly laterally compressed, distinctive greenish-gray or bluish-gray color; lemma obtuse (2–4 mm long), **veins 3; veins villous**; palea equaling or slightly exceeding the lemma, acute, veins 2, puberulent between the veins

glumes: unequal (1.5–3.3 mm long), broad, rounded on the back, often minutely pointed at the apices, glabrous; first usually shorter than the second (1.7–3.3 mm long), veins 1, uniform margins and color

awns: none

VEGETATIVE CHARACTERISTICS

growth habit: cespitose; leaves mostly basal

culms: erect (20–70 cm tall), slender, glabrous, may be scabrous below the nodes, sometimes purplish

sheaths: round, glabrous; margins ciliate

ligules: membranous (0.5–2 mm long), obtuse to truncate, entire to erose-dentate

blades: filiform (5–20 cm long, 1–2 mm wide), involute, often strongly flexuous, glabrous or scabrous; margins scabrous

GROWTH CHARACTERISTICS: starts growth in late June or early July, completes growth in September; reproduces from seeds and tillers; generally comprises only a small portion of the vegetation

FORAGE VALUES: palatability and quality of young plants is good for all classes of livestock, quality rapidly declines with maturity, neglected or only slightly grazed after maturity; generally, the grass growing in timbered areas with the highest forage quality

HABITATS: glades, open slopes, and in dry woodlands at medium to high elevations; adapted to a broad range of soils; most abundant in rocky, moderately dry soils

Prairie sandreed
Calamovilfa longifolia (Hook.) Scribn.

SYN = *Ammophila longfolia* (Hook.) Benth. *ex* Vasey, *Calamagrostis longifolia*
    Hook.

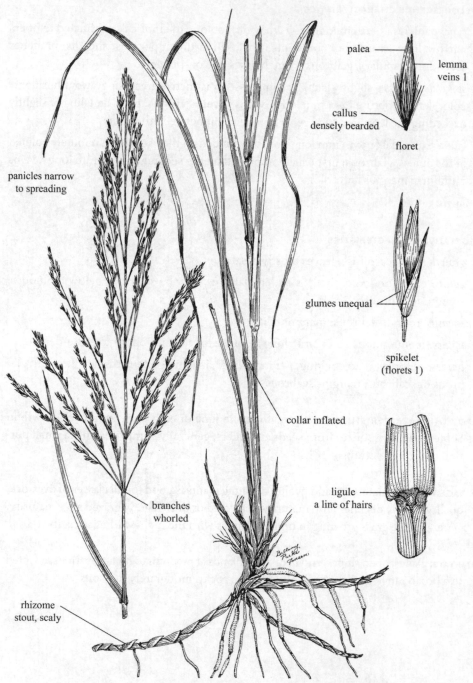

palea

lemma
veins 1

callus
densely bearded

floret

panicles narrow
to spreading

glumes unequal

spikelet
(florets 1)

collar inflated

ligule
a line of hairs

branches
whorled

rhizome
stout, scaly

TRIBE:	ERAGROSTEAE
SPECIES:	*Calamovilfa longifolia* (Hook.) Scribn.
COMMON NAME:	Prairie sandreed (sandgrass)
LIFE SPAN:	Perennial
ORIGIN:	Native
SEASON:	Warm

INFLORESCENCE CHARACTERISTICS

type: panicles (15–70 cm long, 1–6 cm wide), narrow to spreading, shiny; branches whorled, ascending to slightly spreading

spikelets: florets 1; lemma acute (4–7 mm long), veins 1, usually shorter than the second glume, glabrous; **callus densely bearded with hairs usually one-half the length of the lemma**; palea firm, slightly shorter than the lemma

glumes: unequal, first (4–7 mm long) shorter than the second (5–8 mm long), acute to acuminate, rigid, veins 1

awns: none

VEGETATIVE CHARACTERISTICS

growth habit: rhizomatous; rhizomes spreading, stout, scaly

culms: erect to ascending (0.5–2.2 m tall), solitary, robust, glabrous; **internodes short**

sheaths: round, glabrous to pubescent; throats and collars pilose (hairs 2–3 mm long); collars inflated

ligules: line of hairs (0.5–2.5 mm long)

blades: flat below (10–70 cm long, 3–12 mm wide), involute above, basal portions keeled, glabrous, margins scabrous

GROWTH CHARACTERISTICS: grows rapidly in late spring and throughout the summer, remains green until frost, reproduces from seeds and rhizomes; drought tolerant and may increase during dry years when associated grasses decline; tolerates relatively heavy grazing; may form large, dense colonies, especially when grazed only in winter

FORAGE VALUES: fair for cattle, horses, and wildlife throughout the summer; cures well and provides good standing winter forage, produces hay with acceptable quality if it is not cut too late

HABITATS: prairies, plains, dunes, and open woodlands; most abundant in sand and sandy soils, although it may be locally abundant in deep, medium-textured soils; rarely found on subirrigated or other moist sites

Saltgrass
Distichlis spicata (L.) Greene

SYN = *Distichlis stricta* (Torr.) Rydb.

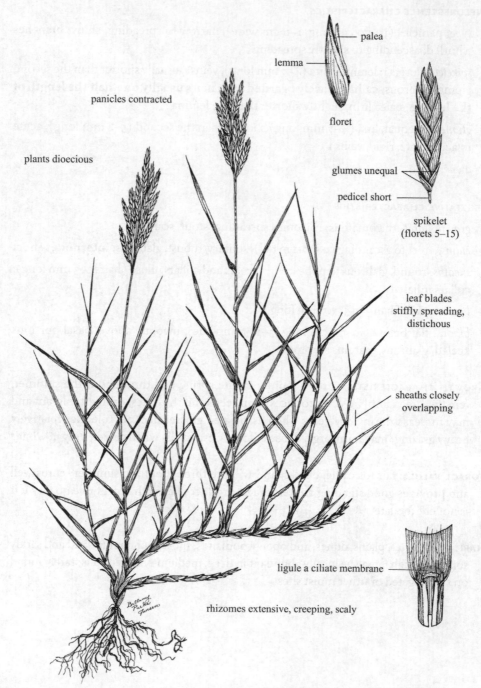

palea

lemma

panicles contracted

floret

plants dioecious

glumes unequal

pedicel short

spikelet
(florets 5–15)

leaf blades
stiffly spreading,
distichous

sheaths closely
overlapping

ligule a ciliate membrane

rhizomes extensive, creeping, scaly

TRIBE:	ERAGROSTEAE
SPECIES:	*Distichlis spicata* (L.) Greene
COMMON NAME:	Saltgrass (zacate salado, inland saltgrass)
LIFE SPAN:	Perennial
ORIGIN:	Native
SEASON:	Warm

INFLORESCENCE CHARACTERISTICS

type: panicles (1–8 cm long), contracted; rarely racemes; branches appressed; pedicels short

spikelets: unisexual (5–20 mm long), laterally compressed, florets 5–15; lemmas acute (3–6 mm long), margins yellowish and coarse; paleas soft, narrowly winged; **staminate spikelets stramineous, pistillate spikelets greenish, otherwise not conspicuously different**

glumes: unequal (3–7 mm long), acute, glabrous, first veins 3–9; second veins 5–11, lateral veins often indistinct

awns: none

other: plants dioecious

VEGETATIVE CHARACTERISTICS

growth habit: rhizomatous; rhizomes extensive, creeping, scaly, stout

culms: decumbent to erect (10–60 cm tall), internodes short and numerous, glabrous; nodes glabrous

sheaths: round, closely overlapping

ligules: ciliate membranes (0.1–0.5 mm long), truncate, often flanked with long hairs

blades: flat to involute (2–12 cm long, 1–4 mm wide), **conspicuously 2-ranked (distichous)**, sharply pointed, tightly involute at the apices, glabrous, **stiffly spreading**

GROWTH CHARACTERISTICS: starts growth in early summer, rate of growth is relatively slow, remains green until autumn, few seeds produced, reproduction mostly from rhizomes, highly resistant to trampling, may aggressively increase when competition from other plants is reduced

LIVESTOCK LOSSES: rumen compaction may develop if cattle are allowed to only graze dried saltgrass in autumn or winter; sharply pointed blades may discourage grazing

FORAGE VALUES: poor for livestock and wildlife, seldom grazed if other grasses are available, heavy grazing pressure will force use

HABITATS: moist alkaline or saline marshes, wetlands, and seashores; frequently the dominant species; can grow in soils crusted with salt, often occurs in pure stands

Weeping lovegrass
Eragrostis curvula (Schrad.) Nees

SYN = *Eragrostis chloromelas* Steud., *Eragrostis robusta* Stent

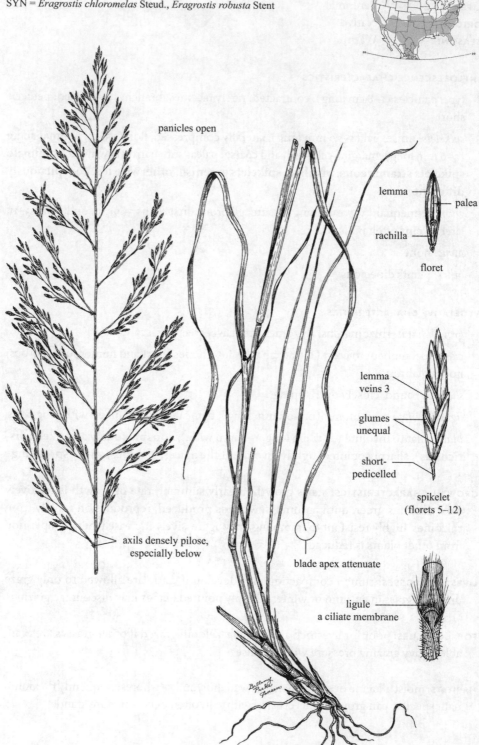

panicles open

lemma

palea

rachilla

floret

lemma
veins 3

glumes
unequal

short-
pedicelled

spikelet
(florets 5–12)

axils densely pilose,
especially below

blade apex attenuate

ligule
a ciliate membrane

TRIBE:	ERAGROSTEAE
SPECIES:	*Eragrostis curvula* (Schrad.) Nees
COMMON NAME:	Weeping lovegrass (zacate llorón, zacate del amor)
LIFE SPAN:	Perennial
ORIGIN:	Introduced (from South Africa)
SEASON:	Warm

INFLORESCENCE CHARACTERISTICS

type: panicles (15–40 cm long, 7–12 cm wide), open, oblong to ovate, eventually drooping or nodding, most often **plumbeous**; branches solitary or in pairs, naked at the bases; **densely pilose in the axils**, especially on the lower axils (compare to *Eragrostis trichodes*)

spikelets: 6–10 mm long, 1.4–2 mm wide, compressed, florets 5–12; lemmas obtuse (2.2–2.7 mm long), membranous, veins 3, lateral veins conspicuous; paleas scabridulous; pedicels shorter than the spikelets

glumes: unequal; first acute to acuminate (1.5–2 mm long); second longer (2–3 mm long), membranous; veins 1

awns: none

VEGETATIVE CHARACTERISTICS

growth habit: cespitose; leaves mainly basal

culms: erect (0.6–1.5 m tall), glabrous or glandular; bases often geniculate

sheaths: keeled, shorter than the internodes; basal sheaths hairy abaxially; upper sheaths pilose only at the throat

ligules: ciliate membranes (0.5–1 mm long), white, backed by longer hairs

blades: involute (culm leaves 15–60 cm long, 1–3 mm wide, basal leaves much longer), flexuous and arching toward the ground, glabrous to scabrous; apices attenuate

GROWTH CHARACTERISTICS: starts growth in spring, produces seeds in late spring and autumn, reproduces from seeds and tillers, withstands dry conditions, does not tolerate extended periods of temperature below -10°C, tolerates moderate to heavy grazing, planted as a forage grass

FORAGE VALUES: fair for livestock and relatively poor for wildlife; most palatable in spring, may be grazed very little from flowering through dormancy, less palatable than many other seeded species

HABITATS: pastures, sandy fields, waste areas, disturbed sites, and roadsides; sometimes seeded on burned and disturbed rangelands in the Southwest

Sand lovegrass
Eragrostis trichodes (Nutt.) Alph. Wood

SYN = *Eragrostis pilifera* Scheele, *Poa trichodes* Nutt.

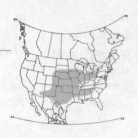

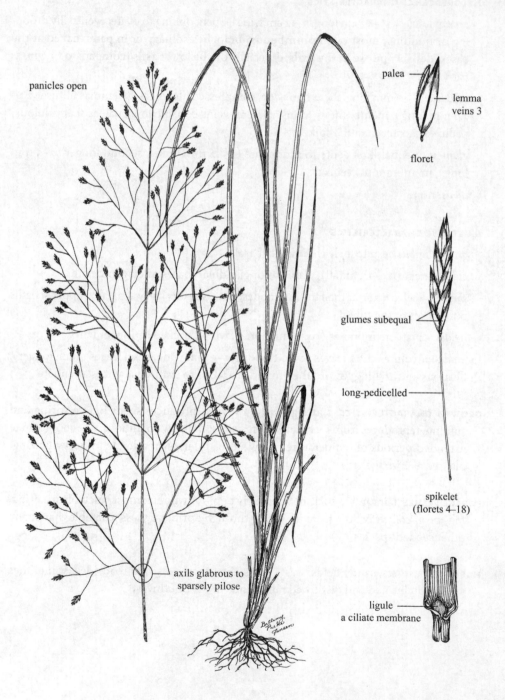

panicles open

palea

lemma
veins 3

floret

glumes subequal

long-pedicelled

spikelet
(florets 4–18)

axils glabrous to
sparsely pilose

ligule
a ciliate membrane

TRIBE:	ERAGROSTEAE
SPECIES:	*Eragrostis trichodes* (Nutt.) Alph. Wood
COMMON NAME:	Sand lovegrass
LIFE SPAN:	Perennial
ORIGIN:	Native
SEASON:	Warm

INFLORESCENCE CHARACTERISTICS

type: panicles (35–70 cm long, 7–30 cm wide), open, diffuse, oblong to ovate, **often purplish or reddish**, may be one-half the length of the plant; branches in groups of 3–4, capillary, sometimes flexuous, glabrous to **sparsely pilose in the axils** (compare to *Eragrostis curvula*)

spikelets: 4–13 mm long, 1.5–3.5 mm wide, compressed, purplish- to reddish-tinged, long-pedicelled, florets 4–18; lemmas acute (2.2–3.5 mm long), veins 3, veins conspicuous; paleas shorter than the lemmas, scabrous on the veins

glumes: subequal (first 1.6–2.6 mm long, second 1.8–4 mm long), narrowly ovate to linear lanceolate, acuminate, veins 1

awns: none

VEGETATIVE CHARACTERISTICS

growth habit: cespitose

culms: erect (0.3–1.6 m tall), glabrous

sheaths: round, imbricate, occasionally hairy on the back or margins; throats pilose

ligules: ciliate membranes (0.2–0.5 mm long), obtuse to truncate

blades: flat (15–55 cm long, 2–8 mm wide), involute apically, elongate, midveins prominent, usually scabrous adaxially but may be pilose near the ligules

GROWTH CHARACTERISTICS: starts growth as much as 2 weeks earlier than most other warm-season grasses, remains green into the autumn if moisture is available; reproduces from seeds and tillers

FORAGE VALUES: good to excellent for all classes of livestock and wildlife during summer, fair to good after maturity, cures well, furnishes good grazing in autumn and winter, sometimes cut for hay

HABITATS: prairies, open woodlands, dunes, roadsides, and disturbed sites; most abundant on deep sands and sandy loam soils, although it is sometimes found on fine-textured soils; sometimes included in seeding mixtures to furnish quick cover and for forage production

Green sprangletop
Leptochloa dubia (Kunth) Nees

SYN = *Chloris dubia* Kunth, *Diplachne dubia* (Kunth) Scribn., *Disakispermia dubium* (Kunth) P.M. Peterson & N. Snow

panicles of 2–15 alternate,
spicate primary unilateral branches

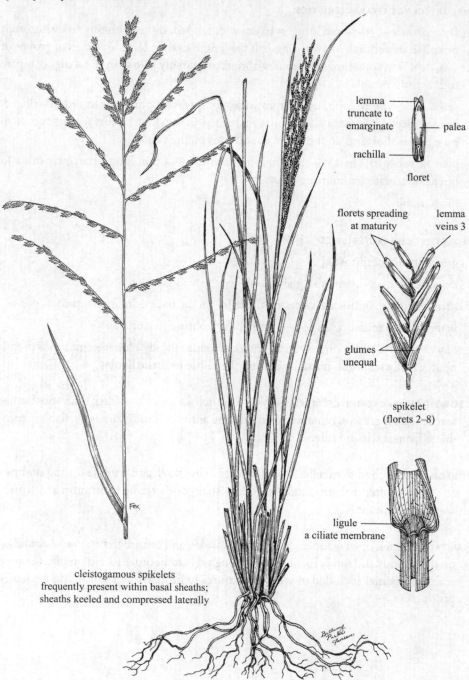

lemma
truncate to
emarginate

palea

rachilla

floret

florets spreading
at maturity

lemma
veins 3

glumes
unequal

spikelet
(florets 2–8)

ligule
a ciliate membrane

cleistogamous spikelets
frequently present within basal sheaths;
sheaths keeled and compressed laterally

TRIBE:	ERAGROSTEAE
SPECIES:	*Leptochloa dubia* (Kunth) Nees
COMMON NAME:	Green sprangletop (zacate gigante)
LIFE SPAN:	Perennial
ORIGIN:	Native
SEASON:	Warm

INFLORESCENCE CHARACTERISTICS

type: panicles (5–25 cm long) of **2–15 alternate, spicate primary unilateral branches**; branches (4–16 cm long) loosely erect or spreading, widely spaced on the central axes

spikelets: 5–12 mm long, nearly sessile, loosely to closely imbricate, spreading at maturity, florets 2–8; **lemmas truncate to emarginate** (3–6 mm long), broad, glabrous, veins 3; veins with appressed pubescence; light to dark olive green; cleistogamous spikelets frequently present within basal sheaths

glumes: unequal; second (4–6 mm long) slightly longer than the first (2.5–4.5 mm long), lanceolate, acute, translucent with greenish scabrous veins

awns: lemmas mucronate or awnless

VEGETATIVE CHARACTERISTICS

growth habit: cespitose

culms: erect to ascending (0.3–1.2 m tall), unbranched above the bases, round or compressed basally, mostly glabrous; nodes dark brown to black, wiry; bases firm

sheaths: lower sheaths keeled, compressed laterally, pilose, often purplish-tinged, throats pilose; upper sheaths glabrous

ligules: ciliate membranes (0.5–1 mm long), truncate, erose

blades: flat or folded (15–30 cm long, 4–10 mm wide), involute on drying, glabrous or slightly scabrous, sometimes drooping; midveins prominent adaxially

GROWTH CHARACTERISTICS: starts growth in April, continued growth depends on available moisture, flowers for most of the growing season; reproduces from seeds, cleistogamous spikelets, and tillers; relatively short life span, present in mixed stands and seldom is dominant

FORAGE VALUES: good for livestock and fair for wildlife when green, cures relatively well and furnishes fair forage during the dormant season, occasionally cut for hay

HABITATS: rocky hills, canyons, roadsides, and prairies; most abundant on well-drained rocky or sandy soils, seldom found on clay soils

Mountain muhly
Muhlenbergia montana (Nutt.) Hitchc.

SYN = *Muhlenbergia trifida* Hack.

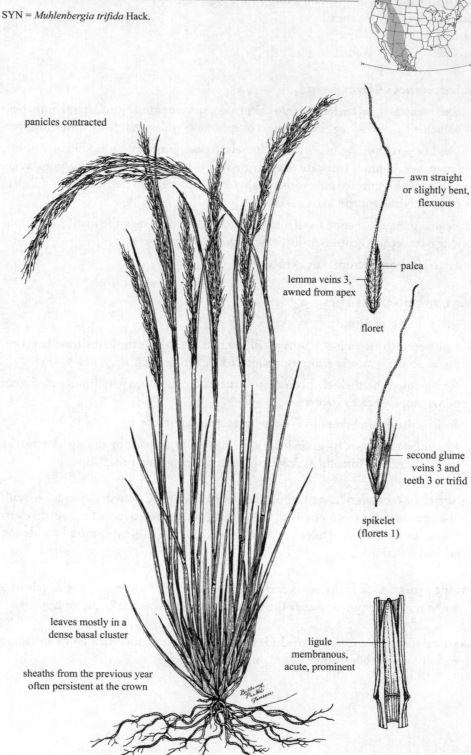

panicles contracted

awn straight or slightly bent, flexuous

palea

lemma veins 3, awned from apex

floret

second glume veins 3 and teeth 3 or trifid

spikelet (florets 1)

leaves mostly in a dense basal cluster

sheaths from the previous year often persistent at the crown

ligule membranous, acute, prominent

TRIBE:	ERAGROSTEAE
SPECIES:	*Muhlenbergia montana* (Nutt.) Hitchc.
COMMON NAME:	Mountain muhly (liendrilla de la montaña)
LIFE SPAN:	Perennial
ORIGIN:	Native
SEASON:	Warm

INFLORESCENCE CHARACTERISTICS

type: panicles (5–25 cm long, 1–3 cm wide), contracted, rather loosely flowered, interrupted below; branches ascending, primary panicle branches usually floriferous to near the base (within 1 cm)

spikelets: florets 1; lemma 3–4.5 mm long, veins 3, greenish or yellowish to grayish with dark green or purple blotches or bands, scabrous above, with fine hairs on veins; palea shorter than lemma; pedicel shorter than spikelet

glumes: subequal, slightly keeled, thin, scabrous to nearly glabrous; first acute to obtuse (1.5–3 mm long), ovate, veins 1, scabrous; second longer (2–3.3 mm long), veins 3, **veins ending as short-trifid or teeth** 3, scaberulous

awns: lemmas awned from apices, **flexuous** (4–25 mm long), straight or slightly bent, scabrous; glumes mucronate

VEGETATIVE CHARACTERISTICS

growth habit: cespitose; leaves mostly in a dense basal cluster, sheaths from the previous year often persistent at the crowns

culms: erect (15–80 cm tall), stout, glabrous

sheaths: round, lower ones often becoming flat and spreading, longer than the internodes, glabrous, may become coiled with maturity

ligules: membranous (5–20 mm long), hyaline, acute, entire, prominent

blades: flat to involute (5–30 cm long, 1–3 mm wide), arcuate, acuminate, abaxially scabrous, adaxially hirsute

GROWTH CHARACTERISTICS: starts growth in late spring, matures August to September, reproduces from seeds and tillers

FORAGE VALUES: good for cattle, horses, and elk; fair for sheep, deer, and bighorn sheep while immature; palatability rapidly declines with maturity

HABITATS: open woodlands, grasslands, hillsides, canyons, and draws up to 3000 m in elevation; most abundant in dry loam to clay soils, but will grow in sandy and gravelly soils

Bush muhly
Muhlenbergia porteri Scribn. *ex* Beal

SYN = *Podosemum porteri* (Scribn. *ex* Beal) Bush

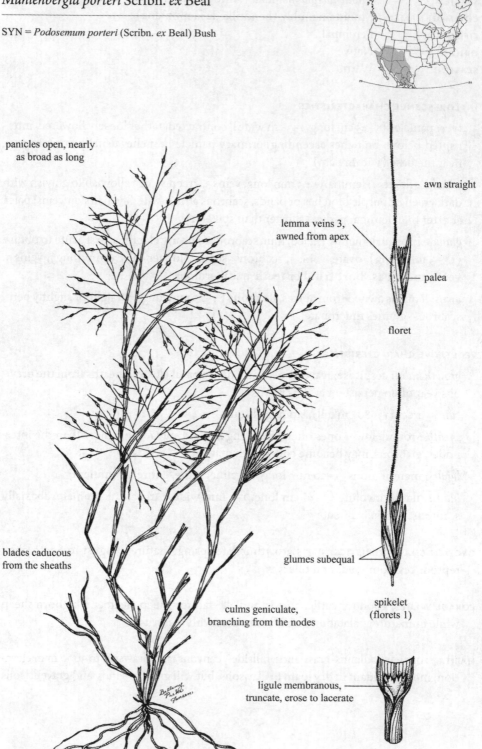

panicles open, nearly
as broad as long

awn straight

lemma veins 3,
awned from apex

palea

floret

blades caducous
from the sheaths

glumes subequal

spikelet
(florets 1)

culms geniculate,
branching from the nodes

ligule membranous,
truncate, erose to lacerate

TRIBE:	ERAGROSTEAE
SPECIES:	*Muhlenbergia porteri* Scribn. *ex* Beal
COMMON NAME:	Bush muhly (zacate araña, mesquitegrass)
LIFE SPAN:	Perennial
ORIGIN:	Native
SEASON:	Warm

INFLORESCENCE CHARACTERISTICS

type: panicles nearly as broad as long (4–10 cm long, 3–9 cm wide), open, diffuse, terminal; branches slender, purplish, stiff; **bearing few spikelets**

spikelets: florets 1, widely spaced; lemma lanceolate (3–4 mm long), acuminate, veins 3, sparsely pubescent, often purplish; palea nearly the same length as the lemma, acuminate, glabrous; long-pedicelled (5–20 mm long)

glumes: subequal (2–3 mm long), narrow, acute to acuminate, glabrous, keels scabrous; veins 1, occasionally mucronate (< 0.4 mm)

awns: lemmas awned (5–15 mm long) from the apices; awns delicate, straight

VEGETATIVE CHARACTERISTICS

growth habit: loosely cespitose, from woody and knotty bases

culms: spreading to ascending (0.3–1 m tall), wiry, **geniculate**, branching from the nodes, scaberulous or finely puberulent below and glabrous above

sheaths: round, spreading away from the internodes, mostly shorter than the internodes, glabrous

ligules: membranous (1–3 mm long), truncate, erose to lacerate, with lateral lobes

blades: flat, becoming involute (2–8 cm long, 1–2.5 mm wide), thin, acuminate, scabrous, caducous from the sheath

other: originally existed in extensive stands but now occurs in the protection of shrubs and often ascends through the shrubs

GROWTH CHARACTERISTICS: growth starts at nodes and crowns, culms do not die back each year; flowers from June to November, reproduces from seeds and tillers; sensitive to heavy grazing but responds favorably in rotational grazing systems

FORAGE VALUES: excellent for domestic livestock, deer, and pronghorn; remains green year-long if moisture is adequate, which makes it especially palatable in winter and early spring before other grasses initiate growth

HABITATS: dry mesas, hills, canyons, arroyos, and rocky deserts; most abundant on calcareous soils

Ring muhly
Muhlenbergia torreyi (Kunth) Hitchc. *ex* Bush

SYN = *Agrostis torreyi* Kunth, *Muhlenbergia gracillima* Torr.

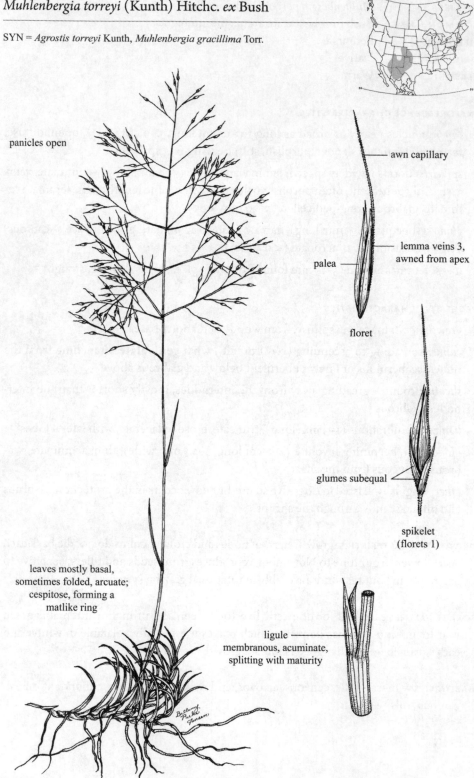

panicles open

awn capillary

lemma veins 3,
awned from apex

palea

floret

glumes subequal

spikelet
(florets 1)

leaves mostly basal,
sometimes folded, arcuate;
cespitose, forming a
matlike ring

ligule
membranous, acuminate,
splitting with maturity

TRIBE:	ERAGROSTEAE
SPECIES:	*Muhlenbergia torreyi* (Kunth) Hitchc. *ex* Bush
COMMON NAME:	Ring muhly (zacate anillo, ringgrass, ringgrass muhly)
LIFE SPAN:	Perennial
ORIGIN:	Native
SEASON:	Warm

INFLORESCENCE CHARACTERISTICS

type: panicles (7–25 cm long, 4–15 cm wide), open, diffuse, usually reddish to purplish; branchlets and pedicels appressed, may be spreading at maturity

spikelets: florets 1; lemma blackish to purplish (2–3.5 mm long), narrowly elliptic to lanceolate, veins 3, minutely bifid, scabrous above; palea equaling or exceeding lemma in length; pedicel equaling or exceeding spikelet length

glumes: subequal (1.5–3 mm long), **shorter than the lemmas** (compare to *Muhlenbergia porteri*), lanceolate, acute or acuminate to mucronate or irregularly toothed, glabrous or scaberulous, veins 1

awns: lemmas awned from the apices, capillary (1–3 mm long); glumes mucronate

VEGETATIVE CHARACTERISTICS

growth habit: cespitose, forming a **matlike ring** as the colony expands outward and dies in the center; leaves mostly basal, crowded

culms: erect (10–30 cm tall) from decumbent bases, somewhat spreading, branching below

sheaths: round, scabrous, puberulent below; margins hyaline; shorter than internodes

ligules: membranous (2–7 mm long), acuminate, entire (splitting with maturity), often with lateral lobes

blades: becoming involute (1–4 cm long, 0.3–1 mm wide), sometimes folded, arcuate, sharply pointed, forming a cushion

GROWTH CHARACTERISTICS: starts growth in late spring or early summer, flowers in midsummer, seeds mature August to September; reproduces from seeds and tillers; usually an indicator of abused rangelands or poor sites

FORAGE VALUES: fair to good for cattle when green, poor to fair for wildlife, quality quickly declines with maturity, rated poor for cattle by midsummer, low forage production

HABITATS: canyons, mesas, rocky slopes, open woodlands, and plains; most abundant on sandy to clay loam soils, but it will grow on gravelly soils

Blowoutgrass
Redfieldia flexuosa (Thurb. *ex* A. Gray) Vasey

SYN = *Graphephorum flexuosum* Thurb. *ex* A. Gray, *Muhlenbergia ammophila*
P.M. Peterson, *Muhlenbergia multiflora* Columbus

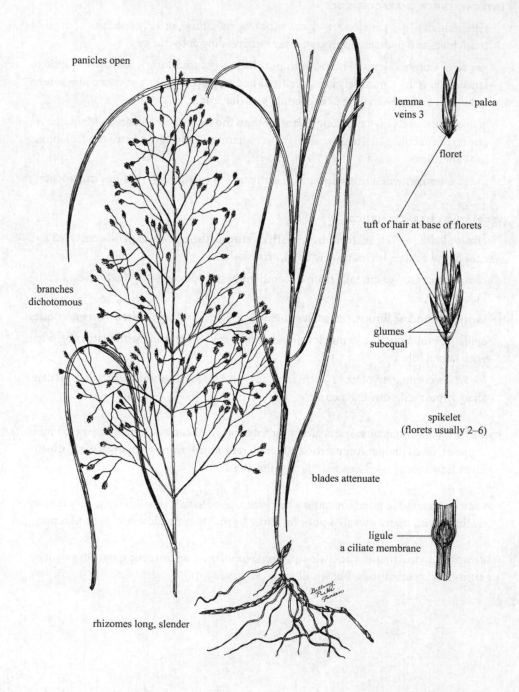

panicles open

lemma
veins 3 —— palea

floret

tuft of hair at base of florets

branches
dichotomous

glumes
subequal

spikelet
(florets usually 2–6)

blades attenuate

ligule ——
a ciliate membrane

rhizomes long, slender

TRIBE:	ERAGROSTEAE
SPECIES:	*Redfieldia flexuosa* (Thurb. *ex* A. Gray) Vasey
COMMON NAME:	Blowoutgrass
LIFE SPAN:	Perennial
ORIGIN:	Native
SEASON:	Warm

INFLORESCENCE CHARACTERISTICS

type: panicles (22–60 cm long, 2–25 cm wide), open, one-third to one-half as long as the culms, ovate to oblong; branches capillary (11–20 cm long), **dichotomous;** lower branches in whorls

spikelets: 5–8 mm long, 2.5–5 mm wide, V-shaped, widely spreading at maturity, florets (1)2–6; lemmas acute to acuminate (4–6 mm long), veins 3; **tuft of hair at base of florets**

glumes: subequal, narrow, lanceolate, acuminate, glabrous; first 3–4 mm long, veins 1; second 3.5–4.5 mm long, veins 3

awns: none

VEGETATIVE CHARACTERISTICS

growth habit: rhizomatous; rhizomes long, slender, sharply pointed, branching

culms: erect to ascending (0.5–1.3 m tall), coarse, glabrous; bases usually buried in the sand, rooting at the nodes

sheaths: nearly round, shorter than the internodes, smooth, open, glabrous or lower ones appressed-pubescent near the bases, with evenly spaced narrow furrows; lower sheaths becoming fibrous with age; collars slightly expanded

ligules: ciliate membranes (1–1.5 mm long), membranes minute; truncate to rounded

blades: involute at maturity (15–75 cm long, 2–8 mm wide), **elongate, attenuate, flexuous**, glabrous, evenly spaced narrow furrows on both surfaces; often fibrous and windtorn

GROWTH CHARACTERISTICS: starts growth in spring, flowers July to October, reproduces from seeds and rhizomes; drought tolerant, resistant to heavy grazing; occurs in large colonies, especially on medium-textured soils and when it is grazed only in winter

FORAGE VALUES: fair for cattle and horses in the summer, but is not readily grazed where other grasses are present, cures rather well and furnishes limited forage in autumn and winter

HABITATS: plains, rolling sand hills, and blowouts; most abundant in loose sandy soils that are subject to movement by wind, also occurs in finely textured soils; one of the most important species in stabilizing sand and reducing wind erosion

Burrograss
Scleropogon brevifolius Phil.

SYN = *Scleropogon longisetus* Beetle

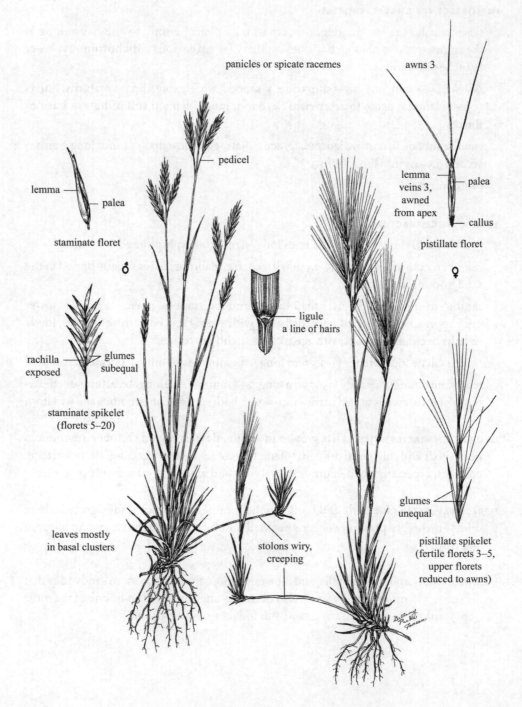

panicles or spicate racemes

awns 3

pedicel

lemma
palea

staminate floret

♂

lemma
veins 3,
awned
from apex

palea

callus

pistillate floret

♀

ligule
a line of hairs

rachilla
exposed

glumes
subequal

staminate spikelet
(florets 5–20)

leaves mostly
in basal clusters

stolons wiry,
creeping

glumes
unequal

pistillate spikelet
(fertile florets 3–5,
upper florets
reduced to awns)

TRIBE:	ERAGROSTEAE
SPECIES:	*Scleropogon brevifolius* Phil.
COMMON NAME:	Burrograss (zacate burrero)
LIFE SPAN:	Perennial
ORIGIN:	Native
SEASON:	Warm

INFLORESCENCE CHARACTERISTICS

type: staminate panicles or spicate racemes (2–4 cm long), contracted; pistillate panicles or spicate racemes longer (2.5–6 cm long, excluding awns), contracted

spikelets: unisexual; staminate spikelets 2–3 cm long, stramineous, rachillas exposed, florets 5–20; lemmas persistent (3.5–7 mm long) similar to glumes; **pistillate spikelets** 2–3 cm long excluding awns; **fertile florets 3**–5, below; veins 3, lemmas firm and rounded; calluses sharply pointed, bearded; upper florets 1 to many, reduced to awns, stramineous to purplish

glumes: staminate spikelet glumes subequal (0.5–5 mm long), thin, lanceolate, veins 1–3, first and second glumes separated by a short internode; pistillate spikelet glumes unequal (first 1–2 cm long, second 1.6–3 cm long), thin, lanceolate, subtended by glumelike bracts (1–2 cm long)

awns: pistillate spikelet awns 3 from lemma apices; awns lightly twisted (6–10 cm long), divergent; reduced florets usually only represented by awns; staminate spikelet lemmas mucronate

other: plants usually dioecious, less frequently monoecious

VEGETATIVE CHARACTERISTICS

growth habit: stoloniferous; stolons wiry, creeping (internodes 5–15 cm long), forming mats; leaves mostly in basal clusters

culms: erect (10–40 cm tall), may be decumbent at the bases

sheaths: round, short, strongly veined, upper sheaths glabrous, lower sheaths hispid or villous; collars hispid

ligules: line of hairs, minute (0.2–1 mm long)

blades: flat or folded (2–9 cm long, 1–2.5 mm wide), sharply pointed, twisted, often reflexed, coarse

GROWTH CHARACTERISTICS: starts growth in May or June, flowers mostly in late summer and autumn but occasionally in the spring; reproduces from seeds and stolons; often develops into large dense stands

LIVESTOCK LOSSES: awns may cause eye irritation of grazing animals and contaminate wool

FORAGE VALUES: poor for livestock and wildlife

HABITATS: open hills, plains, abused rangelands, flats, and disturbed sites; generally on fine-textured, calcareous soils

Alkali sacaton
Sporobolus airoides (Torr.) Torr.

SYN = *Agrostis airoides* Torr., *Sporobolus tharpii* Hitchc., *Vilfa airoides* (Torr.) Trin. *ex* Steud.

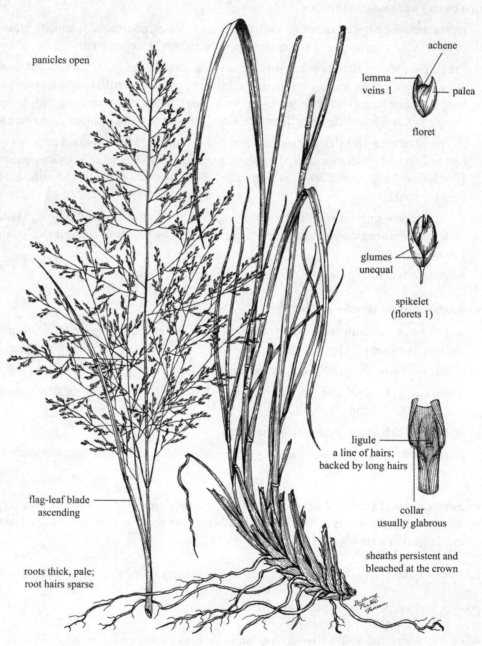

panicles open

achene

lemma
veins 1

palea

floret

glumes
unequal

spikelet
(florets 1)

ligule
a line of hairs;
backed by long hairs

collar
usually glabrous

flag-leaf blade
ascending

sheaths persistent and
bleached at the crown

roots thick, pale;
root hairs sparse

TRIBE:	ERAGROSTEAE
SPECIES:	*Sporobolus airoides* (Torr.) Torr.
COMMON NAME:	Alkali sacaton (zacatón alcalino)
LIFE SPAN:	Perennial
ORIGIN:	Native
SEASON:	Warm

INFLORESCENCE CHARACTERISTICS

type: panicles, highly variable in size (20–50 cm long, 15–25 cm wide), **subpyramidal to pyramidal**, open, diffuse, exserted or only lower portions enclosed in the sheaths, often purplish; branches without spikelets near the base; lower branches single or paired

spikelets: 1.3–2.8 mm long, mostly on spreading pedicels (0.5–2 mm long), imbricate, florets 1; lemma (1.1–2.6 mm long) about equaling the length of the second glume, acute, glabrous, veins 1, purplish or greenish; palea (1–2.4 mm long) nearly equal to the lemma

glumes: unequal; first 0.5–2 mm long; second 1–2.8 mm long, lanceolate to ovate, acute, veins usually 1

awns: none

VEGETATIVE CHARACTERISTICS

growth habit: densely cespitose; roots thick, pale; root hairs sparse

culms: erect (0.3–1.5 m tall), stout, firm, glabrous, shiny

sheaths: round, coarse, margins glabrous; collars usually glabrous, rarely sparingly hairy; **lower sheaths persistent and usually bleached, persistent at the crowns**

ligules: line of hairs (to 0.5 mm long), minute, backed and/or flanked with longer hairs (1–3 mm long)

blades: flat or becoming involute (5–60 cm long, 2–6 mm wide), firm, pointed, glabrous, rarely long-pilose near the base; prominently ridged adaxially; flag-leaf blades ascending

GROWTH CHARACTERISTICS: starts growth in midspring, flowers June until frost, reproduces from seeds and tillers; withstands flooding and considerable soil deposition, may occur in nearly pure stands

FORAGE VALUES: fair to good for cattle and horses, poor for sheep and wildlife while actively growing, poor for all animals when dry, makes fair hay when cut before flowering or earlier

HABITATS: alkaline or saline soils in meadows, flats, floodplains, and valleys; sandy soils of desert foothills or roadsides, and dry and gravelly slopes; most abundant on moderately moist alkaline soils of bottomlands where other species are not as adapted

Tall dropseed
Sporobolus compositus (Poir.) Merr.

SYN = *Agrostis asper* Michx, *Sporobolus asper* (P. Beauv.) Kunth, *Sporobolus canovirens* Nash, *Sporobolus macer* (Trin.) Hitchc., *Sporobolus pilosus* Vasey, *Vilfa composita* (Poir.) P. Beauv.

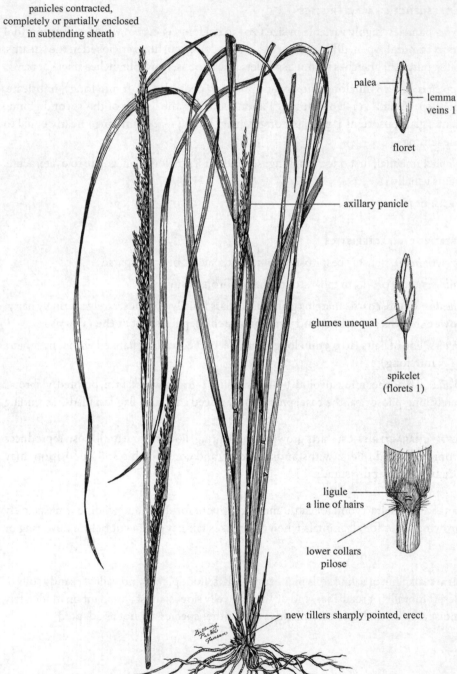

panicles contracted,
completely or partially enclosed
in subtending sheath

palea

lemma
veins 1

floret

axillary panicle

glumes unequal

spikelet
(florets 1)

ligule
a line of hairs

lower collars
pilose

new tillers sharply pointed, erect

TRIBE:	ERAGROSTEAE
SPECIES:	*Sporobolus compositus* (Poir.) Merr.
COMMON NAME:	Tall dropseed (zacatón, meadow dropseed, rough dropseed, zacate alcalino espigado)
LIFE SPAN:	Perennial
ORIGIN:	Native
SEASON:	Warm

INFLORESCENCE CHARACTERISTICS

type: panicles (5–30 cm long, 4–25 mm wide) contracted, terminal and axillary, **completely or partially enclosed in the subtending sheaths**; panicle branches appressed

spikelets: florets 1; lemma compressed, veins 1, somewhat rounded at apex (2.5–6.5 mm long), usually longer than the second glume, keeled, glabrous, stramineous to purplish; palea conspicuous, equal to the lemma; cleistogamous spikelets often within axillary panicles

glumes: unequal; first 1.5–4.5 mm long; second 2–5.5 mm long, acute to blunt, keeled, glabrous or pubescent, green to purplish with bright green midveins

awns: none

VEGETATIVE CHARACTERISTICS

growth habit: usually cespitose, some varieties with short rhizomes; new tillers are sharply pointed, erect

culms: erect (0.6–1.2 m tall), slender, solitary or in small tufts, glabrous

sheaths: round, open, glabrous or lower ones pilose near the collar and at the throat

ligules: line of hairs (0.1–0.5 mm long), minute, truncate

blades: flat or folded (10–70 cm long, 2–8 mm wide), involute on drying, tapered to filiform apices, glabrous or pilose abaxially, glabrous or scabridulous adaxially; flag-leaf blades ascending

other: a variable species with several varieties

GROWTH CHARACTERISTICS: starts growth in late spring, flowers in August, some leaves in dense bunches remain green in winter; reproduces from seeds, tillers, and occasionally rhizomes; drought tolerant, increases during dry periods and with excessive grazing

FORAGE VALUES: fair for livestock and poor for wildlife, most palatable in spring, palatability rapidly declines with maturity

HABITATS: prairies, desert grasslands, foothills, terraces, alluvial fans on dry clayey to silty soils, most abundant on soils that are intermittently wet and dry, rarely grows on deep sandy soils or soils with a high water table

Sand dropseed
Sporobolus cryptandrus (Torr.) A. Gray

SYN = *Agrostis cryptandra* Torr., *Sporobolus subinclusus* Phil., *Vilfa triniana* Steud.

panicles contracted, completely enclosed in subtending sheath;
or panicles open, partially enclosed in subtending sheath

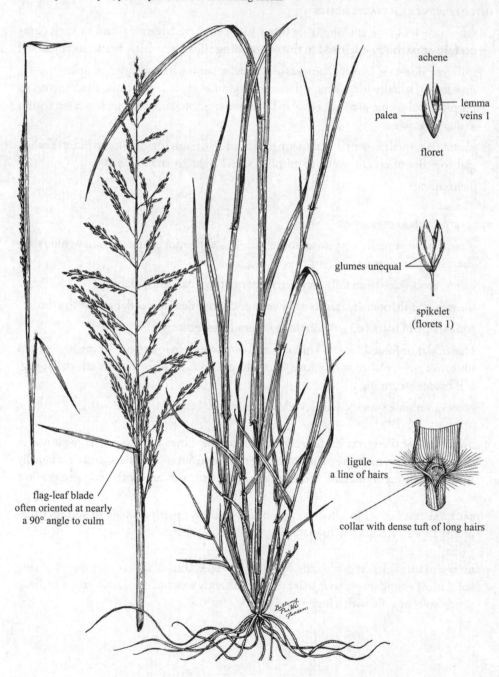

achene

lemma
veins 1

palea

floret

glumes unequal

spikelet
(florets 1)

flag-leaf blade
often oriented at nearly
a 90° angle to culm

ligule
a line of hairs

collar with dense tuft of long hairs

TRIBE:	ERAGROSTEAE
SPECIES:	*Sporobolus cryptandrus* (Torr.) A. Gray
COMMON NAME:	Sand dropseed (zacatón arenoso, zacate encubierto)
LIFE SPAN:	Perennial
ORIGIN:	Native
SEASON:	Warm

INFLORESCENCE CHARACTERISTICS

type: panicles (15–40 cm long, 2–15 cm wide), terminal, contracted to open above, **completely or partially enclosed in the subtending sheaths**; primary branches distant, occasionally pubescent near panicle central axes; pedicels short

spikelets: florets 1, densely crowded on upper portion of panicle branches, imbricate, plumbeous to purplish; lemma ovate to lanceolate (1.4–2.5 mm long), acute, veins 1; palea equaling or slightly shorter than the lemma; spikelets within enclosed portion of the inflorescences often cleistogamous

glumes: unequal; first 0.7–1.9 mm long; second 1.4–2.5 mm long, veinless or veins 1; second glumes equaling or slightly shorter than lemmas, thin, linear lanceolate to ovate, acute

awns: none

VEGETATIVE CHARACTERISTICS

growth habit: cespitose

culms: erect to ascending (0.3–1.2 m tall) to geniculate below, compressed to sulcate on one side, glabrous

sheaths: round, glabrous or scabridulous; margins often ciliate; throats and collars with **dense tufts of long hairs** (2–4 mm long)

ligules: line of hairs (0.3–1 mm long), rounded to truncate

blades: flat to involute (4–25 cm long, 2–6 mm wide), tapering to long and slender apices; adaxially glabrous, abaxially scabridulous to scabrous; margins slightly scabrous; flag-leaf blades oriented at nearly a 90° angle to the culm

GROWTH CHARACTERISTICS: starts growth in early spring, seeds mature June to August, produces an abundance of seeds, reproduces readily from seeds and tillers; increases with abusive grazing or after drought

FORAGE VALUES: fair to good for livestock and poor for wildlife while actively growing, declines rapidly with maturity

HABITATS: dunes, alluvial fans, washes, rocky slopes, roadsides, open woodlands and disturbed sites; most common on sandy soils, but occurs on rocky and silty soils, not tolerant of wet soils

Oniongrass
Melica bulbosa Geyer *ex* Porter & J.M. Coult.

SYN = *Bromelica bulbosa* (Geyer *ex* Porter & J.M. Coult.) W.A. Weber,
Melica bella Piper, *Melica inflata* (Bol.) Vasey

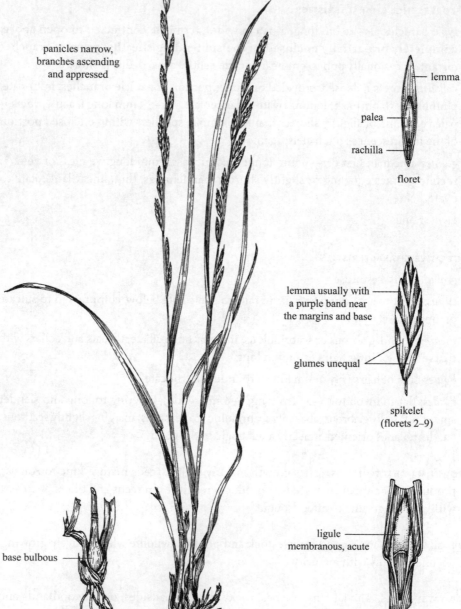

panicles narrow,
branches ascending
and appressed

lemma

palea

rachilla

floret

lemma usually with
a purple band near
the margins and base

glumes unequal

spikelet
(florets 2–9)

ligule
membranous, acute

base bulbous

TRIBE:	MELICEAE
SPECIES:	*Melica bulbosa* Geyer *ex* Porter & J.M. Coult.
COMMON NAME:	Oniongrass (bulbous oniongrass, onion melic)
LIFE SPAN:	Perennial
ORIGIN:	Native
SEASON:	Cool

INFLORESCENCE CHARACTERISTICS

type: panicles (8–30 cm long), narrow, contracted, dense, elongate; branches ascending, appressed, short

spikelets: 7–20 mm long, florets 2–9, perfect florets usually 3, upper florets sterile; lemmas papery (6–12 mm long), broadly acute to obtuse and occasionally emarginate, veins 7–11, usually with a **purple band near the margins and base**; palea veins ciliate

glumes: unequal, ovate, acute to obtuse, scaberulous; first 5–11 mm long, veins 1–5; second 6–13 mm long, veins 5–7; veins sometimes scabrous

awns: none

VEGETATIVE CHARACTERISTICS

growth habit: cespitose, often with short rhizomes

culms: erect (0.3–1 m tall), clustered, **bases bulbous** (bulblike corms); internodes scabrous above the nodes

sheaths: round, glabrous to scabrous (infrequently pubescent), margins connate nearly the full length

ligules: membranous (2–6 mm long), acute, erose to deeply lacerate

blades: flat (10–30 cm long, 2–5 mm wide), becoming involute; glabrous to scabrous, occasionally sparsely pubescent adaxially

GROWTH CHARACTERISTICS: starts growth in early spring, flowers in late spring or early summer, seeds mature in July, low seed viability; reproduces from rhizomes, seeds, and tillers from bulblike corms

FORAGE VALUES: excellent for cattle, sheep, horses, elk, bighorn sheep, and deer; many species of small mammals eat the seeds and bulblike corms; generally not abundant enough to produce high quantities of forage

HABITATS: meadows, alluvial fans, open woodlands, hills, and along streams; occurs on all exposures but is most abundant on north and east exposures; often in mesic sites; most abundant in rich sandy loams or clay loams from midalpine to subalpine elevations

Arizona cottontop
Digitaria californica (Benth.) Henrard

SYN = *Trichachne californica* (Benth.) Chase

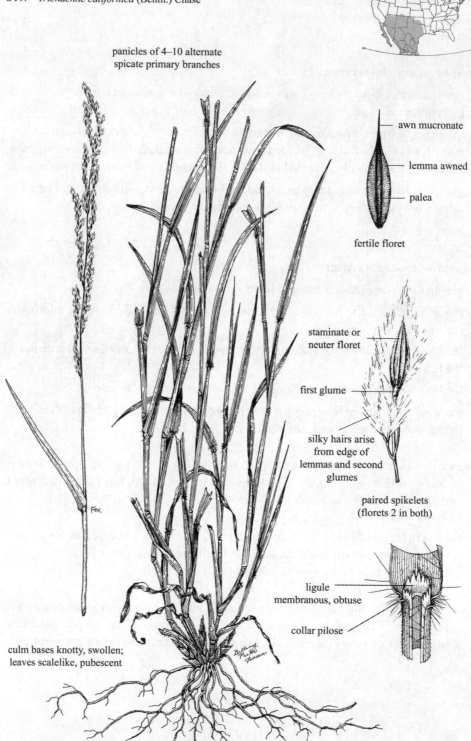

panicles of 4–10 alternate
spicate primary branches

awn mucronate

lemma awned

palea

fertile floret

staminate or
neuter floret

first glume

silky hairs arise
from edge of
lemmas and second
glumes

paired spikelets
(florets 2 in both)

ligule
membranous, obtuse

collar pilose

culm bases knotty, swollen;
leaves scalelike, pubescent

TRIBE:	PANICEAE
SPECIES:	*Digitaria californica* (Benth.) Henrard
COMMON NAME:	Arizona cottontop (zacate punta blanca, zacate punta del algodón, cottongrass, California cottontop)
LIFE SPAN:	Perennial
ORIGIN:	Native
SEASON:	Warm

INFLORESCENCE CHARACTERISTICS

type: panicles (8–20 cm long, 4–20 mm wide) of 4–10 generally alternate spicate primary branches (3–15 cm long), narrow, densely flowered, erect; branches appressed, mostly simple, not winged

spikelets: in pairs (3–4 mm long), both with florets 2; lower floret staminate or neuter, broad; upper floret perfect, brown to dark brown, covered and **exceeded by silky hairs** that arise from edge of lemmas and second glumes; hairs (2–6 mm long) white or silver to purple-tinged; lower lemmas veins 7

glumes: first minute (0.3–0.6 mm long) or absent; second narrow (3–4 mm long), veins 3, densely villous; hairs white to purplish (2–4 mm long)

awns: fertile lemmas mucronate

other: can be confused with *Bothriochloa laguroides*; compare arrangements, spikelets, and awns

VEGETATIVE CHARACTERISTICS

growth habit: cespitose

culms: erect (0.4–1 m tall), stiff; freely branching at bases, often purplish; bases knotty, swollen; scalelike leaves densely pubescent

sheaths: round, longer than the internodes, lower sheaths lanate pubescent to sparsely hairy; upper sheaths glabrous to sparsely hairy; collars pilose

ligules: membranous (2–6 mm long), obtuse, erose to lacerate

blades: flat to somewhat folded (2–18 cm long, 2–7 mm wide), involute upon drying, pubescent to glabrous, often bluish-green, occasionally with glandular hairs abaxially, midveins prominent

GROWTH CHARACTERISTICS: starts growth in late spring or early summer, responds quickly to moisture, reproduces primarily from seeds; seldom occurs in pure stands; responds well to summer rest

FORAGE VALUES: good to fair for livestock and fair for wildlife; palatable throughout the year, cures well and some culms remain green in winter furnishing valuable winter forage; frequently over-utilized because of its relatively high palatability in winter

HABITATS: plains, prairies, shrublands, dry open ground, steep slopes, chaparral, and semidesert grasslands; most abundant in well-drained sandy and gravelly loam soils

Hall panicum
Panicum hallii Vasey

SYN = *Panicum filipes* Scribn., *Panicum lepidulum* Hitchc. & Chase

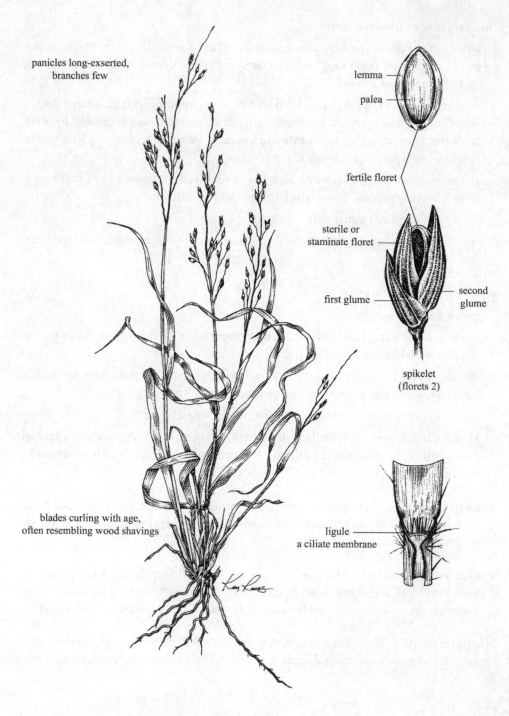

panicles long-exserted,
branches few

lemma

palea

fertile floret

sterile or
staminate floret

first glume

second
glume

spikelet
(florets 2)

blades curling with age,
often resembling wood shavings

ligule
a ciliate membrane

TRIBE:	PANICEAE
SPECIES:	*Panicum hallii* Vasey
COMMON NAME:	Hall panicum (panzio, Hall panicgrass, zacate rizado)
LIFE SPAN:	Perennial
ORIGIN:	Native
SEASON:	Warm

INFLORESCENCE CHARACTERISTICS

type: panicles (6–20 cm long, 8–12 cm wide), terminal, long-exserted, pyramidal to ovate; branches few, spreading; spikelets pedicellate; pedicels short, appressed to slightly spreading

spikelets: 2.2–3.9 mm long, 1–1.5 mm wide, usually ovoid, florets 2; lower floret sterile or staminate; upper floret fertile; glabrous, acute to acuminate, about as long as the spikelet; lemma elliptical, smooth, dark brown, shiny, strongly veined; veins 5–7

glumes: unequal; first (0.8–1.9 mm long) one-third to two-thirds length of spikelet, **acute**; second about as long as spikelet (2.2–3.9 mm long), veins 5–7

awns: none

VEGETATIVE CHARACTERISTICS

growth habit: cespitose; leaves basal and cauline

culms: erect or ascending (15–90 cm tall), simple or sparingly branched from lower nodes, glaucous; nodes glabrous to appressed-pubescent

sheaths: round, lower overlapping, upper about as long as internodes, glabrous to sparsely papillose-hispid

ligules: ciliate membranes (membrane 0.8–1.5 mm long, hairs about 1 mm long), truncate

blades: flat (10–30 cm long, 4–10 mm wide), stiffly ascending or lax, glabrous to **sparsely papillose-hirsute**, occasionally glaucous, **curling with age**, often resembling wood shavings when mature

GROWTH CHARACTERISTICS: starts growth in early spring, flowers April to November; reproduces from seeds and tillers

FORAGE VALUES: fair to good for livestock and fair for wildlife, palatability rapidly declines with maturity

HABITATS: dry prairies, rocky to gravelly hills, canyons, oak and pine savannas, and bottomlands; sandy to clayey soils, particularly calcareous soils, weakly salt tolerant; one variety occurs in moist soils, sometimes found in irrigated fields

Vinemesquite
Panicum obtusum Kunth

SYN = *Brachiaria obtusa* (Kunth) Nash, *Hopia obtusa* (Kunth) Zuloaga & Morrone

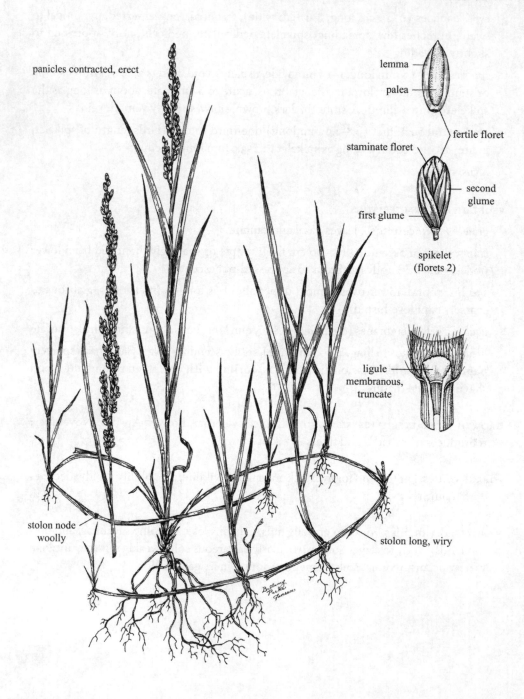

panicles contracted, erect

lemma

palea

fertile floret

staminate floret

second glume

first glume

spikelet (florets 2)

ligule membranous, truncate

stolon node woolly

stolon long, wiry

TRIBE:	PANICEAE
SPECIES:	*Panicum obtusum* Kunth
COMMON NAME:	Vinemesquite (zacate guía, grapevine mesquite)
LIFE SPAN:	Perennial
ORIGIN:	Native
SEASON:	Warm

INFLORESCENCE CHARACTERISTICS

type: panicles (3–14 cm long, 5–13 mm wide), terminal, contracted, erect; primary branches usually unbranched, distant, ascending to appressed

spikelets: 3.2–4.5 mm long, **oblong or obovate, blunt**, glabrous, indurate, florets 2; lower floret usually staminate; lemma veins 5–9; palea often exceeding the lemma; upper floret fertile (2.6–3.6 mm long), elliptic, glabrous, minutely reticulate

glumes: subequal (3–4.3 mm long), obtuse to rounded, brown at maturity; first nearly as long as spikelets, veins 3–5; second with veins 5–9

awns: none

VEGETATIVE CHARACTERISTICS

growth habit: stoloniferous or less frequently rhizomatous; **stolons long (to 3 m), wiry, nodes woolly**

culms: erect (20–80 cm tall), from hairy nodes or rhizomes or stolons, glaucous, base knotty, internodes glabrous; nodes pubescent

sheaths: round, covers one-half to three-fourths the length of the internodes, glabrous to hispid at collars; basal sheaths usually villous; upper sheaths glaucous

ligules: membranous (1–2 mm long), truncate, erose

blades: flat to involute (5–20 cm long, 2–7 mm wide), firm, ascending, elongate, glaucous; margins scabrous; pilose adaxially at collars; midveins white, prominent abaxially; stolon leaves may be strongly villous

GROWTH CHARACTERISTICS: starts growth in April or May; flowers May to October; seeds are slow to disseminate; reproduces from seeds, rhizomes, and stolons; may grow in pure stands

FORAGE VALUES: fair to good for livestock and fair for wildlife, withstands heavy grazing; produces fair hay, but is seldom abundant; doves and quail eat the seeds in autumn and winter

HABITATS: moist depressions that periodically dry out, banks of rivers and irrigation ditches, and lowland pastures; adapted to a wide range of soil textures, but most abundant on sandy to sandy loam soils

Switchgrass
Panicum virgatum L.

SYN = *Chasea virgata* (L.) Nieuwl., *Milium virgatum* (L.) Lunell

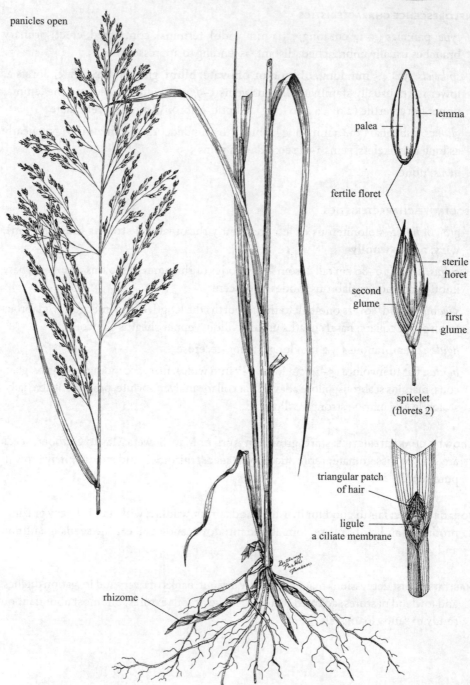

panicles open

lemma

palea

fertile floret

sterile floret

second glume

first glume

spikelet (florets 2)

triangular patch of hair

ligule a ciliate membrane

rhizome

TRIBE:	PANICEAE
SPECIES:	*Panicum virgatum* L.
COMMON NAME:	Switchgrass (panzio)
LIFE SPAN:	Perennial
ORIGIN:	Native
SEASON:	Warm

INFLORESCENCE CHARACTERISTICS

type: panicles (15–60 cm long, 10–35 cm wide), terminal, open, pyramidal to ovate; spikelets clustered toward ends of long branches; lower branches in whorls, pairs, or single; inflorescences variable

spikelets: 3–6 mm long, turgid to slightly compressed, florets 2; lower floret sterile or staminate, glabrous; upper floret fertile; lemma of fertile floret smooth, shiny, acute, inrolled at base, clasping the palea

glumes: unequal, first 2.3–5 mm long, shorter than the second, veins 5–9; second 3.3–6 mm long, **narrowly acute or acuminate**, veins 7–11; base of the second glume enclosed by the first glume

awns: none

VEGETATIVE CHARACTERISTICS

growth habit: rhizomatous, closely overlapping scales

culms: erect to ascending (0.5–3 m tall), glabrous, robust, seldom branching above base; nodes glabrous

sheaths: round, glabrous to ciliate at margins, often purplish to reddish at bases

ligules: ciliate membranes (2–4 mm long), mostly hairs, obtuse

blades: flat (10–60 cm long, 3–15 mm wide), elongate, firm; **triangular patch of hairs adaxially just above the ligules**; margins weakly scabrous

GROWTH CHARACTERISTICS: starts growth in April or May, flowers in summer, seeds mature in late summer to early autumn; reproduces from seeds, tillers, and rhizomes

FORAGE VALUES: good for all types of livestock and fair for wildlife; as the plants begin to mature in midsummer, nutrient content and palatability decline rapidly; produces high yields of good hay if cut before maturity; potentially valuable for biofuels production

HABITATS: prairies, dunes, meadows, open ground, open oak woodlands, brackish marshes, and pine woodlands; adapted to a broad range of soil textures; tolerates flooding for short periods

Knotgrass
Paspalum distichum L.

SYN = *Digitaria paspalodes* Michx., *Paspalum paspalodes* (Michx.) Scribn.,
    Paspalum vaginatum Sw.

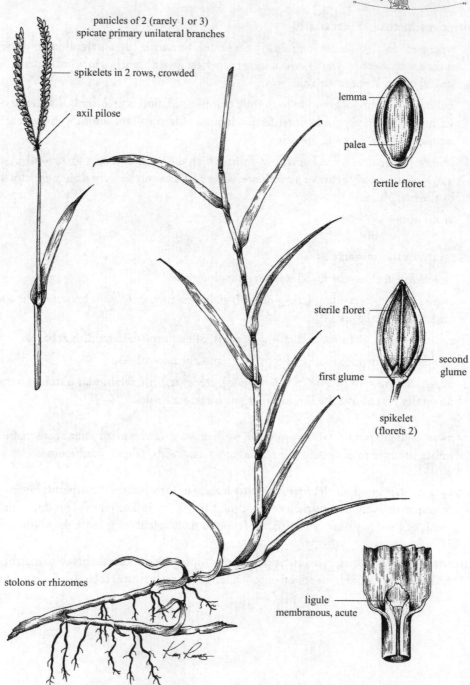

panicles of 2 (rarely 1 or 3)
spicate primary unilateral branches

spikelets in 2 rows, crowded

axil pilose

lemma

palea

fertile floret

sterile floret

first glume

second glume

spikelet
(florets 2)

stolons or rhizomes

ligule
membranous, acute

TRIBE:	PANICEAE
SPECIES:	*Paspalum distichum* L.
COMMON NAME:	Knotgrass (camalote saladillo, grama dulce, zacate nudoso, zacate de arena, jointgrass)
LIFE SPAN:	Perennial
ORIGIN:	Native
SEASON:	Warm

INFLORESCENCE CHARACTERISTICS

type: **panicles of 2 (rarely 1 or 3) spicate primary unilateral branches**; branches usually not more than 1 cm apart at the culm apices, erect to somewhat spreading (2–7 cm long), compressed (1–2 cm wide), arcuate, not winged; pilose in the axils

spikelets: solitary (2.5–3.3 mm long), dorsally compressed, **crowded in 2 rows on lower side of branches**, florets 2; lower floret sterile; lower lemma similar in size, shape, and texture to second glume, glabrous; upper floret perfect, elliptic, blunt to broadly acute; lemma ovate (2.2–3.1 mm long), greenish, indurate, glabrous; veins 3, faint

glumes: unequal; first usually present, minute (to 1 mm long), triangular, truncate to lanceolate; second broadly lanceolate (2.5–3.3 mm long), veins 3, apices obtuse to acute, pubescence appressed

awns: none

VEGETATIVE CHARACTERISTICS

growth habit: stoloniferous or rhizomatous; extensively creeping

culms: ascending to decumbent (0.3–1 m long), often rooting at the lower nodes, compressed; nodes glabrous or pubescent; internodes glabrous, longer than the sheaths

sheaths: round to somewhat keeled, cross-septate; pilose at base and on upper margins, sometimes pilose throughout

ligules: membranous (1–2.5 mm long), acute, erose to lacerate

blades: flat or folded (3–20 cm long 3–10 mm wide), glabrous, occasionally hairs on adaxial surface; margins scabrous

GROWTH CHARACTERISTICS: flowers June to October; reproduces by seed, stolons, rhizomes, and rooting at lower culm nodes

FORAGE VALUES: while actively growing, good for livestock and fair for wildlife, becomes less palatable as it matures

HABITATS: edges of woodlands, open areas, fields, and roadsides; moist to wet places along streams, canals, and ditches; important species for holding soil along streams, although has been known to restrict the water flow in irrigation ditches

Buffelgrass
Pennisetum ciliare (L.) Link

SYN = *Cenchrus ciliaris* L., *Pennisetum cenchroides* Rich. *ex* Pers.

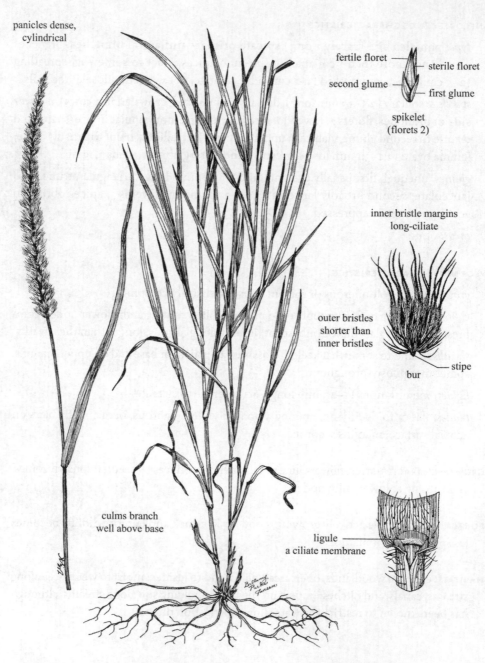

panicles dense,
cylindrical

fertile floret

sterile floret

second glume

first glume

spikelet
(florets 2)

inner bristle margins
long-ciliate

outer bristles
shorter than
inner bristles

stipe

bur

culms branch
well above base

ligule
a ciliate membrane

156

TRIBE:	PANICEAE
SPECIES:	*Pennisetum ciliare* (L.) Link
COMMON NAME:	Buffelgrass (zacate buffel)
LIFE SPAN:	Perennial
ORIGIN:	Introduced (from India and Africa)
SEASON:	Warm

INFLORESCENCE CHARACTERISTICS

type: panicles (3–18 cm long, 1–2.6 cm wide), spicate, contracted, dense, cylindrical, exserted from the sheaths, erect to nodding; branches highly reduced; inflorescences usually purplish to blackish; burs on minute stipes; stipes pilose

spikelets: in clusters (2–6 mm long), florets 2, lower floret sterile; **subtended and enclosed by numerous (15–70) bristles**, forming a bur; bristles unequal (4–20 mm long), united at base, erect or spreading, outer bristles shorter than inner bristles, long-ciliate on inner bristle margins, purplish, bleached at maturity, flexuous, terete, connate or free at base; **bristles disarticulate with the spikelets**

glumes: unequal (1–3.4 mm long); first veins 0–1; second veins 1–3

awns: none

VEGETATIVE CHARACTERISTICS

growth habit: cespitose, from a **knotty base**

culms: erect or geniculate to spreading (0.5–1.2 m tall), branched well above the bases, glabrous, sometimes scabrous beneath the panicles

sheaths: laterally compressed and keeled, open, glabrous to sparsely pilose; margins scabrous

ligules: ciliate membranes (1–3 mm long), truncate

blades: flat (8–35 cm long, 2.5–8 mm wide), scabrous or slightly pilose, tapering to acuminate apices

GROWTH CHARACTERISTICS: starts growth in late winter to early spring, flowers from spring through autumn, reproduces from seeds; will not tolerate extended subfreezing temperatures, commonly seeded on rangelands following mechanical removal of brush; tends to spread and crowd out more desirable plants

FORAGE VALUES: good for cattle and horses and poor forage for sheep, and deer; most valuable for spring and early summer grazing, forage quality declines in summer; produces hay of fair quality

HABITATS: seeded pastures, roadsides, old fields, gardens, and disturbed sites; occurs in all soil textures but is most common in sandy soils, does not tolerate extended flooding

Plains bristlegrass
Setaria leucopila (Scribn. & Merr.) K. Schum.

SYN = *Chaetochloa leucopila* Scribn. & Merr.

panicles contracted,
cylindrical

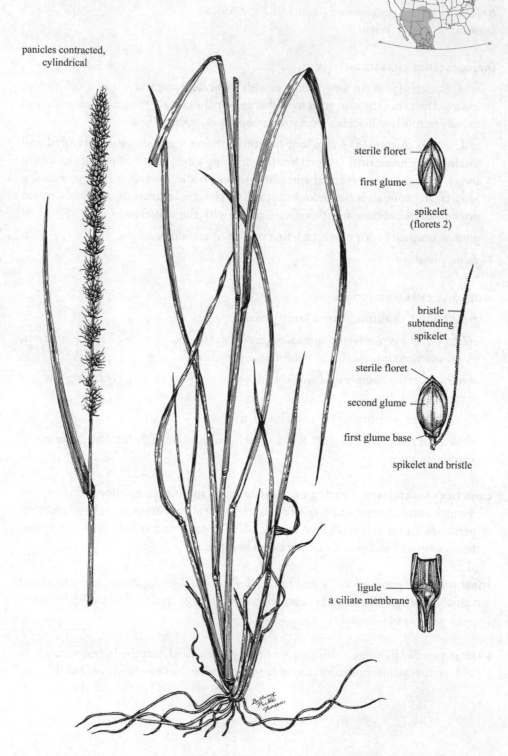

sterile floret

first glume

spikelet
(florets 2)

bristle
subtending
spikelet

sterile floret

second glume

first glume base

spikelet and bristle

ligule
a ciliate membrane

TRIBE:	PANICEAE
SPECIES:	*Setaria leucopila* (Scribn. & Merr.) K. Schum.
COMMON NAME:	Plains bristlegrass (zacate tempranero, zacate erizado, streambed bristlegrass)
LIFE SPAN:	Perennial
ORIGIN:	Native
SEASON:	Warm

INFLORESCENCE CHARACTERISTICS

type: panicles (6–20 cm long, 6–15 mm wide), contracted, cylindrical, erect, densely flowered, bristly; bristles represent reduced branches or pedicels; central axis scabrous to pilose

spikelets: 2–3 mm long, florets 2; usually **subtended by a single bristle** (4–15 mm long), antrorsely barbed, green; **bristles persistent on inflorescence**; lower floret sterile, palea of lower floret one-half to three-fourths as long as the lemma; upper floret perfect, lemma and palea rugose

glumes: unequal, first reduced (1–1.4 mm long), veins 3; second as long as spikelet (2–3 mm long), veins 5

awns: none

VEGETATIVE CHARACTERISTICS

growth habit: cespitose

culms: stiffly erect to geniculate below (0.2–1.2 m tall), infrequently branched above, scabrous, often pubescent below the nodes

sheaths: keeled, glabrous, often villous on upper margins

ligules: ciliate membranes (1.5–2.5 mm long), obtuse to rounded

blades: flat or folded (8–25 cm long, 2–6 mm wide), glabrous to scabrous or infrequently pubescent on both surfaces, pale brownish or glaucous

GROWTH CHARACTERISTICS: starts growth midspring, flowers from May to September, may produce more than one seed crop depending on available moisture, good seed producer; reproduces from seeds and tillers; cannot withstand heavy grazing; usually does not occur in dense stands

FORAGE VALUES: good for cattle and horses, fair to for wildlife and sheep, usually not abundant enough to be cut for hay

HABITATS: prairies, dry woodlands, and rocky slopes; in open shade of brush and small trees where it is protected from livestock; most abundant in well-drained alkaline soils along gullies or streams or other areas with occasional abundant moisture

Orchardgrass
Dactylis glomerata L.

SYN = *Bromus glomeratus* (L.) Scop., *Festuca glomerata* (L.) All., *Limnetis glomerata* (L.) Eaton

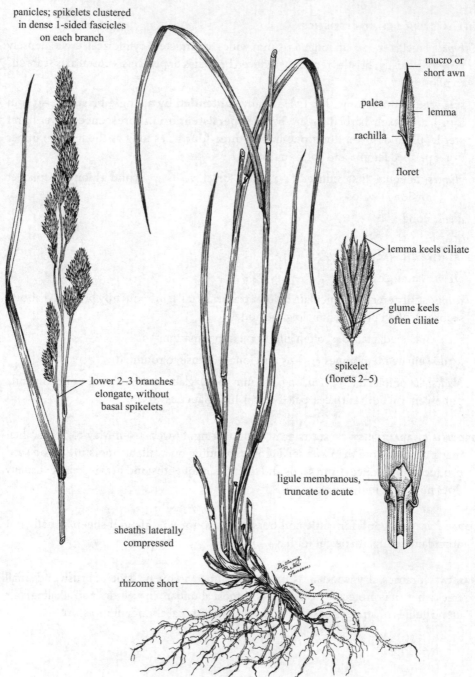

panicles; spikelets clustered
in dense 1-sided fascicles
on each branch

mucro or
short awn

palea

lemma

rachilla

floret

lemma keels ciliate

glume keels
often ciliate

spikelet
(florets 2–5)

lower 2–3 branches
elongate, without
basal spikelets

ligule membranous,
truncate to acute

sheaths laterally
compressed

rhizome short

TRIBE:	POEAE
SPECIES:	*Dactylis glomerata* L.
COMMON NAME:	Orchardgrass (alpiste, zacate dactilo, zacate ovillo, cocksfoot)
LIFE SPAN:	Perennial
ORIGIN:	Introduced (from Europe)
SEASON:	Cool

INFLORESCENCE CHARACTERISTICS

type: panicles (3–20 cm long, 3–10 cm wide); lower 2–3 branches elongate, without basal spikelets, lower branches spreading; upper branches short, floriferous to the base, appressed; spikelets clustered in dense 1-sided fascicles on each branch

spikelets: 5–9 mm long, oval to elliptic, laterally compressed, nearly sessile florets 2–5; lemmas acute (3–7 mm long), veins 5, keeled; **keels ciliate**; margins hyaline

glumes: equal to subequal (2.5–7 mm long), either may be reduced, acute, veins 1, keeled; keel ciliate or glabrous

awns: glumes and lemmas mucronate to short-awned (up to 2 mm long)

VEGETATIVE CHARACTERISTICS

growth habit: cespitose, rarely with short rhizomes

culms: erect (0.3–1 m tall), glabrous

sheaths: laterally compressed, keeled, glabrous to slightly scabrous, most shorter than the internodes, open at least one-half their length

ligules: membranous (5–9 mm long), truncate to acute, erose to lacerate

blades: flat or folded (10–40 cm long, 2–11 mm wide), elongate, lax; midveins prominent, scabrous; margins scabrous

GROWTH CHARACTERISTICS: starts growth in early spring, reproduces from seeds, rhizomes, and tillers; may be injured in areas with dry, cold winters and no snow cover or if subjected to warm temperatures in January or February followed by a period of extremely cold temperatures; does not tolerate extended periods of drought; shade tolerant; responds to nitrogen fertilizer and irrigation

FORAGE VALUES: good to excellent for livestock and wildlife, especially relished by deer, provides early spring forage, cures well as hay, sometimes mixed with alfalfa or other legumes to improve hay quality; quality of standing forage rapidly declines with maturity

HABITATS: hay fields, meadows, pastures, lawns, along ditch banks, and waste places; on fine or coarse soils; only slightly salt tolerant; commonly seeded in mixtures on pastures and haylands

Rough fescue
Festuca campestris Rydb.

SYN = *Festuca scabrella* Torr. in part

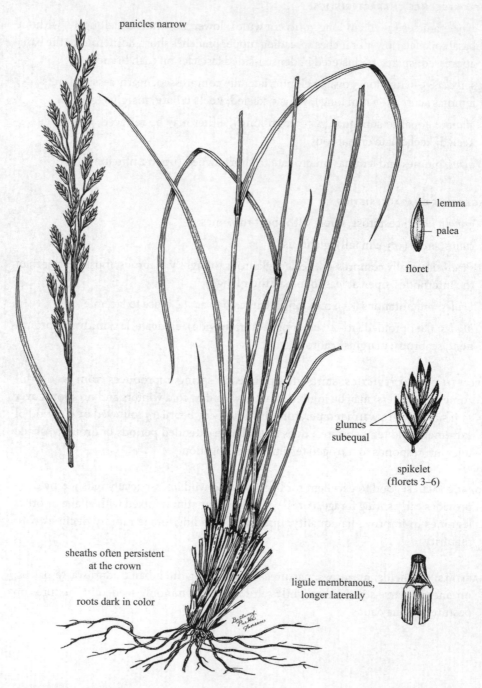

panicles narrow

lemma
palea

floret

glumes
subequal

spikelet
(florets 3–6)

sheaths often persistent
at the crown

roots dark in color

ligule membranous,
longer laterally

TRIBE:	*POEAE*
SPECIES:	*Festuca campestris* Rydb.
COMMON NAME:	Rough fescue (buffalo bunchgrass)
LIFE SPAN:	Perennial
ORIGIN:	Native
SEASON:	Cool

INFLORESCENCE CHARACTERISTICS

type: panicles (5–20 cm long, 1–2.5 cm wide), narrow, loosely contracted; branches solitary or in pairs above, occasionally in groups of 3 in the lower portion, appressed or stiffly spreading

spikelets: 9–15 mm long, often purplish, florets 3–6; lemmas stout (7–10 mm long); veins 5, faint; lemma backs rounded below the middle, acute, scabrous

glumes: subequal, acute; first lanceolate (4–7.5 mm long), veins 1; second 5.5–8 mm long, veins 3

awns: **lemmas awnless** (compare to *Festuca idahoensis*), mucronate, or short-awned (< 2 mm long)

VEGETATIVE CHARACTERISTICS

growth habit: cespitose in large tufts (to 60 cm in diameter), rarely with short rhizomes; leaves mostly basal; **roots dark in color**

culms: erect (0.3–1 m tall), glabrous, scabrous and naked below panicles, stout, purplish at the bases

sheaths: compressed, enlarged at bases, glabrous to scabrous; margins broad, hyaline, open more than one-half their length; often persistent at the crown; collars glabrous

ligules: membranous (< 1 mm long), truncate, longer on lateral edges

blades: folded (30–70 cm long, 2.5–4.5 mm wide), becoming involute, erect, stiff, pointed; abaxial surfaces usually scabrous; adaxially scabrous or puberulent

GROWTH CHARACTERISTICS: growth period is June to August, reproduces from seeds, tillers, and occasionally from short rhizomes; does not withstand continuous heavy grazing

FORAGE VALUES: during all growth stages, excellent for cattle and horses and good for sheep and wildlife; valuable for winter grazing because it retains much of its protein and palatability, cures well and makes good hay

HABITATS: prairies, open woodlands, foothills, mountains to near timberline, benches, and valleys; most abundant on dry, deep sandy loam soils

Idaho fescue
Festuca idahoensis Elmer

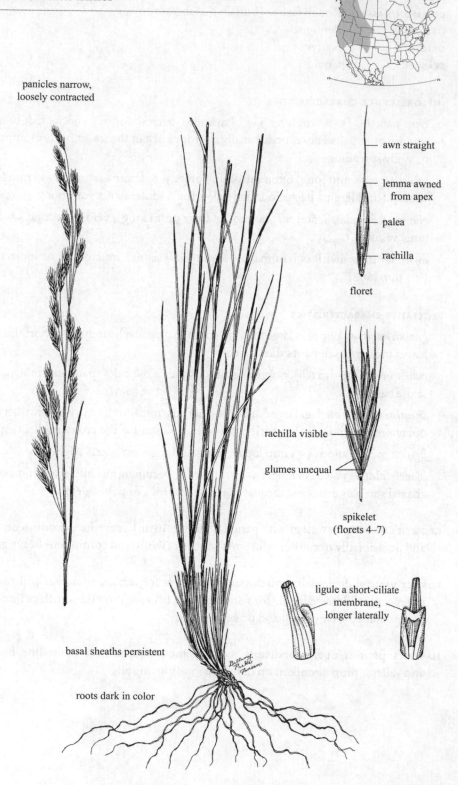

panicles narrow,
loosely contracted

awn straight

lemma awned
from apex

palea

rachilla

floret

rachilla visible

glumes unequal

spikelet
(florets 4–7)

ligule a short-ciliate
membrane,
longer laterally

basal sheaths persistent

roots dark in color

TRIBE:	POEAE
SPECIES:	*Festuca idahoensis* Elmer
COMMON NAME:	Idaho fescue (blue bunchgrass)
LIFE SPAN:	Perennial
ORIGIN:	Native
SEASON:	Cool

INFLORESCENCE CHARACTERISTICS

type: panicles (7–15 cm long, 1.5–2.5 cm wide), narrow, loosely contracted; branches ascending to narrowly spreading, lower branches spreading

spikelets: 8–14 mm long, florets 4–7; lemmas (5–9 mm long) somewhat laterally compressed at maturity, scabrous to glabrous; rachilla usually visible

glumes: unequal, ovate-lanceolate to lanceolate, acute; first 2.5–5 mm long, veins 1; second 4–6 mm long, veins 3 (faint)

awns: **lemmas awned from apices** (compare to *Festuca campestris*); awns straight (2–6 mm long)

VEGETATIVE CHARACTERISTICS

growth habit: cespitose; leaves mostly basal; **roots dark in color**

culms: erect to ascending (0.3–1 m tall), bases slightly geniculate, glabrous to scaberulous, glaucous

sheaths: compressed, keeled, glabrous or scabrous, glaucous or green; basal sheaths short, open more than one-half their length, wider than the blades, persistent; collars indistinct; auricles small or absent

ligules: membranous (< 2 mm long), slightly longer on lateral edges, usually short-ciliate

blades: involute (5–25 cm long, about 1.5 mm wide), **filiform**, firm, elongate, scabrous, often glaucous, adaxially pubescent to scabrous, abaxially glabrous to scabrous

GROWTH CHARACTERISTICS: starts growth in early spring, seeds mature by midsummer, reproduces from seeds and tillers; withstands some excessive grazing

FORAGE VALUES: excellent for livestock and wildlife, especially important late in the growing season because it remains green longer than the associated species; not readily grazed upon drying unless more desirable forage is not available

HABITATS: prairies, foothills, open woodlands, moist parks, and rocky slopes; grows on all exposures and on many soil types, most abundant on well-drained loams with a neutral to slightly alkaline pH, occurs at elevations from 300 m to nearly 4000 m

Muttongrass
Poa fendleriana (Steud.) Vasey

SYN = *Eragrostis fendleriana* Steud., *Poa longiligula* Scribn. & T.A. Williams

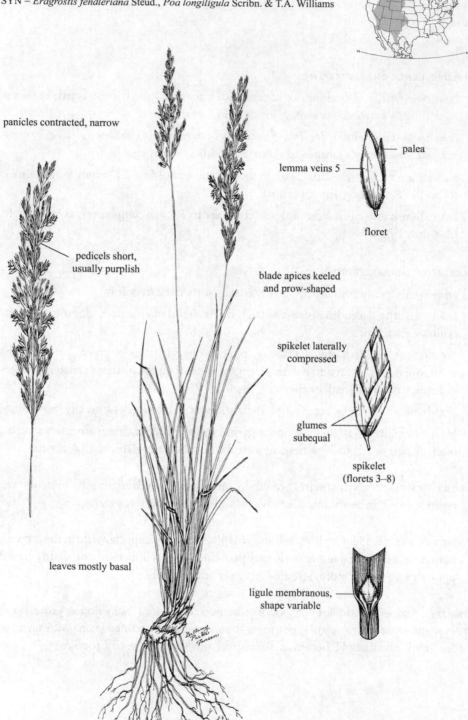

panicles contracted, narrow

palea

lemma veins 5

floret

pedicels short,
usually purplish

blade apices keeled
and prow-shaped

spikelet laterally
compressed

glumes
subequal

spikelet
(florets 3–8)

leaves mostly basal

ligule membranous,
shape variable

TRIBE:	POEAE
SPECIES:	*Poa fendleriana* (Steud.) Vasey
COMMON NAME:	Muttongrass (zacate azúl borreguero, mutton bluegrass)
LIFE SPAN:	Perennial
ORIGIN:	Native
SEASON:	Cool

INFLORESCENCE CHARACTERISTICS

type: panicles (2–12 cm long, 1–2.5 cm wide), **contracted**, narrow, erect, densely flowered, oblong, exserted above the basal leaves, tan to purplish; branches 2–3 per node, short, erect or erect-spreading; pedicels short, **usually purplish**

spikelets: 4–10 mm long, 1.8–3 mm wide, mostly unisexual; usually twice as long as wide, lanceolate, **laterally compressed**, papery, florets 3–8; lemmas compressed-keeled (3–6 mm long), blunt, veins 5, marginal veins pubescent below; bases usually not webbed (compare to *Poa pratensis*)

glumes: subequal, usually one-half to two-thirds as long as lowest lemmas; first 2.8–4.5 mm long, veins 1; second 3–5.5 mm long, papery, strongly keeled, veins 3

awns: none

other: incompletely dioecious, most plants pistillate, few staminate; occasional plants with functional, perfect flowers

VEGETATIVE CHARACTERISTICS

growth habit: cespitose, rarely with rhizomes; rhizomes short, slender, inconspicuous; leaves mostly basal

culms: erect to ascending (20–70 cm tall), bases decumbent, glabrous, scabrous below inflorescences, **weakly compressed at the base**

sheaths: round, short, glabrous to scabrous; sheath bases bleached, expanded, and persistent for several years; margins hyaline, open two-thirds of their length

ligules: membranous, highly variable (usually 0.5 mm long, occasionally up to 3–7 mm long); obtuse, truncate, or acuminate; erose

blades: folded or involute (10–28 cm long, 1–4 mm wide), stiff, erect; glabrous to scabrous, glaucous, often remaining green after drying; double midveins; apices keeled, prow-shaped; flag leaf blades absent or vestigial

GROWTH CHARACTERISTICS: starts growth in early spring and matures in June or July, reproduces from seeds and tillers and rarely from rhizomes

FORAGE VALUES: excellent for cattle and horses, good for sheep, bighorn sheep, elk, pronghorn, and deer; value declines rapidly with maturity

HABITATS: prairies, mesas, foothills, mountains, open woodlands, cold deserts, and rocky hills; grows on a broad range of soils

Kentucky bluegrass
Poa pratensis L.

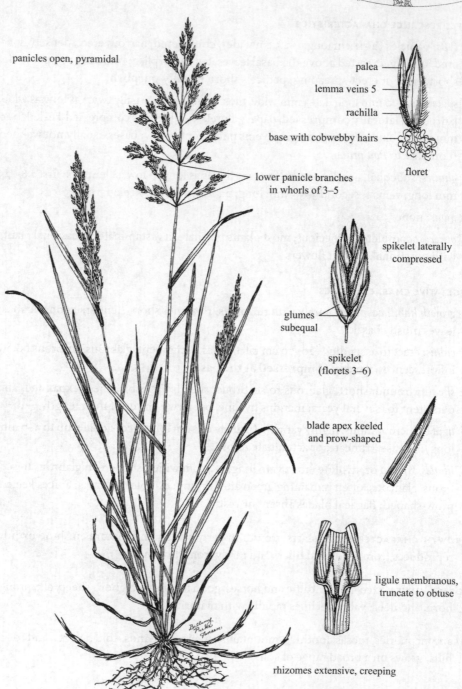

panicles open, pyramidal

palea

lemma veins 5

rachilla

base with cobwebby hairs

floret

lower panicle branches
in whorls of 3–5

spikelet laterally
compressed

glumes
subequal

spikelet
(florets 3–6)

blade apex keeled
and prow-shaped

ligule membranous,
truncate to obtuse

rhizomes extensive, creeping

TRIBE:	POEAE
SPECIES:	*Poa pratensis* L.
COMMON NAME:	Kentucky bluegrass (zacate azúl de Kentucky, zacate azúl de las paderas)
LIFE SPAN:	Perennial
ORIGIN:	Introduced (from Europe)
SEASON:	Cool

INFLORESCENCE CHARACTERISTICS

type: panicles (3–13 cm long, 3–8 cm wide), open, **pyramidal; branches long, flexuous, lower branches in whorls of 3–5** (commonly 5); spikelets along upper one-half of branches

spikelets: 3–6 mm long, laterally compressed, often nearly as wide as long, ovate; florets 3–6; lemmas keeled (2.5–4 mm long), acute or obtuse, veins 5; veins pubescent with tufts of long, cobwebby hairs at the bases; paleas slightly shorter than the lemmas

glumes: subequal, acute, keeled, scabrous on keels; first 1.5–3.5 mm long, veins 1–3; second 2–4 mm long, veins 3; margins usually scarious

awns: none

VEGETATIVE CHARACTERISTICS

growth habit: rhizomatous; rhizomes extensive, creeping, forming a dense sod

culms: erect to ascending (0.2–1 m tall), occasionally decumbent at bases, slender, wiry, somewhat flattened, not branching at the bases; bases not bulblike

sheaths: round, glabrous to scabrous, veins distinct; margins connate from one-fourth to one-half of their length

ligules: membranous (1–2 mm long), truncate to obtuse, minutely erose to entire

blades: flat or folded (5–40 cm long, 1–5 mm wide), elongated, glabrous to slightly pubescent, **double midveins**; apices keeled, prow-shaped

GROWTH CHARACTERISTICS: growth from rhizomes initiated in spring or autumn; initiates aerial culms in early spring and summer, becomes dormant during summer if moisture is limiting; reproduces from seeds, tillers, and rhizomes; not tolerant of drought; able to withstand continuous heavy grazing

FORAGE VALUES: good for livestock and wildlife in early spring when other plants are not growing; undesirable in hay meadows and tallgrass prairies because of its low growth form, poor yield, and early maturity

HABITATS: meadows, pastures, open woodlands, open ground, roadsides, lawns, and disturbed sites; commonly planted for lawns and some pastures, adapted to a broad range of soil textures; most common on sites with abundant soil moisture

Sandberg bluegrass
Poa secunda J. Presl

SYN = *Festuca patagonica* Phil., *Poa ampla* Merr., *Poa canbyi* (Scribn.) Howell,
Poa fallens Pilg., *Poa gracillima* Vasey, *Poa incurva* Scribn. & T.A. Williams,
Poa juncifolia Scribn., *Poa nevadensis* Vasey *ex* Scribn., *Poa orcuttiana*
Vasey, *Poa sandbergii* Vasey, *Poa scabrella* (Thurb.) Benth. *ex* Vasey,
Poa tenuifolia Nutt. *ex* S. Watson

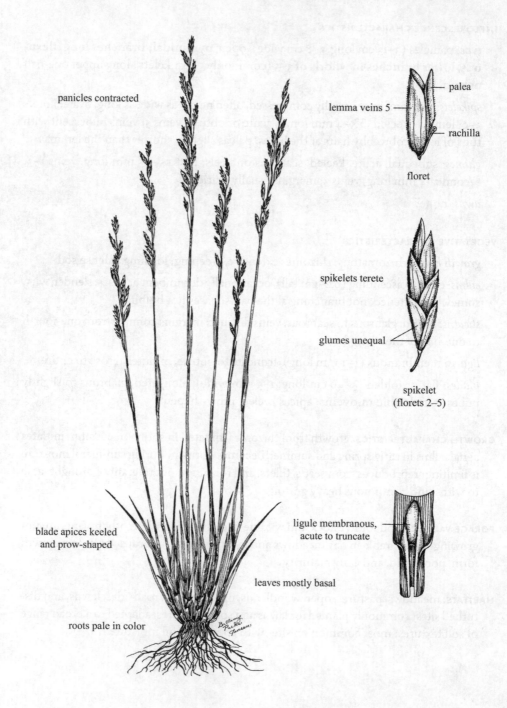

panicles contracted

palea

lemma veins 5

rachilla

floret

spikelets terete

glumes unequal

spikelet
(florets 2–5)

blade apices keeled
and prow-shaped

ligule membranous,
acute to truncate

leaves mostly basal

roots pale in color

TRIBE:	POEAE
SPECIES:	*Poa secunda* J. Presl
COMMON NAME:	Sandberg bluegrass
LIFE SPAN:	Perennial
ORIGIN:	Native
SEASON:	Cool

INFLORESCENCE CHARACTERISTICS

type: panicles (2–15 cm long, 1–4 cm wide), **contracted**, erect to somewhat lax, not densely flowered, **yellowish-green to purplish**; branches per node 2–3, short, appressed, rarely spreading below; pedicels yellowish-green (compare to *Poa fendleriana*); spikelets along most of the branches

spikelets: 4–9 mm long, 0.9–2 mm wide, terete, acute, much longer than wide, florets 2–5; lemmas convex on back (3–5.5 mm long), veins 5, purplish, apices acute; bases short-pubescent to scabrous, not webbed (compare to *Poa pratensis*)

glumes: unequal, papery, lanceolate, acute; first 2.2–5 mm long, veins 1–3; second 3–5 mm long, veins 3, shorter than lowermost lemmas; keels indistinct

awns: none

VEGETATIVE CHARACTERISTICS

growth habit: strongly cespitose, rarely with rhizomes; leaves mostly basal; roots pale

culms: erect (10–90 cm tall), decumbent at bases, wiry, glabrous; bases not bulblike; nodes occasionally reddish; leaves 1–3 per culm

sheaths: round to slightly keeled, glabrous to scabrous, veins prominent, persistent; margins hyaline, connate one-half to three-fourths of their length

ligules: membranous (2–7 mm long), acute to truncate, usually entire

blades: flat or folded or involute (3–16 cm long, 1–3 mm wide), glabrous, double midveins, apices keeled, prow-shaped; margins slightly barbed

GROWTH CHARACTERISTICS: one of the first plants to start growth in early spring, seeds mature in early summer, reproduces from seeds, tillers, and rarely from rhizomes; drought tolerant; withstands moderate to heavy grazing; forage yield is generally low, but it can be an important forage species

FORAGE VALUES: good for cattle and fair for sheep, bighorn sheep, deer, and pronghorn in spring and early summer; with adequate moisture, remains green and furnishes good forage throughout the summer

HABITATS: plains, meadows, dry woodlands, and rocky slopes; adapted to a wide variety of soils, most abundant on deep sandy to silt loam soils

Sixweeksgrass
Vulpia octoflora (Walter) Rydb.

SYN = *Festuca gracilenta* Buckley, *Festuca octoflora* Walter, *Festuca tenella* Willd.

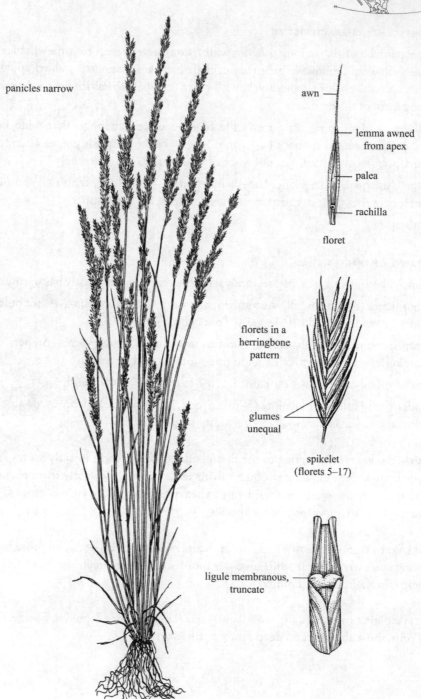

panicles narrow

awn

lemma awned from apex

palea

rachilla

floret

florets in a herringbone pattern

glumes unequal

spikelet (florets 5–17)

ligule membranous, truncate

TRIBE:	POEAE
SPECIES:	*Vulpia octoflora* (Walter) Rydb.
COMMON NAME:	Sixweeksgrass (cañuela anual, sixweeks fescue, vulpia)
LIFE SPAN:	Annual
ORIGIN:	Native
SEASON:	Cool

INFLORESCENCE CHARACTERISTICS

type: panicles (1–10 cm long), narrow, erect, exserted; branches short, appressed to slightly spreading

spikelets: 4–10 mm long (excluding the awns), glabrous or scabrous, or pubescent, florets 5–17; **florets closely arranged in a herringbone pattern**; lemmas (3–6 mm long) lanceolate, veins 5 (obscure), convex, rounded on the back; apices acute to acuminate

glumes: unequal, subulate-lanceolate; first 1.7–4.5 mm long, veins 1; second 2.7–6.7 mm long, veins 3

awns: lemmas awned (0.3–10 mm long) from apices; glumes awnless or occasionally with a short awn

VEGETATIVE CHARACTERISTICS

growth habit: cespitose or rarely solitary

culms: erect (5–60 cm tall), occasionally from decumbent bases, slender and weak, glaucous, glabrous or pubescent

sheaths: round, ridged, smooth, glabrous or sparingly pubescent; margins open

ligules: membranous (0.5–2 mm long), truncate, erose or ciliate

blades: involute (2–10 cm long, 0.5–2 mm wide), glabrous to finely pubescent adaxially

GROWTH CHARACTERISTICS: starts growth in early spring and matures about 6 weeks later, reproduces from seeds; extreme variation in plant height is directly related to available moisture; abundance is often an indicator of improper grazing; forage production is usually low

FORAGE VALUES: little forage value for livestock or wildlife except during 2–3 weeks in early spring; livestock commonly pull the roots from the soil when grazing

HABITATS: prairies, plains, mesas, woodland borders, open ground, waste areas, roadsides, and disturbed sites; adapted to a broad range of soils, most common on coarse-textured soils

Indian ricegrass
Achnatherum hymenoides (Roem. & Schult.) Barkworth

SYN = *Oryzopsis hymenoides* (Roem. & Schult.) Ricker, *Stipa hymenoides*
Roem. & Schult.

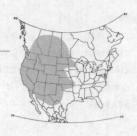

panicles open, diffuse,
branches dichotomous

awn straight to wavy, caducous

lemma indurate, brown to black,
pilose with whitish hairs, awned

callus pilose
with whitish hairs

floret

glumes
subequal

spikelet
(florets 1)

ligule membranous,
acuminate

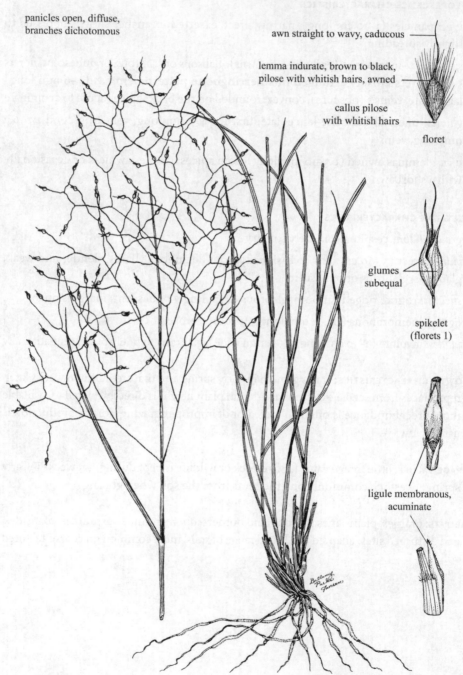

TRIBE:	STIPEAE
SPECIES:	*Achnatherum hymenoides* (Roem. & Schult.) Barkworth
COMMON NAME:	Indian ricegrass (arrocillo, zacate arroz de indio, sand bunchgrass)
LIFE SPAN:	Perennial
ORIGIN:	Native
SEASON:	Cool

INFLORESCENCE CHARACTERISTICS

type: panicles (7–20 cm long, 4–11 cm wide), diffuse; **branches spreading, dichotomous**, terminating in 2 pedicellate spikelets, flexuous; pedicels curved (5–30 mm long)

spikelets: 5–8.2 mm long, excluding awn; florets 1; lemma indurate (3–5 mm long), **brown to black at maturity, usually covered with whitish hairs** (2.5–4.5 mm long); callus blunt (0.4–1 mm long), pilose with whitish hairs; palea slightly shorter than the lemma

glumes: subequal; first 5–8.2 mm long, veins 3–5; second 4.2–7.5 mm long, glabrous to puberulent, ovate to acuminate, veins 5–7, prominent; margins papery

awns: lemmas awned (3–6 mm long), straight to wavy, caducous; glumes sometimes mucronate

VEGETATIVE CHARACTERISTICS

growth habit: cespitose

culms: erect to ascending (30–90 cm tall), slender, glabrous or partially scabridulous

sheaths: round, glabrous to ciliate on the overlapping margins, papery, shorter than the internodes, open, persistent; collars with tufts of hair (to 1 mm long) on the margins

ligules: membranous (4–7 mm long), acuminate, erose to lacerate or deeply notched

blades: involute (5–40 cm long, 1–3 mm wide), filiform, scabrous to hirsute adaxially, glabrous abaxially with prominent midveins

GROWTH CHARACTERISTICS: starts growth in early spring, flowers in late spring, reproduces from seeds and tillers; drought tolerant

FORAGE VALUES: good for cattle, sheep, horses, and wildlife; especially valuable for winter grazing because the plants cure well and lower plant parts remain somewhat green; seeds are high in protein; some Native Americans used the seeds to make flour

HABITATS: prairies, plains, dunes, deserts, foothills, mesas, alluvial fans, and disturbed sites; most abundant on sandy soils but will occur on well-drained silty and calcareous soils; moderately salt and alkali tolerant; frequently a codominant with *Atriplex confertifolia*.

Columbia needlegrass
Achnatherum nelsonii (Scribn.) Barkworth

SYN = *Stipa columbiana* Macoun, *Stipa nelsonii* (Scribn.) Barkworth

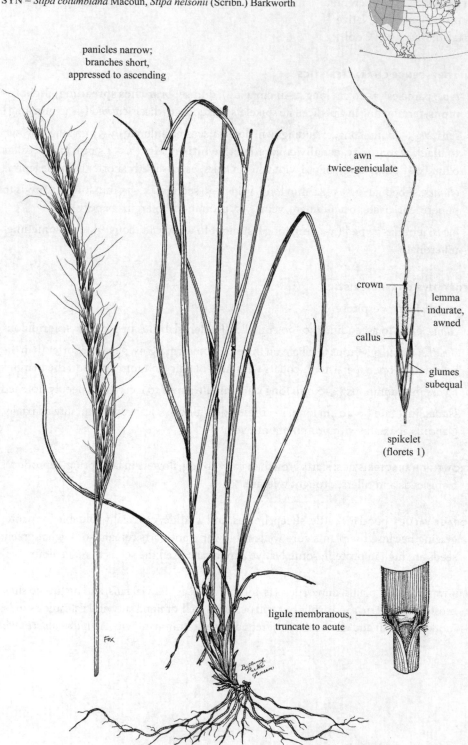

panicles narrow;
branches short,
appressed to ascending

awn
twice-geniculate

crown

lemma
indurate,
awned

callus

glumes
subequal

spikelet
(florets 1)

ligule membranous,
truncate to acute

Fox

TRIBE:	STIPEAE
SPECIES:	*Achnatherum nelsonii* (Scribn.) Barkworth
COMMON NAME:	Columbia needlegrass (subalpine needlegrass, Nelson needlegrass)
LIFE SPAN:	Perennial
ORIGIN:	Native
SEASON:	Cool

INFLORESCENCE CHARACTERISTICS

type: panicles (7–30 cm long, 1–2.5 cm wide), narrow, dense, **sometimes purplish**; branches short, appressed to ascending

spikelets: florets 1; lemma slender (5–7 mm long), indurate, slightly pubescent; crowns with tufts of hair (longest 0.7–1 mm); callus usually sharp (0.4–1 mm long), occasionally blunt

glumes: subequal (7–11 mm long), first longest, exceeds lemmas, acuminate to mucronate, veins 3–5, glabrous, papery, stramineous to purplish

awns: **lemmas awned** (1.8–4 cm long), twice-geniculate; lower segments minutely scabrous and twisted; upper segments not twisted (often 1.5 cm or longer)

VEGETATIVE CHARACTERISTICS

growth habit: cespitose; leaves mostly basal

culms: erect (0.3–1.3 m tall), stout, glabrous, straight; **nodes glabrous to pubescent**, may be purplish

sheaths: round to slightly keeled, prominently veined, open, glabrous to scabrous; margins occasionally pubescent at apices

ligules: membranous (0.5–2 mm long), usually longest laterally, truncate to acute, erose to entire

blades: flat to mostly involute (10–30 cm long, 1–5 mm wide), glabrous; margins barbed

GROWTH CHARACTERISTICS: starts growth in midspring, matures by September, reproduces from seeds and tillers; may regrow in the autumn if moisture is adequate

LIVESTOCK LOSSES: the sharply pointed calluses may work into the ears, eyes, nostrils, and tongues of grazing animals; sheep are especially susceptible to injury; awns may contaminate fleece

FORAGE VALUES: fair to good for cattle and horses; fair for sheep, deer, and elk; becomes nearly unpalatable at maturity

HABITATS: prairies, plains, meadows, and open woodlands in foothills and mountains; adapted to soils ranging from sandy loams to clays, most abundant in fine-textured soils

Needleandthread
Hesperostipa comata (Trin. & Rupr.) Barkworth

SYN = *Stipa comata* Trin. & Rupr.

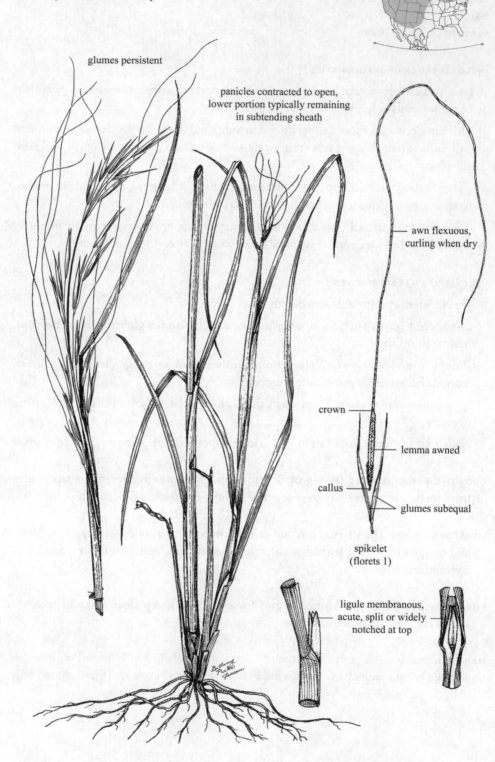

glumes persistent

panicles contracted to open,
lower portion typically remaining
in subtending sheath

awn flexuous,
curling when dry

crown

lemma awned

callus

glumes subequal

spikelet
(florets 1)

ligule membranous,
acute, split or widely
notched at top

TRIBE:	STIPEAE
SPECIES:	*Hesperostipa comata* (Trin. & Rupr.) Barkworth
COMMON NAME:	Needleandthread
LIFE SPAN:	Perennial
ORIGIN:	Native
SEASON:	Cool

INFLORESCENCE CHARACTERISTICS

type: panicles (10–40 cm long), contracted to open, **lower portions typically remaining in subtending sheaths**; branches slender and ascending

spikelets: florets 1, large, drooping at anthesis and thereafter; lemma (1–1.5 cm long, excluding the awn) slender, indurate, pale to brown at maturity, veins 5, lightly pubescent; crown hairs minute; callus sharp (3 mm long), **bearded with stiff hairs** (0.2–0.5 mm long)

glumes: subequal (1.8–3.5 cm long); first longer, narrow, acuminate, veins 3–7, margins hyaline

awns: **lemmas awned (10–23 cm long), flexuous, twisted**; curling when dry; lower segment short-pubescent; terminal segment glabrous or scabrous, not twisted

VEGETATIVE CHARACTERISTICS

growth habit: densely cespitose; leaves mostly basal

culms: erect or ascending (0.3–1.1 m tall); **nodes glabrous** to puberulent

sheaths: round, usually longer than the internodes, glabrous to scabrous, prominently veined, open

ligules: membranous (3–6 mm long), acute, split or widely notched at top

blades: flat or folded or involute (10–40 cm long, 1–3 mm wide), scabrous adaxially, glabrous to scaberulous abaxially

GROWTH CHARACTERISTICS: starts growth in early spring or when moisture is available, seeds mature in early summer, reproduces from seeds and tillers

LIVESTOCK LOSSES: sharply pointed calluses and long awns may cause injury by working into the eyes, tongues, nostrils, and ears of grazing animals; sheep are especially susceptible to injury, awns may contaminate the fleece and carcass

FORAGE VALUES: fair to good for livestock and poor to fair for wildlife, extensively utilized by elk in winter and deer in spring; cures well to provide autumn and winter forage for livestock

HABITATS: prairies, plains, dry hills, foothills, alluvial fans, and sandy benches; most abundant on excessively drained soils

Texas wintergrass
Nassella leucotricha (Trin. & Rupr.) R.W. Pohl

SYN = *Stipa leucotricha* Trin. & Rupr.

panicles; upper branches ascending

awn once- to twice-geniculate

crown

lemma awned

callus

glumes subequal

spikelet (florets 1)

nodes pubescent, glabrate with maturity

ligule membranous, truncate

blades usually pubescent with short, stiff hairs

cleistogamous spikelet in basal sheath

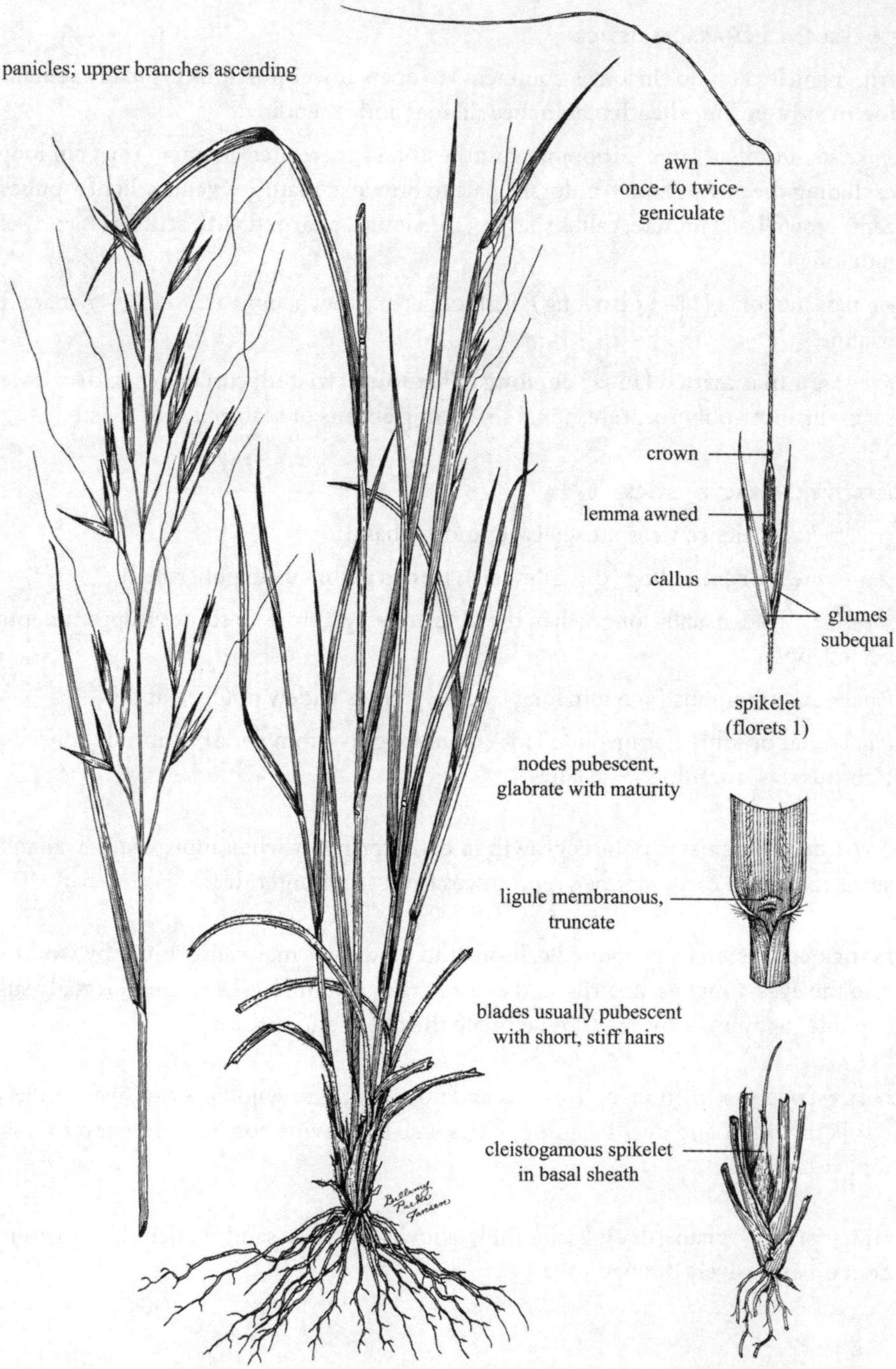

TRIBE:	STIPEAE
SPECIES:	*Nassella leucotricha* (Trin. & Rupr.) R.W. Pohl
COMMON NAME:	Texas wintergrass (zacate flechilla, Texas needlegrass, speargrass)
LIFE SPAN:	Perennial
ORIGIN:	Native
SEASON:	Cool

INFLORESCENCE CHARACTERISTICS

type: panicles (6–50 cm long, usually less than 10 cm wide), lower portions occasionally enclosed in subtending sheaths; lower branches flexuous, slender, spreading; upper branches ascending

spikelets: florets 1; lemma 9–12 mm long, veins 3, coriaceous to indurate, light brown, appressed pubescence below, rugose on body above the base, rounded, summits white; crown hairs stiff, minute (0.5–2 mm long); callus sharp (2–5 mm long) with silky hairs (to 3 mm long)

glumes: subequal (1.2–2 cm long), lanceolate; first longest, thin, glabrous, veins 3–5, acuminate to awn-tipped, somewhat hyaline

awns: **lemmas awned (4–10 cm long)**, **stout**, once- or twice-geniculate; lower portions scabrous to pubescent, twisted (2–3.5 cm long); upper portions essentially glabrous, loosely twisted

other: produces cleistogamous spikelets in the basal sheaths

VEGETATIVE CHARACTERISTICS

growth habit: cespitose; leaves mostly basal

culms: ascending to erect (30–90 cm tall), spreading at bases; **nodes pubescent**, glabrate with maturity

sheaths: round, pubescent to nearly glabrous; collars with hairs (0.5–1 mm long) on sides

ligules: membranous (< 1 mm long), variable, sometimes absent, truncate, entire to erose

blades: flat, becoming involute (5–40 cm long, 1–5 mm wide), usually pubescent; abaxial surfaces with short and stiff hairs, adaxial surfaces glabrous

GROWTH CHARACTERISTICS: starts growth in late autumn, remains green through winter and spring; reproduces from seeds, cleistogamous spikelets, and tillers

LIVESTOCK LOSSES: sharply pointed calluses and awns may cause injury to livestock and contaminate fleece

FORAGE VALUES: fair to good for both livestock and wildlife, most valuable for early spring forage; also valuable as winter forage because it remains partially green

HABITATS: prairies, open woodlands, disturbed sites, and heavily grazed areas

Purple needlegrass
Nassella pulchra (Hitchc.) Barkworth

SYN = *Stipa pulchra* Hitchc.

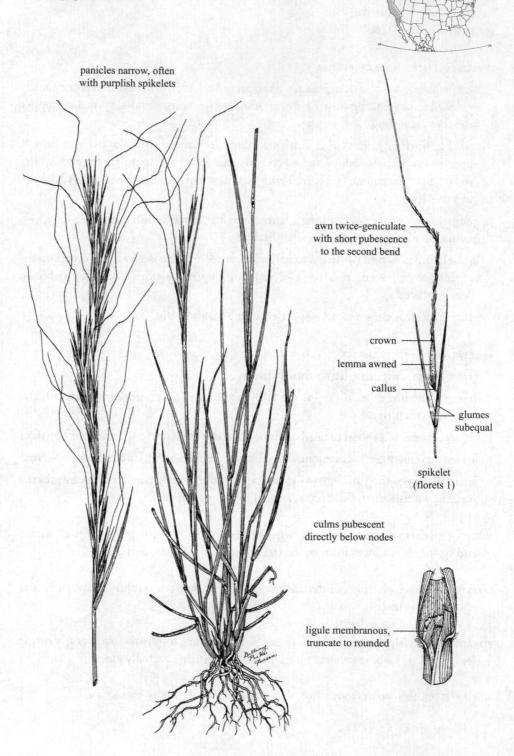

panicles narrow, often
with purplish spikelets

awn twice-geniculate
with short pubescence
to the second bend

crown

lemma awned

callus

glumes
subequal

spikelet
(florets 1)

culms pubescent
directly below nodes

ligule membranous,
truncate to rounded

TRIBE:	STIPEAE
SPECIES:	*Nassella pulchra* (Hitchc.) Barkworth
COMMON NAME:	Purple needlegrass (flechilla púrpura, purple tussockgrass)
LIFE SPAN:	Perennial
ORIGIN:	Native
SEASON:	Cool

INFLORESCENCE CHARACTERISTICS

type: panicles (15–40 cm long), nodding, narrow, loose; lower branches spreading (2.5–9 cm long); often pilose at the axils

spikelets: florets 1, **often strongly purplish**; lemma narrow (7–13 mm long), fusiform, sparingly pilose, summit smooth; crown ciliate; callus sharp (2–3 mm long), with white hairs (1 mm long)

glumes: subequal (1.2–2 cm long, 1–2.2 mm wide); first longest, narrow, long-acuminate, glabrous, veins 3

awns: **lemmas awned (5–9 cm long)**, twice-geniculate; **first segment twisted** (1.5–2 cm long), second segment shorter and loosely twisted, third segment straight (4–6 cm long); short pubescence to the second bend

VEGETATIVE CHARACTERISTICS

growth habit: densely cespitose; leaves mostly basal

culms: erect (0.4–1 m tall), erect or geniculate below, **pubescent only directly below the nodes**

sheaths: round, mostly glabrous; margins ciliate and open; collars pilose, hairs (0.5–0.8 mm long) at the sides

ligules: membranous (to 1.2 mm long), truncate to rounded, erose

blades: flat to involute (10–25 cm long, 1–4 mm wide), lightly pubescent

GROWTH CHARACTERISTICS: starts growth in late autumn or early spring, flowers April to June, reproduces from seeds and tillers; productivity reduced if grazed during period of maximum growth

LIVESTOCK LOSSES: calluses may cause injury to livestock by working into the eyes, tongues, and ears; sheep are especially susceptible to injury; may contaminate fleece

FORAGE VALUES: good for cattle, sheep, and horses; fair for wildlife; basal leaves remain green for 9–10 months of the year

HABITATS: prairies, open timbered areas of foothills and valleys, waste places, and disturbed sites; adapted to a broad range of soils, most abundant on sandy loams

Crested wheatgrass
Agropyron cristatum (L.) Gaertn.

SYN = *Agropyron desertorum* Fisch. *ex* Link, *Agropyron imbricatum*
Roem. & Schult., *Agropyron pectiniforme* Roem. & Schult.,
Triticum cristatum (L.) Schreb.

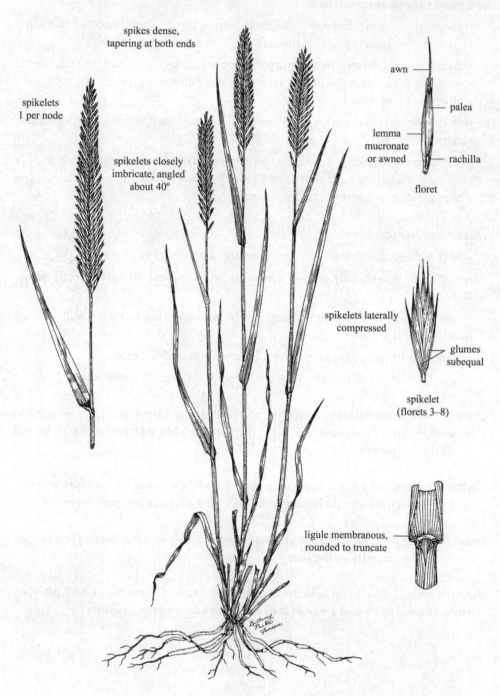

spikes dense,
tapering at both ends

spikelets
1 per node

spikelets closely
imbricate, angled
about 40°

awn

palea

lemma
mucronate
or awned

rachilla

floret

spikelets laterally
compressed

glumes
subequal

spikelet
(florets 3–8)

ligule membranous,
rounded to truncate

TRIBE:	TRITICEAE
SPECIES:	*Agropyron cristatum* (L.) Gaertn.
COMMON NAME:	Crested wheatgrass (triguillo crestado)
LIFE SPAN:	Perennial
ORIGIN:	Introduced (from Eurasia)
SEASON:	Cool

INFLORESCENCE CHARACTERISTICS

type: spikes (2–9 cm long, 7–25 mm wide), dense, tapering at both ends; spikelets 1 per node, **closely imbricate**, several times longer than rachis internodes, spreading to ascending, angled about 40°; rachises scabrous to pilose, occasionally wavy, not readily disarticulating

spikelets: 5–15 mm long, **laterally compressed**, florets 3–8; lemmas 4–7 mm long, firm, acute, veins 5, margins ciliate; upper florets sterile

glumes: subequal (3–6 mm long); second longest, lanceolate, not bowed at the base, glabrous to pilose, midveins prominent and ciliate

awns: glumes and lemmas mucronate or tapering to short awns (1–6 mm long)

VEGETATIVE CHARACTERISTICS

growth habit: cespitose, rarely with rhizomes

culms: erect to ascending (0.2–1 m tall), glabrous; bases occasionally geniculate

sheaths: round, glabrous (sometimes pubescent below); margins overlapping, open; auricles slender (1 mm long)

ligules: membranous (0.5–1.5 mm long), rounded to truncate, erose

blades: flat or folded (5–20 cm long, 2–6 mm wide), glabrous to puberulent, veins raised adaxially, smooth abaxially; margins weakly scabrous

GROWTH CHARACTERISTICS: starts growth in early spring, flowers in late spring, reproduces from seeds, tillers, and rarely rhizomes; drought and cold tolerant, may regrow in the autumn if moisture is sufficient

FORAGE VALUES: good for livestock and fair for wildlife; cures well for use as winter forage, produces good hay if cut early, most valuable for early spring grazing

HABITATS: planted in pastures, hay meadows, and roadsides; most abundant on dry, medium-textured soils, less adapted to heavy clays and sands; relatively salt tolerant; a major species for reseeding in sagebrush areas

Canada wildrye
Elymus canadensis L.

SYN = *Elymus wiegandii* Fernald

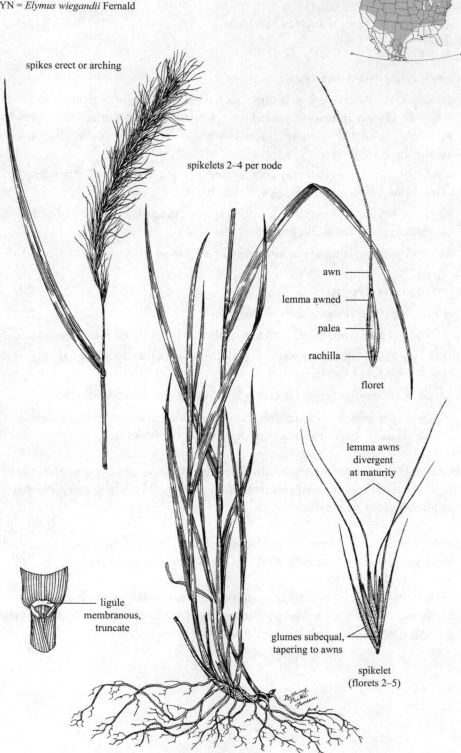

spikes erect or arching

spikelets 2–4 per node

awn

lemma awned

palea

rachilla

floret

lemma awns divergent at maturity

ligule membranous, truncate

glumes subequal, tapering to awns

spikelet (florets 2–5)

TRIBE:	TRITICEAE
SPECIES:	*Elymus canadensis* L.
COMMON NAME:	Canada wildrye (centeno silvestre)
LIFE SPAN:	Perennial
ORIGIN:	Native
SEASON:	Cool

INFLORESCENCE CHARACTERISTICS

type: spikes (8–25 cm long), **erect or arching**, totally exserted, somewhat linear, thick, bristly, occasionally interrupted below; **spikelets 2–4 per node**, imbricate, slightly spreading; rachises not readily disarticulating

spikelets: 1.2–1.5 cm long, florets 2–5; lemmas broad at bases (8–12 mm long), scabrous to hirsute

glumes: subequal (1–2.5 cm long, 0.8–1.5 mm wide), subulate, veins 3–7; veins glabrous, scabrous, or ciliate

awns: **lemmas awned (1.5–5 cm long), flexuous**; awns divergent at maturity; glumes tapering to awns (1–3 cm long); awns arching to somewhat straight

VEGETATIVE CHARACTERISTICS

growth habit: cespitose, rarely with rhizomes (to 4 cm long)

culms: erect or ascending from decumbent bases (1–1.5 m tall), coarse; nodes glabrous

sheaths: round, glabrous, rarely pubescent; auricles well developed (1–4 mm long), clasping, fingerlike

ligules: membranous (0.5–2 mm long), truncate, erose or rarely ciliate

blades: flat or folded (5–40 cm long, 7–15 mm wide), elongate, ascending, tapering to fine points, scabrous adaxially, midveins prominent abaxially; margins finely toothed

GROWTH CHARACTERISTICS: starts growth in autumn and makes some growth in winter in the southern portions of its range, plants mature by late spring to early summer, reproduces from seeds and tillers

LIVESTOCK LOSSES: inflorescences may become infested with ergot and be potentially dangerous to livestock

FORAGE VALUES: good for cattle and horses and fair for sheep and wildlife during the spring when green and growing, value decreases sharply when the plants mature

HABITATS: prairies, stream banks, disturbed areas, open woodlands, fence lines, and ditches; usually in open areas, adapted to a broad range of dry and moist soils

Squirreltail
Elymus elymoides (Raf.) Swezey

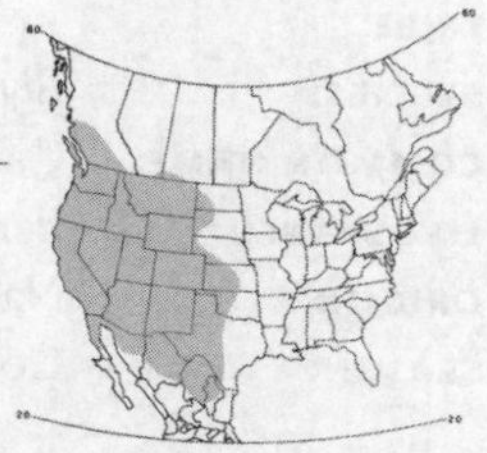

SYN = *Elymus glaber* (J.G. Sm.) Burtt Davy, *Elymus pubiflorus* (J.G. Sm.)
Burtt Davy, *Sitanion hystrix* (Nutt.) J.G. Sm., *Sitanion longifolium*
J.G. Sm., *Sitanion pubiflorum* J.G. Sm.

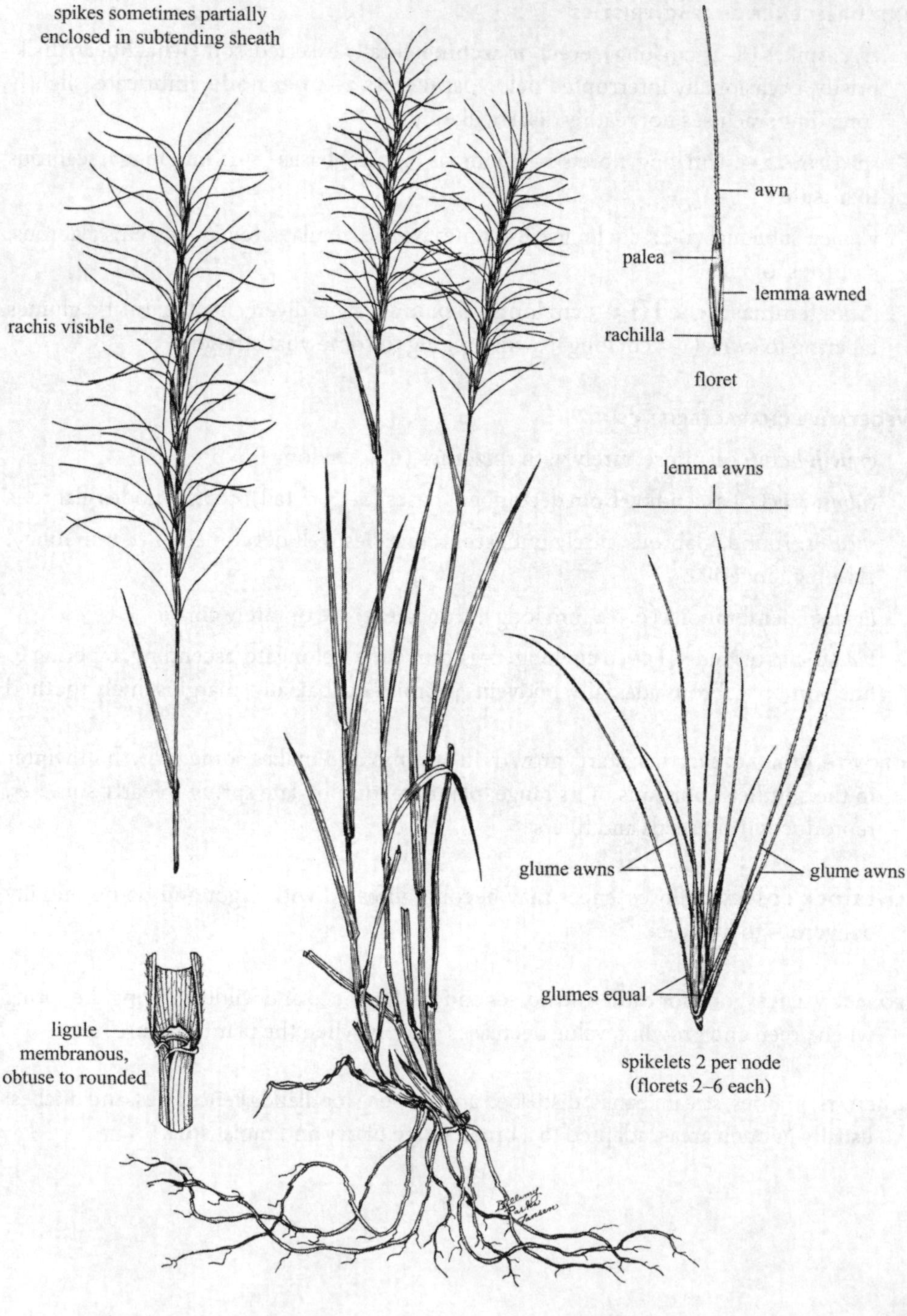

spikes sometimes partially
enclosed in subtending sheath

rachis visible

awn

palea

lemma awned

rachilla

floret

lemma awns

glume awns   glume awns

glumes equal

spikelets 2 per node
(florets 2–6 each)

ligule
membranous,
obtuse to rounded

TRIBE:	TRITICEAE
SPECIES:	*Elymus elymoides* (Raf.) Swezey
COMMON NAME:	Squirreltail (zacate triguillo, triguillo desértico, cola de zorra, bottlebrush squirreltail)
LIFE SPAN:	Perennial
ORIGIN:	Native
SEASON:	Cool

INFLORESCENCE CHARACTERISTICS

type: spikes (2–20 cm long excluding awns), sometimes partially enclosed in subtending sheaths, erect; **spikelets usually 2 per node; rachises disarticulating readily, visible**

spikelets: 1–2 cm long, divergent, florets 2–6; apical 1–2 florets maybe reduced to awn-like structures; fertile lemmas convex (8–12 mm long), glabrous to lightly pubescent, veins 3–5

glumes: equal (6–9 mm long), narrow, straight, attenuate, subulate to lanceolate, veins 1–3, entire or bifid

awns: **lemmas awned** (5–15 mm long); glumes awned (2–10 cm long), widely spreading, **stiff**, scabrous, green to purplish

VEGETATIVE CHARACTERISTICS

growth habit: cespitose

culms: erect to geniculate (10–60 cm tall), stiff, glabrous to puberulent

sheaths: round, pubescent or glabrous; margins open, translucent; auricles small (about 1 mm long), often purplish

ligules: membranous (0.6–1 mm long), obtuse to rounded, erose

blades: flat to involute (5–20 cm long, 1–5 mm wide), stiffly ascending, tapering to fine points, glabrous to lightly pubescent, prominently veined

GROWTH CHARACTERISTICS: starts growth in early spring, flowers in late spring, may regrow and flower a second time with favorable moisture, reproduces from seeds and tillers

LIVESTOCK LOSSES: sharply pointed calluses and awns may cause injury to soft tissue and may contaminate fleece

FORAGE VALUES: fair for cattle and horses and poor for sheep before inflorescences develop, unpalatable by midsummer because of troublesome awns, important winter forage because leaves remain partially green

HABITATS: disturbed rangelands, plains, open woodlands, and rocky slopes of deserts; most abundant on disturbed sites on either deep or shallow soils; may grow in saline or alkaline soils

Slender wheatgrass
Elymus trachycaulus (Link) Gould *ex* Shinners

SYN = *Agropyron caninum* (L.) P. Beauv., *Agropyron pauciflorum* Hitchc.,
Agropyron subsecundum (Link) Hitchc., *Agropyron trachycaulum*
(Link) Malte *ex* H.F. Lewis

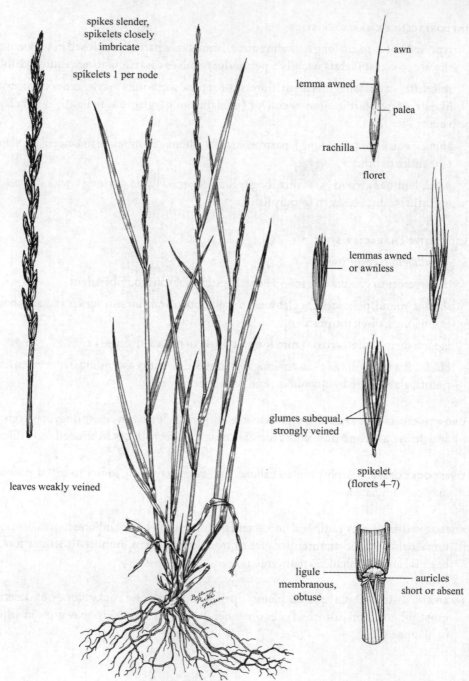

spikes slender,
spikelets closely
imbricate

spikelets 1 per node

awn

lemma awned

palea

rachilla

floret

lemmas awned
or awnless

glumes subequal,
strongly veined

spikelet
(florets 4–7)

leaves weakly veined

ligule
membranous,
obtuse

auricles
short or absent

TRIBE:	TRITICEAE
SPECIES:	*Elymus trachycaulus* (Link) Gould *ex* Shinners
COMMON NAME:	Slender wheatgrass (triguillo largo, agropiro delgado, bearded wheatgrass)
LIFE SPAN:	Perennial
ORIGIN:	Native
SEASON:	Cool

INFLORESCENCE CHARACTERISTICS

type: spikes (5–25 cm long), slender, erect, distinctly bilateral, sometimes purplish-tinged; spikelets 1 per node, **closely imbricate** (usually about one-half of each spikelet overlaps); rachises not readily disarticulating, rachis segments (4–11 mm long) scabrous

spikelets: 1–1.5 cm long, florets 4–7; lemmas acute (8–13 mm long), veins 3–5, glabrous to scabrous to occasionally hirsute on margins; rachillas pubescent

glumes: subequal (6–14 mm long); second longest, symmetrical (compare to *Pascopyrum smithii*), broad, not bowed out at the base, lanceolate, acute, nearly enclosing the floret, **strongly veined; veins 5–7**, veins dark green; margins hyaline, scabrous

awns: lemmas awnless or awned (1–12 mm long); glumes may taper to short awns

VEGETATIVE CHARACTERISTICS

growth habit: cespitose

culms: erect or from decumbent bases (0.5–1.5 m tall), slender, green or glaucous, glabrous

sheaths: round, glabrous to pilose, open; auricles short (0.3–1 mm long) or absent, 1 often rudimentary

ligules: membranous (0.4–1 mm long), obtuse, erose to ciliate

blades: flat or folded (5–25 cm long, 2–8 mm wide), slender with pointed apices, elongate, scabrous to glabrous, **weakly veined** (compare to *Pascopyrum smithii*); margins white

GROWTH CHARACTERISTICS: starts growth in midspring, seeds mature by August to September, reproduces from seeds and tillers

FORAGE VALUES: good for cattle and fair for sheep when green, good when mature; good to excellent for elk and bighorn sheep

HABITATS: prairies, pastures, meadows, riverbanks, open woodlands, mountain slopes, and rolling hills; most abundant in well-drained medium to finely textured soils; tolerant of moderate drought as well as long, wet periods

Foxtail barley
Hordeum jubatum L.

SYN = *Critesion jubatum* (L.) Nevski

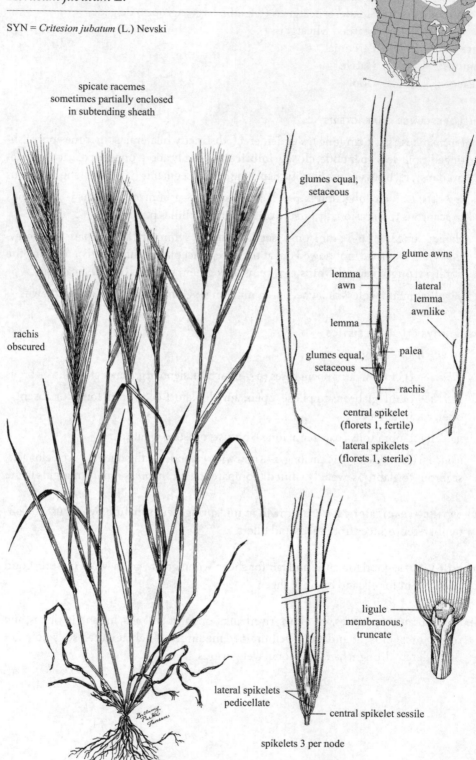

spicate racemes
sometimes partially enclosed
in subtending sheath

glumes equal,
setaceous

glume awns

lemma
awn

lateral
lemma
awnlike

lemma

palea

glumes equal,
setaceous

rachis

rachis
obscured

central spikelet
(florets 1, fertile)

lateral spikelets
(florets 1, sterile)

ligule
membranous,
truncate

lateral spikelets
pedicellate

central spikelet sessile

spikelets 3 per node

TRIBE:	TRITICEAE
SPECIES:	*Hordeum jubatum* L.
COMMON NAME:	Foxtail barley (cola de zorra)
LIFE SPAN:	Perennial
ORIGIN:	Native
SEASON:	Cool

INFLORESCENCE CHARACTERISTICS

type: spicate racemes (4–15 cm long, to 1.5 cm wide, excluding awns), sometimes partially enclosed in subtending sheaths, nodding, purplish; **spikelets 3 per node**; central spikelet sessile; 2 lateral spikelets pedicellate, reduced; rachises obscured, readily disarticulating at maturity

spikelets: central spikelet florets 1, fertile lemma bodies narrow (4–8 mm long, not including the awns); lateral spikelets florets 1, sterile, of similar length (0.7–1.2 mm long), reduced to little more than a series of 3 awns

glumes: equal, setaceous, spreading or straight; central spikelet glumes awnlike (2.5–7 cm long, not including the awns), narrow, scabrous; lateral spikelet glumes slightly shorter, similar in shape

awns: central spikelet lemmas awned (1.2–9 cm long), lateral spikelet lemmas awned (2–15 mm long), **glumes awned** (2–6 cm long), **thin**; scabrous

VEGETATIVE CHARACTERISTICS

growth habit: cespitose

culms: erect or ascending (30–75 cm tall), decumbent below, slender; nodes dark, glabrous to soft-pubescent

sheaths: round, glabrous to lightly pilose; auricles small (to 0.5 mm long) or absent, sometimes present on some leaves of some plants but not on others

ligules: membranous (0.2–1 mm long), truncate, erose to ciliolate

blades: flat (5–15 cm long, 2–5 mm wide), tapering to fine points, scabrous, glabrous to occasionally pilose abaxially

GROWTH CHARACTERISTICS: starts growth in late April to May, matures June to August, reproduces from seeds and tillers, short-lived

LIVESTOCK LOSSES: awns may cause sores in and around the noses, eyes, and mouths of grazing livestock and contaminate fleece

FORAGE VALUES: poor for all classes of livestock and wildlife, but it may be lightly grazed before inflorescence development, presence in hay greatly reduces hay value

HABITATS: meadows, waste places, disturbed pastures, ditches, roadsides, and alkaline and saline sites; adapted to a broad range of soil types, most abundant where water occasionally accumulates

Little barley
Hordeum pusillum Nutt.

SYN = *Critesion pusillum* (Nutt.) Á. Löve

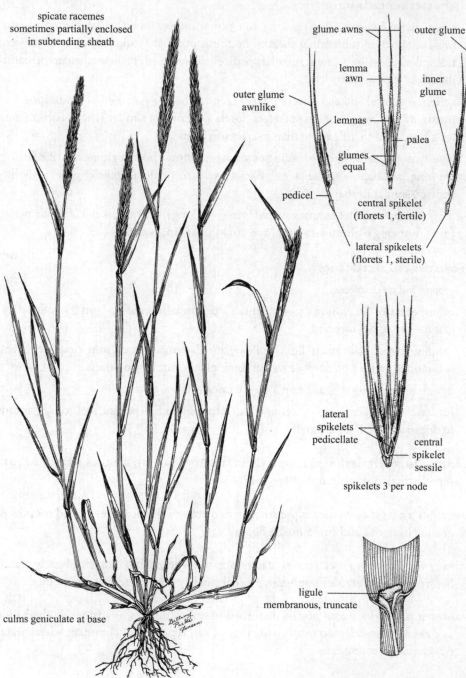

spicate racemes
sometimes partially enclosed
in subtending sheath

outer glume
awnlike

pedicel

glume awns

lemma
awn

lemmas

glumes
equal

outer glume

inner
glume

palea

central spikelet
(florets 1, fertile)

lateral spikelets
(florets 1, sterile)

lateral
spikelets
pedicellate

central
spikelet
sessile

spikelets 3 per node

culms geniculate at base

ligule
membranous, truncate

TRIBE:	TRITICEAE
SPECIES:	*Hordeum pusillum* Nutt.
COMMON NAME:	Little barley (cebadilla, cola de ardilla)
LIFE SPAN:	Annual
ORIGIN:	Native
SEASON:	Cool

INFLORESCENCE CHARACTERISTICS

type: spicate racemes (4–8 cm long, 3–8 mm wide, not including awns), erect, stiff, sometimes partially enclosed in subtending sheaths; **spikelets 3 per node**, central spikelets 1, sessile; lateral spikelets 2, pedicellate (pedicels 0.3–0.7 mm long), reduced; **rachises readily disarticulating at maturity**

spikelets: central spikelet florets 1, fertile; lemma body narrow (5–7 mm long), scabrous; lateral spikelets florets 1, sterile, smaller; lemma bodies acuminate (1.5–3.5 mm long)

glumes: equal; central spikelet glumes lanceolate (3.4–5.5 mm long, not including awns), broadened above bases; veins 3, veins faint; scabrous; lateral glumes shorter, flattened toward bases; inner 2 glumes similar to central spikelet glumes, outer 2 glumes awn-like, scabrous

awns: lemmas of central spikelets awned (2–7 mm long); lemmas of lateral spikelets short-awned; outer glumes of lateral spikelets awnlike; other glumes awned (7–15 mm long)

VEGETATIVE CHARACTERISTICS

growth habit: weakly cespitose or solitary

culms: erect from geniculate bases (to 40 cm tall), glabrous, nodes dark

sheaths: round, glabrous or with short-spreading pubescence, inflated; auricles absent or present, triangular

ligules: membranous (0.4–0.7 mm long), truncate, erose to short ciliate

blades: flat (to 12 cm long, 2–5 mm wide), erect to lax; margins weakly scabrous, glabrous to pubescent

GROWTH CHARACTERISTICS: starts growth in early spring, matures May to June, reproduces from seeds; especially frequent during years with abundant winter and spring moisture

FORAGE VALUES: essentially no value for either livestock or wildlife, although it is sometimes lightly grazed in the spring

HABITATS: plains, open ground, cultivated land, deteriorated pastures, disturbed sites, waste places, roadsides, and borders of marshes; most abundant on dry or alkaline soils of formerly cultivated land and deteriorated rangelands

Basin wildrye
Leymus cinereus (Scribn. & Merr.) Á. Löve

SYN = *Aneuroledipedium piperi* (Bowden) B.R. Baum, *Elymus cinereus*
Scribn. & Merr., *Elymus condensatus* J. Presl, *Elymus piperi* Bowden

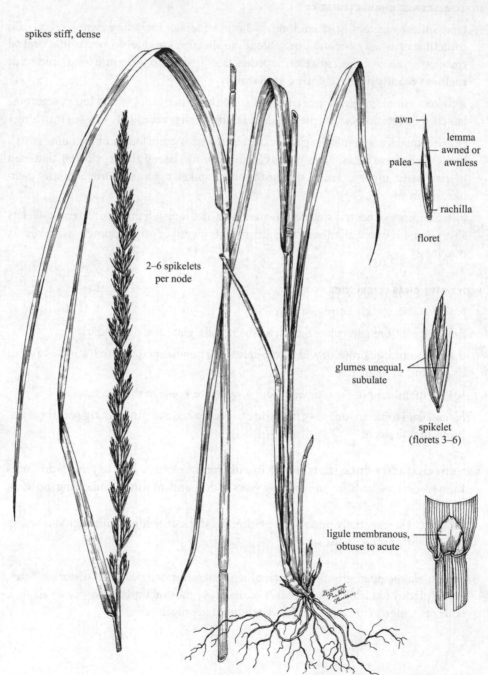

spikes stiff, dense

2–6 spikelets
per node

awn

lemma
awned or
awnless

palea

rachilla

floret

glumes unequal,
subulate

spikelet
(florets 3–6)

ligule membranous,
obtuse to acute

TRIBE:	TRITICEAE
SPECIES:	*Leymus cinereus* (Scribn. & Merr.) Á. Löve
COMMON NAME:	Basin wildrye (giant wildrye, Great Basin wildrye)
LIFE SPAN:	Perennial
ORIGIN:	Native
SEASON:	Cool

INFLORESCENCE CHARACTERISTICS

type: spikes or rarely spicate racemes (10–25 cm long, 1–2 cm wide), dense, generally erect, stiff, occasionally interrupted below, rarely develop short branches with up to 6 nodes; **spikelets 2–6 per node** (1 may be pedicellate), highly imbricate, ascending or appressed

spikelets: 9–20 mm long, florets 3–6; lemmas acute to blunt (7–12 mm long), glabrous to sparsely strigose (hairs < 0.7 mm long), apices acute or awned, margins hyaline

glumes: unequal (2–16 mm long, 0.2–0.8 mm wide), subulate, stiff, keeled, central portions thicker than the margins, mucronate, scabrous, veins usually obscure

awns: lemmas awned (1–4 mm long) or awnless; glumes mucronate

VEGETATIVE CHARACTERISTICS

growth habit: cespitose, rarely with short rhizomes

culms: erect (1–2.5 m tall), coarse, robust, glabrous to harshly puberulent; nodes glabrous or puberulent

sheaths: round, glabrous to puberulent, margins open; auricles well developed (1–2 mm long) to lacking, fingerlike

ligules: membranous (3–7 mm long), obtuse to acute, entire to erose

blades: flat to involute (20–60 cm long, 5–15 mm wide), firm, narrowing to acute apices, glabrous to harshly puberulent, strongly veined

GROWTH CHARACTERISTICS: starts growth in early spring, seeds mature by August, reproduces from seeds, tillers, and rarely from rhizomes; not tolerant of heavy grazing because of the relatively high positions of the growing points (10–15 cm above the ground)

FORAGE VALUES: good for cattle and horses, fair for sheep and wildlife, provides abundant forage in early spring, relatively unpalatable in the summer, furnishes important winter feed for most classes of livestock but usually requires a protein supplement

HABITATS: riverbanks, wet meadows, floodplains, ravines, moist or dry slopes, and plains; adapted to a broad range of soils, including moderately saline soils

Western wheatgrass
Pascopyrum smithii (Rydb.) Barkworth & D.R. Dewey

SYN = *Agropyron elmeri* Scribn., *Agropyron smithii* Rydb., *Elymus smithii* (Rydb.) Gould, *Elytrigia dasystachya* (Hook.) Á Löve & D. Löve, *Elytrigia smithii* (Rydb.) Nevski

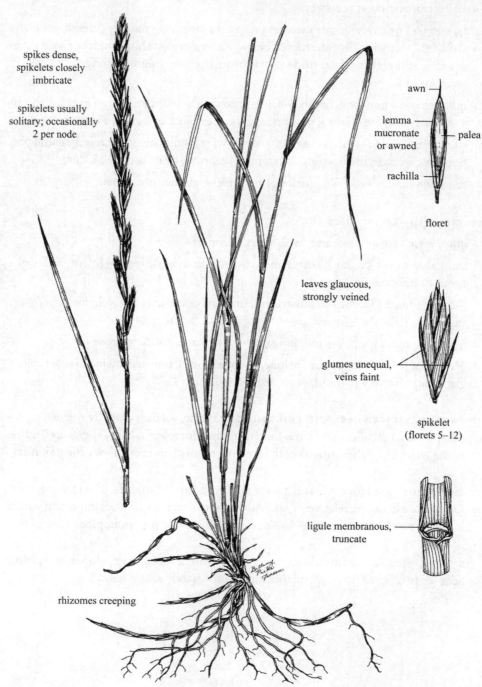

spikes dense, spikelets closely imbricate

spikelets usually solitary; occasionally 2 per node

awn

lemma mucronate or awned

palea

rachilla

floret

leaves glaucous, strongly veined

glumes unequal, veins faint

spikelet (florets 5–12)

ligule membranous, truncate

rhizomes creeping

TRIBE:	TRITICEAE
SPECIES:	*Pascopyrum smithii* (Rydb.) Barkworth & D.R. Dewey
COMMON NAME:	Western wheatgrass (triguillo corto, agropiro del oeste, bluestem)
LIFE SPAN:	Perennial
ORIGIN:	Native
SEASON:	Cool

INFLORESCENCE CHARACTERISTICS

type: spikes (6–20 long), often dense, stiff, erect; spikelets usually solitary or occasionally 2 per node, **closely imbricate (about one-half of each spikelet overlaps)**; rachises not readily disarticulating, rachis segments scabrous (5–6 mm long)

spikelets: 1.5–2.5 cm long, florets 5–12, glaucous; lemmas acute (8–14 mm long), glabrous to pubescent on margins, veins 5

glumes: unequal, first shorter (6–12 mm long) than the second (7–15 mm long), asymmetrical, acute, narrow, rigid; **veins 3–5, faint** (compare to *Thinopyrum intermedium*); glabrous to scabrous

awns: glumes occasionally mucronate; lemmas occasionally mucronate or awned (to 5 mm long)

VEGETATIVE CHARACTERISTICS

growth habit: **rhizomatous**; rhizomes creeping and may form a loose sod

culms: erect (30–90 cm tall) or ascending from a decumbent base, single or in small clusters, glabrous to glaucous

sheaths: round, glaucous, glabrous or scabrous, open, involute on drying; auricles short (1–2 mm long) or absent, often purplish

ligules: membranous (to 1 mm long), truncate, erose to minutely ciliate

blades: flat (10–25 cm long, 2–6 mm wide), involute on drying, rigid, tapering to a short point, **strongly veined** (compare to *Elymus trachycaulus*), **glaucous**, scabrous to rarely pilose adaxially

GROWTH CHARACTERISTICS: growth starts in early spring, dormant in summer, begins growth again in autumn if soil moisture is adequate; reproduces from seeds and rhizomes

FORAGE VALUES: fair to good for all classes of livestock, fair for pronghorn and other wildlife; cures well, making good winter forage

HABITATS: prairies, foothills, swales, alkaline meadows, ditch banks, and roadsides; occurs in all soil textures, but most abundant in finely textured soils

Bluebunch wheatgrass
Pseudoroegneria spicata (Pursh) Á. Löve

SYN = *Agropyron inerme* (Scribn. & J.G. Sm.) Rydb., *Agropyron spicatum*
(Pursh) Scribn. & J.G. Sm., *Agropyron vaseyi* Scribn. & J.G. Sm.,
Elymus spicatus (Pursh) Gould

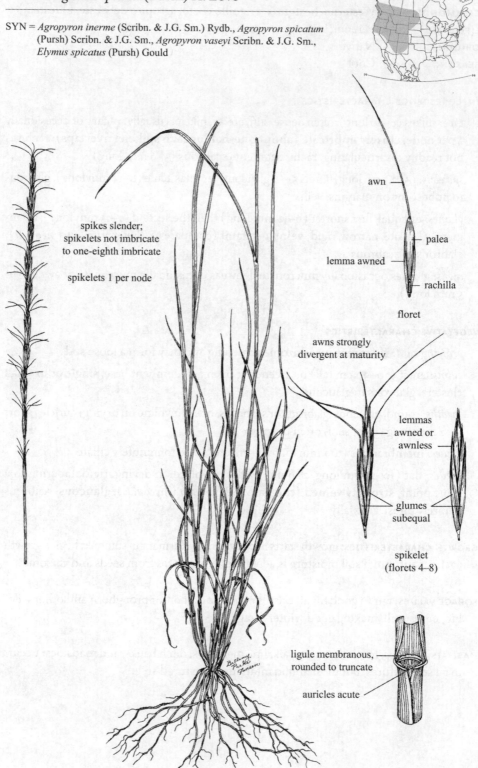

spikes slender;
spikelets not imbricate
to one-eighth imbricate

spikelets 1 per node

awn

palea

lemma awned

rachilla

floret

awns strongly
divergent at maturity

lemmas
awned or
awnless

glumes
subequal

spikelet
(florets 4–8)

ligule membranous,
rounded to truncate

auricles acute

TRIBE:	TRITICEAE
SPECIES:	*Pseudoroegneria spicata* (Pursh) Á. Löve
COMMON NAME:	Bluebunch wheatgrass (triguillo de Arizona, azúl, agropiron, beardless wheatgrass)
LIFE SPAN:	Perennial
ORIGIN:	Native
SEASON:	Cool

INFLORESCENCE CHARACTERISTICS

type: spikes (6–15 cm long), slender, often drooping; spikelets 1 per node, usually only 5–14 per spike, **not imbricate to one-eighth imbricate**; rachises not readily disarticulating, rachis segments 1–2 cm long

spikelets: 1–2 cm long, florets 4–8; lemmas acute (7–11 mm long), glabrous or scabrous; veins 5, faint

glumes: subequal (4.5–11 mm long), not bowed out at the base; second longest, narrow, obtuse to acute, about one-half as long as spikelets; veins 4–5, glabrous or scabrous

awns: **lemmas awnless or awned**; awns strongly divergent at maturity (1–2.5 cm long); glumes awnless, rarely mucronate

VEGETATIVE CHARACTERISTICS

growth habit: cespitose or rarely with short rhizomes, bunches large (up to 15 cm in diameter at the base)

culms: erect to ascending (0.2–1 m tall), slender, glaucous, glabrous or puberulent below the nodes

sheaths: round, glabrous to appressed-puberulent, prominently veined; margins overlapping, open; old sheaths strongly persisting; auricles acute (0.1–1 mm long), clasping, occasionally reddish

ligules: membranous (0.5–2 mm long), rounded to truncate, erose to ciliate

blades: loosely involute to flat (5–25 cm long, 1.5–4.5 mm wide), elongate, pubescent or scabrous adaxially; margins white and weakly barbed

GROWTH CHARACTERISTICS: growth begins in April, plants stay green well into the summer; regrowth occurs following autumn rains; reproduces from seeds, tillers, and rarely by rhizomes

FORAGE VALUES: excellent for cattle and horses; good for sheep, elk, pronghorn, and deer; cures well and makes good standing winter forage

HABITATS: plains, mountain slopes, canyons, open woodlands, and stream banks; most abundant in dry soils but will tolerate moist soils; sometimes seeded alone or in mixtures for grazing or hay

Medusahead rye
Taeniatherum caput-medusae (L.) Nevski

SYN = *Elymus caput-medusae* L., *Taeniatherum asperum* (Simonk.) Nevski

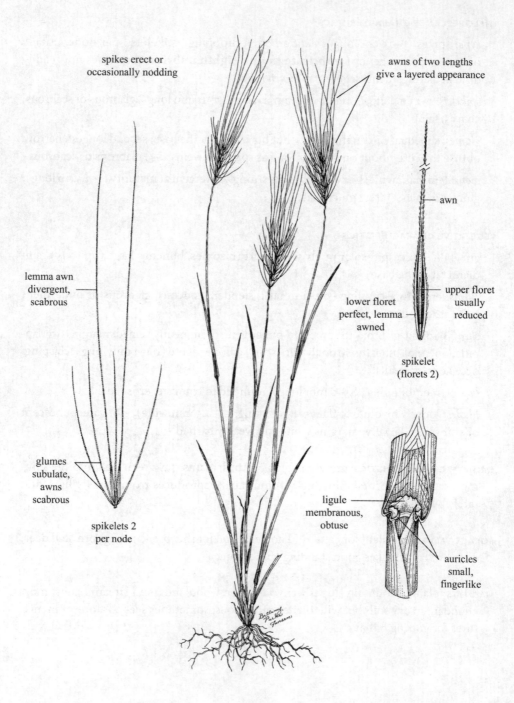

spikes erect or
occasionally nodding

awns of two lengths
give a layered appearance

awn

lemma awn
divergent,
scabrous

upper floret
usually
reduced

lower floret
perfect, lemma
awned

spikelet
(florets 2)

glumes
subulate,
awns
scabrous

spikelets 2
per node

ligule
membranous,
obtuse

auricles
small,
fingerlike

TRIBE:	TRITICEAE
SPECIES:	*Taeniatherum caput-medusae* (L.) Nevski
COMMON NAME:	Medusahead rye
LIFE SPAN:	Annual
ORIGIN:	Introduced (from Europe)
SEASON:	Cool

INFLORESCENCE CHARACTERISTICS

type: spikes (2–6 cm long, excluding awns), erect or occasionally nodding, dense, bristly; spikelets 2 per node

spikelets: 7–12 mm long (excluding awns), florets 2; lower florets perfect, upper florets usually reduced; fertile lemma bodies 5–8 mm long, lanceolate, narrow; veins 3, indistinct, scabrous

glumes: equal (1.5–3.5 cm long), subulate, stiff, indurate below, veinless, glabrous

awns: **glumes awned (1–2.5 cm long)**, scabrous; **lemmas awned (3–10 cm long)**, flat, divergent, scabrous; awns of 2 lengths give a layered appearance

VEGETATIVE CHARACTERISTICS

growth habit: cespitose or solitary

culms: ascending from decumbent and geniculate bases (to 60 cm tall), slender, **weak**, often branching at the base

sheaths: round, slightly inflated; margins open, strigose-puberulent to glabrous; auricles small (0.1–0.5 mm long), fingerlike, rarely absent

ligules: membranous (0.2–0.5 mm long), obtuse, erose

blades: involute (3–10 cm long, 1–3 mm wide), glabrous to puberulent; margins sometimes ciliate

GROWTH CHARACTERISTICS: starts growth in March or April, matures May to June, produces abundant seeds, reproduces from seeds; an aggressive, rapidly spreading weed

LIVESTOCK LOSSES: stiff awns may cause injury to grazing animals by working into the ears, eyes, noses, and tongues; may contaminate fleece

FORAGE VALUES: poor for livestock early in the spring, becomes worthless with production of inflorescences; worthless for wildlife at all times

HABITATS: open ground, disturbed areas, waste places, and deteriorated rangelands; adapted to a broad range of soil types

Intermediate wheatgrass
Thinopyrum intermedium (Host) Barkworth & D.R. Dewey

SYN = *Agropyron intermedium* (Host) P. Beauv., *Agropyron trichophorum*
(Link) K. Richt., *Elymus hispidus* (Opiz) Melderis, *Elytrigia intermedia*
(Host) Nevski

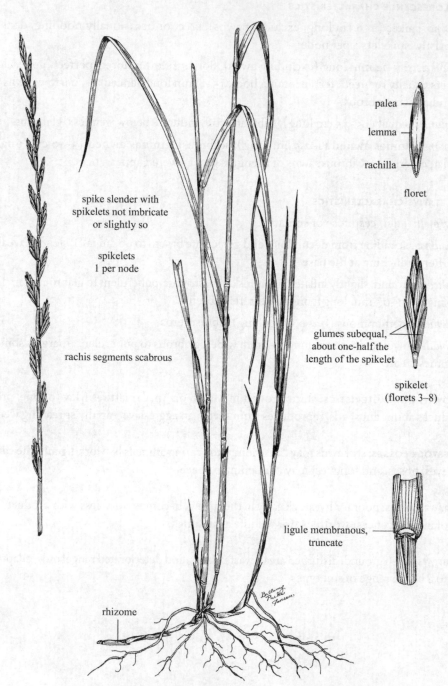

palea

lemma

rachilla

floret

spike slender with
spikelets not imbricate
or slightly so

spikelets
1 per node

glumes subequal,
about one-half the
length of the spikelet

spikelet
(florets 3–8)

rachis segments scabrous

ligule membranous,
truncate

rhizome

TRIBE:	TRITICEAE
SPECIES:	*Thinopyrum intermedium* (Host) Barkworth & D.R. Dewey
COMMON NAME:	Intermediate wheatgrass (triguillo intermedio)
LIFE SPAN:	Perennial
ORIGIN:	Introduced (from Eurasia)
SEASON:	Cool

INFLORESCENCE CHARACTERISTICS

type: spikes (8–25 cm long), slender; spikelets 1 per node, **not imbricate or slightly so**, may be curved away from the rachises at maturity; rachis not readily disarticulating, rachis segments scabrous (6–15 mm long)

spikelets: 1–2 cm long, florets 3–8; lemmas lanceolate (7–11 mm long), broad, blunt, glabrous to hirsute

glumes: subequal (4–9 mm long); second longest, about one-half the length of the spikelets, lanceolate, **apices blunt to acute**, usually distinctively notched, glabrous to hirsute, distinctly veined (compare to *Pascopyrum smithii*)

awns: lemmas may be mucronate

VEGETATIVE CHARACTERISTICS

growth habit: **rhizomatous**, forming a sod

culms: erect (0.4–1.2 m tall), robust, glabrous, glaucous; nodes sometimes hairy

sheaths: round, mostly glabrous; upper margins ciliate, open; auricles acute (0.5–1 mm long)

ligules: membranous (1–2 mm long), truncate, erose to entire

blades: flat to loosely involute (10–40 cm long, 5–15 mm wide), stiff, broad at bases and tapering to points, strongly veined, green or glaucous, glabrous to scabrous to pilose adaxially

GROWTH CHARACTERISTICS: starts growth in early spring; matures June to August; little growth during the summer, even with adequate moisture; reproduces from seeds, tillers, and rhizomes; responds to nitrogen fertilizer

FORAGE VALUES: good for all classes of livestock, fair for wildlife; produces good to excellent hay if cut early; forage cures relatively well and remains palatable; commonly seeded for pasture and hay

HABITATS: meadows, pastures, hills, roadsides, and disturbed sites; dryland and irrigated pastures and hay meadows, adapted to a broad range of soil textures and soil moisture conditions

GRASSLIKE PLANTS

Threadleaf sedge
Carex filifolia Nutt.

SYN = *Carex elyniformis* Porsild, *Olotrema filifolia* (Nutt.) Raf., *Uncina filifolia* (Nutt.) Nees

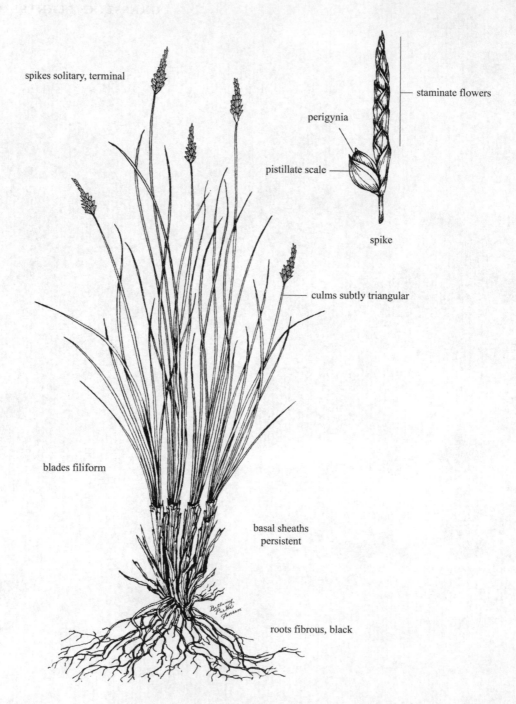

spikes solitary, terminal

staminate flowers

perigynia

pistillate scale

spike

culms subtly triangular

blades filiform

basal sheaths persistent

roots fibrous, black

FAMILY:	CYPERACEAE
SPECIES:	*Carex filifolia* Nutt.
COMMON NAME:	Threadleaf sedge (blackroot)
LIFE SPAN:	Perennial
ORIGIN:	Native
SEASON:	Cool

GROWTH FORM: monoecious, grasslike herbs (5–30 cm tall), erect; densely cespitose with old basal sheaths persistent; starts growth in early spring, flowers April to May, reproduces from seeds and tillers

INFLORESCENCE CHARACTERISTICS

type: spikes (5–30 mm long, 2–6 mm wide), without subtending bracts, comprised of staminate flowers above and pistillate flowers below; solitary, terminal, compact, cylindrical, generally tapering to both ends

flowers: unisexual; staminate flowers 3–25, stamens 3; pistillate scales broadly obovate, apices obtuse; margins hyaline, broad, conspicuous, hiding the perigynia, white to gold, reddish-brown in the center; perigynia 1–15, erect or ascending, obovoid to obovoid-orbicular (3–4 mm long, 2–3 mm wide); ribs 2, obscure; stramineous; abruptly contracted into a short, cylindrical, obliquely cut beaks (0.2–0.5 mm long); stigmas 3

fruits: achenes; obovoid to elliptic (2.3–3 mm long), triangular in cross section, brown to black

VEGETATIVE CHARACTERISTICS

leaves: grasslike; **blades filiform, involute** (3–20 cm long, 0.3–1 mm wide), stiff, mostly basal, generally 2–3 per culm, light green to yellowish-green, glabrous; margins scabrous; sheaths persistent, brown near bases, glabrous, fibrillose when old

culms: erect (5–30 cm tall), subtly triangular, filiform, stiff, wiry

other: roots fibrous, stout, black

HISTORICAL, FOOD, AND MEDICINAL USES: some Native Americans used culm bases for famine food

LIVESTOCK LOSSES: none

FORAGE VALUES: good for cattle, excellent for sheep, horses, and wildlife; extremely valuable early spring forage, maintains high palatability throughout the growing season; retains good forage quality during winter

HABITATS: prairies, hills, ridges, and valleys; adapted to a broad range of soil types; most abundant on dry soils

Elk sedge
Carex geyeri Boott

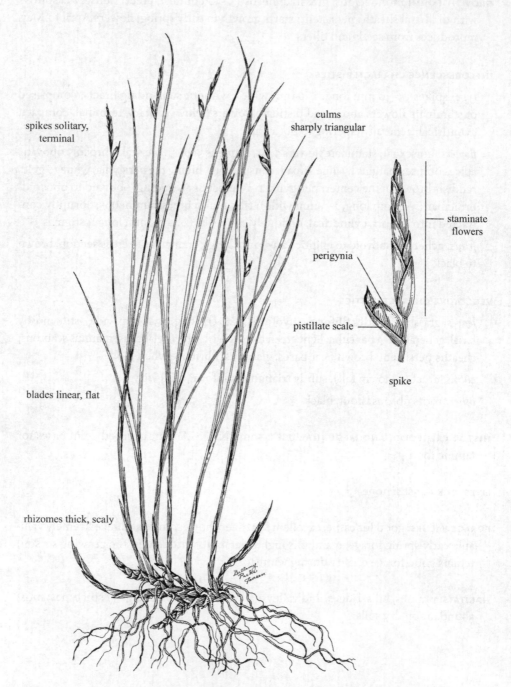

spikes solitary, terminal

culms sharply triangular

staminate flowers

perigynia

pistillate scale

blades linear, flat

spike

rhizomes thick, scaly

FAMLY:	CYPERACEAE
SPECIES:	*Carex geyeri* Boott
COMMON NAME:	Elk sedge (Geyer sedge, pine sedge)
LIFE SPAN:	Perennial
ORIGIN:	Native
SEASON:	Cool

GROWTH FORM: monoecious, grasslike herbs (10–50 cm tall), erect; rhizomatous; rhizomes thick, scaly; starts growth in early spring, flowers April to June, reproduces from seeds and rhizomes

FLORAL AND FRUIT CHARACTERISTICS

type: spikes (5–25 mm long, **1.5–3 mm wide**); solitary, terminal, **cylindrical above**, light brown; staminate portions elevated on short peduncles

flowers: unisexual; staminate flowers 2–15, stamens 3; pistillate scales usually longer and wider than the perigynia, brownish to greenish, often with a lighter midvein, margins hyaline, lower scales short-awned and surpassing the perigynium; perigynia 1 to few, oblong-obovoid (5.5–7 mm long, 2.5–3.5 mm wide), ribs 2, rather abruptly contracted above to the scarcely beaked tip; stigmas 3

fruits: achenes; obovoid (4–5 mm long), sharply triangular in cross section, sides concave, completely filling the perigynia

VEGETATIVE CHARACTERISTICS

leaves: **grasslike**; blades linear (10–50 cm long, 1.5–4 mm wide), short at flowering and then elongating, commonly 2 per culm, veins parallel, flat, coriaceous, margins scabrous, tips often withered; sheaths tight; basal sheaths reddish-brown to dark brown

culms: sharply triangular, rough above, stiffly erect; numerous sterile culms present

HISTORICAL, FOOD, AND MEDICINAL USES: some Native Americans boiled and ate the culm bases

FORAGE VALUES: fair to good for cattle, fair for sheep and deer; grazed in early spring and autumn since it starts growth earlier than most forages and remains green late into the autumn; good for elk in winter

HABITATS: slopes, dry montane and submontane grasslands, open woodlands, and dry meadows at high and mid-elevations; adapted to a broad range of soil types

Nebraska sedge
Carex nebrascensis Dewey

SYN = *Carex jamesii* Schwein.

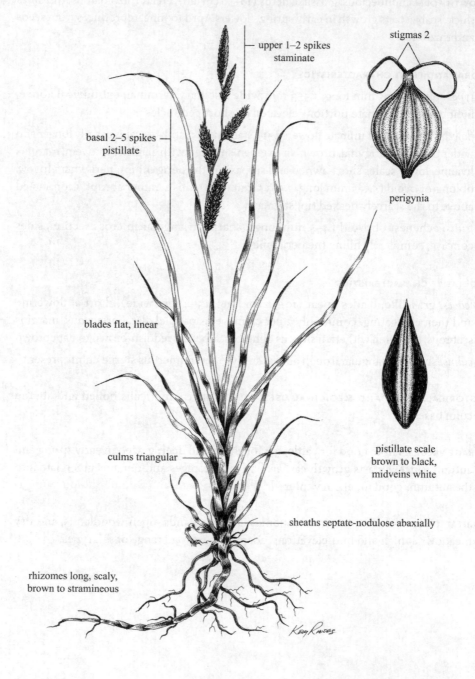

stigmas 2

perigynia

upper 1–2 spikes staminate

basal 2–5 spikes pistillate

blades flat, linear

culms triangular

pistillate scale brown to black, midveins white

sheaths septate-nodulose abaxially

rhizomes long, scaly, brown to stramineous

Kay Rhodes

FAMILY:	CYPERACEAE
SPECIES:	*Carex nebrascensis* Dewey
COMMON NAME:	Nebraska sedge
LIFE SPAN:	Perennial
ORIGIN:	Native
SEASON:	Cool

GROWTH FORM: monoecious, grasslike herbs (25–110 cm tall), erect; rhizomatous (occasionally stoloniferous); rhizomes long, scaly, brown to stramineous; flowers May to July, reproduces from seeds and rhizomes

FLORAL AND FRUIT CHARACTERISTICS

type: spikes (3–6) unisexual or 1–2 androgynous; upper 1–2 spikes staminate (1.5–7 cm long, 3–9 mm wide), broadly linear, sessile; basal 2–5 spikes pistillate (1.5–6 cm long, 5–9 mm wide), occasionally androgynous, cylindrical; lowest bract leaflike, usually exceeding inflorescence, not sheathing

flowers: unisexual; staminate flowers several per spike; **scales brownish- to purplish-black**, margins hyaline; pistillate flowers 30–150 per spike; **scales brown to black**, lanceolate, obtuse to acuminate, about the same size as the perigynia, **midveins white**, excurrent; perigynia stramineous with reddish dots, oblong to obovate (3–4 mm long, 2 mm wide), flattened planoconvex or biconvex, strongly veined, beaks bidentate (0.5–1 mm long); stigmas 2

fruits: achenes; lenticular (1.5–2.2 mm long, about 1.2 mm wide)

VEGETATIVE CHARACTERISTICS

leaves: grasslike; blades flat, linear (10–40 cm long, 3–12 mm wide), 8–15 per culm, mostly on the lower one-third of the culms, **compressed or channeled near bases**, glaucous, thick, firm; sheaths septate-nodulose abaxially, hyaline and yellowish-brown adaxially

culms: triangular, stout, erect, scabrous

HISTORICAL, FOOD, AND MEDICINAL USES: some Native Americans ate raw stem bases as a famine food

FORAGE VALUES: poor to fair for sheep, fair to good for cattle and wildlife; although not as palatable as some sedge species, it is a valuable late-season forage; makes fair to good hay

HABITATS: wet meadows, swamps, streams, marshes, edges of lakes, ponds, and ditches; adapted to a broad range of soil textures

Beaked sedge
Carex utriculata Boott

SYN = *Carex ampullacea* Gooden., *Carex rostrata* Stokes, *Carex vesicaria* L.

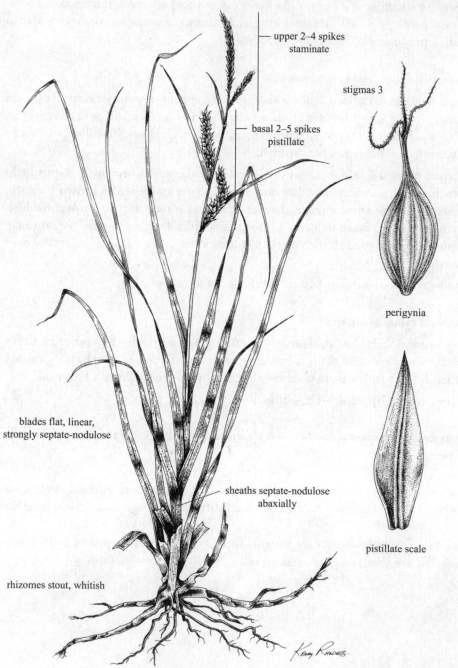

upper 2–4 spikes staminate

basal 2–5 spikes pistillate

stigmas 3

perigynia

blades flat, linear, strongly septate-nodulose

sheaths septate-nodulose abaxially

pistillate scale

rhizomes stout, whitish

FAMILY:	CYPERACEAE
SPECIES:	*Carex utriculata* Boott
COMMON NAME:	Beaked sedge (Northwest Territory sedge)
LIFE SPAN:	Perennial
ORIGIN:	Native
SEASON:	Cool

GROWTH FORM: monoecious, grasslike herb (0.4–1.2 m tall), erect; rhizomatous, sod-forming; rhizomes stout, whitish; flowers June to August; reproduces by seeds and rhizomes

FLORAL AND FRUIT CHARACTERISTICS

type: spikes unisexual or 1–2 androgynous; upper 2–4 spikes staminate (1–6 cm long, 3–4 mm wide), narrowly linear, upper ones on peduncles, lower ones sessile; basal 2–5 spikes pistillate (1–10 cm long, 8–12 mm wide), occasionally androgynous, flowers 40–150, dense, cylindrical, sessile above to short-peduncled below; bracts leaflike, shorter than to slightly exceeding inflorescences

flowers: unisexual; pistillate scale veins 3, margins hyaline; perigynia oval to ovoid (3.5–8 mm long, 2.5–3.5 mm wide), **yellowish-green to brown**, inflated, subterete; veins 8–16, strong; beaks bidentate (1–2 mm long); stigmas 3, styles S-curved toward base

fruits: achenes; trigonous (1.7–2 mm long, 1.2–1.4 mm wide), styles persistent

VEGETATIVE CHARACTERISTICS

leaves: grasslike; blades flat, linear (20–60 cm long, 4–12 mm wide), 4–10 per culm, **strongly septate-nodulose**, thickened at the bases, stiff; margins revolute, more or less channeled at bases; sheaths brown, lightly tinged pinkish-red, abaxially septate-nodulose, adaxially white hyaline

culms: stout, erect, glabrous to scabrous; bluntly triangular below the inflorescence; bases thickened, spongy, reddish

HISTORICAL, FOOD, AND MEDICINAL USES: some Native Americans used stem bases for food

FORAGE VALUES: low palatability, rarely utilized by domestic livestock in North America, but reported to be excellent forage in Iceland and Siberia; good for elk, caribou, and moose

HABITATS: wet meadows, marshes, swamps, fens, lake and pond edges, springs

Hardstem bulrush

Schoenoplectus acutus (Muhl. *ex* Bigelow) Á. Löve & D. Löve

SYN = *Schoenoplectus lacustris* (L.) Palla, *Scirpus acutus* Muhl. *ex* Bigelow

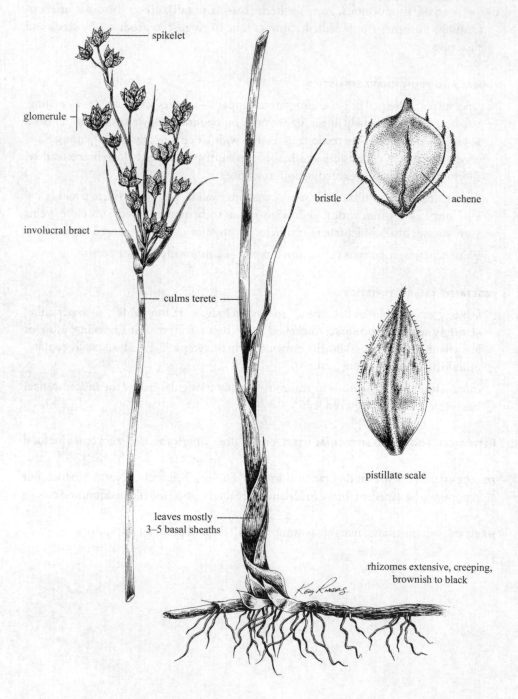

spikelet

glomerule

involucral bract

culms terete

bristle

achene

pistillate scale

leaves mostly
3–5 basal sheaths

rhizomes extensive, creeping,
brownish to black

FAMILY:	CYPERACEAE
SPECIES:	*Schoenoplectus acutus* (Muhl. *ex* Bigelow) Á. Löve & D. Löve
COMMON NAME:	Hardstem bulrush (tule)
LIFE SPAN:	Perennial
ORIGIN:	Native
SEASON:	Cool

GROWTH FORM: grasslike herb (1–3 m tall, 0.8–2.3 cm wide at base), rhizomatous; forming dense colonies from extensive rhizomes; rhizomes creeping, brownish to black; flowers June to mid-August, reproduces from seeds and rhizomes

FLORAL AND FRUIT CHARACTERISTICS

type: umbelliform or paniculiform (3–10 cm long), appearing laterally with up to 160 spikelets, compact to open; spikelets in glomerules of 2–15 (rarely solitary), pendulous with short to long pedicels; principal involucral bract appearing as a continuation of the culm (1.5–10 cm long), solitary or rarely 2–3

spikelets: perfect; flowers 20–50, ovoid to linear (5–15 mm long, 3–5 mm wide), grayish-brown, acute; scales pale-brown with reddish spots (2–8 mm long, 2–2.6 mm wide), scarious, **lower ones puberulent abaxially**; apices acute to cleft, mucronate; perianths of 4–6 bristles, retrorsely barbed, shorter than or equaling achenes; stamens 3, often persisting; style clefts 2–3

fruits: achenes; trigonous (1.8–2.9 mm long, 1.2–1.9 mm wide), light green to dark brown, planoconvex or unequally trigonous, beaks to 0.5 mm long

VEGETATIVE CHARACTERISTICS

leaves: consisting mostly of 3–5 basal sheaths, upper sheaths occasionally with tapering blades to 25 cm long; blades dark olive green

culms: **erect, terete, stout**, dark green, firm, unyielding to crushing when green

HISTORICAL, FOOD, AND MEDICINAL USES: some Native Americans ate young shoots raw or boiled, pollen was used in bread, starchy root was boiled or ground into meal; leaves were woven into mats

FORAGE VALUES: poor for livestock, usually unpalatable; poor to fair for wildlife

HABITATS: emergent in shallow to deep waters of sloughs, ponds, lakes, ditches, and roadsides; in fresh to brackish water

Wire rush
Juncus balticus Willd.

SYN = *Juncus arcticus* Willd.

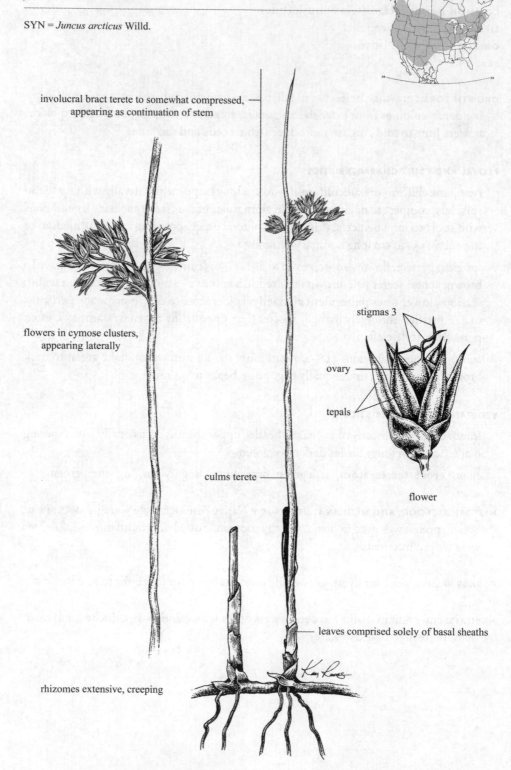

involucral bract terete to somewhat compressed, appearing as continuation of stem

flowers in cymose clusters, appearing laterally

culms terete

stigmas 3

ovary

tepals

flower

leaves comprised solely of basal sheaths

rhizomes extensive, creeping

FAMILY:	JUNCACEAE
SPECIES:	*Juncus balticus* Willd.
COMMON NAME:	Wire rush (Baltic rush)
LIFE SPAN:	Perennial
ORIGIN:	Native
SEASON:	Cool

GROWTH FORM: grasslike herbs (to 1.1 m tall), erect, wiry; from extensively creeping rhizomes, often forming a sod; flowers May to August, reproduces from seeds and rhizomes

FLORAL AND FRUIT CHARACTERISTICS

type: cymose clusters (1–7 cm long and wide); appearing laterally near the top of the plant, dense to spreading; **involucral bracts terete** to somewhat compressed (to 20 cm long), **erect, appearing as a continuation of the stem**, obtuse to mucronate; flowers borne singly, subsessile or on brownish pedicels (3–4 mm long)

flowers: perfect; tepals 6, greenish becoming stramineous or dark brown (3–6 mm long), lanceolate, acute to ovate, margins hyaline, inner tepals shorter than outer tepals; stamens 6; stigmas 3; ovary locules 3, parietal placentation; paired bracteoles below each flower

fruits: capsules; narrowly ovoid, ranging from shorter to longer than perianth, acute, mucronate, dark amber; seeds ovoid-ellipsoid (0.3–0.8 mm long), grayish brown, ends apiculate

VEGETATIVE CHARACTERISTICS

leaves: with the exception of the involucre, comprised solely of basal sheaths, blades and auricles absent; sheaths rounded, clustered near plant bases (2–15 cm long), light brown to reddish-brown

culms: erect, terete, stout (1.5–4 mm thick), crowded, arising from thick, dark brown, creeping rhizomes

other: variable species with several varieties recognized in North America

HISTORICAL, FOOD, AND MEDICINAL USES: some Native Americans made mats and baskets from the stems

FORAGE VALUES: fair to good for cattle and sheep when actively growing, often more palatable in hay than when growing, worthless when mature

HABITATS: marshes, wet meadows, ditches, shallow streams, seeps, lake and pond shores; often growing in alkaline soils; frequently abundant

FORBS AND WOODY PLANTS

Skunkbrush
Rhus aromatica Aiton

SYN = *Rhus trilobata* Nutt., *Schmalzia trilobata* (Nutt.) Greene

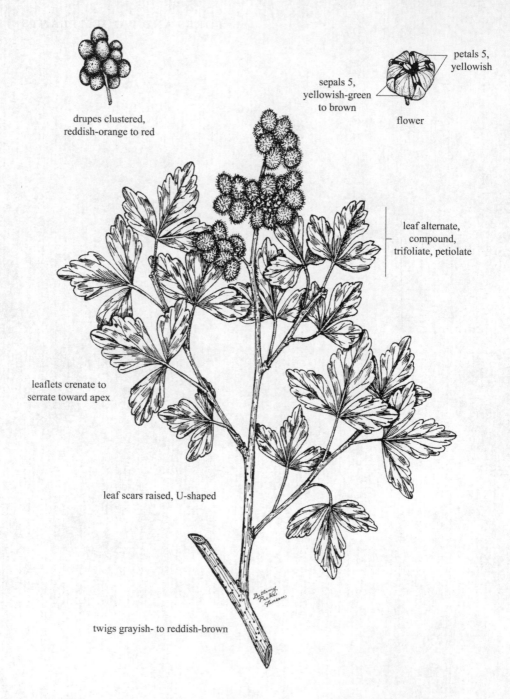

drupes clustered,
reddish-orange to red

sepals 5,
yellowish-green
to brown

petals 5,
yellowish

flower

leaf alternate,
compound,
trifoliate, petiolate

leaflets crenate to
serrate toward apex

leaf scars raised, U-shaped

twigs grayish- to reddish-brown

FAMILY:	ANACARDIACEAE
SPECIES:	*Rhus aromatica* Aiton
COMMON NAME:	Skunkbrush (skunkbrush sumac, fragrant sumac, lemita, lantrisco, agrillo)
LIFE SPAN:	Perennial
ORIGIN:	Native
SEASON:	Cool

GROWTH FORM: dioecious (polygamo-dioecious) shrubs (to 2.5 m tall), dense, branches from several trunks; branches ascending; flowers March to May, reproduces from seeds

FLORAL AND FRUIT CHARACTERISTICS

inflorescences: globular clusters or catkinlike spikes (6 cm long, 3 cm wide), terminal

flowers: regular, unisexual or perfect, usually opening before the leaves emerge; sepals 5, obtuse (1 mm long), united at base, yellowish-green to brown; petals 5, obtuse (1–2 mm long), spreading at maturity, yellowish; stamens 5

fruits: drupes (4–7 mm in diameter); **globose, reddish-orange to red, lightly to densely hirsute**, hairs simple to glandular (1–1.3 mm long); occurring in globular clusters, often confluent; persistent in winter; seeds 1

other: flowers formed late in the previous season and before, or as leaves appear in early spring

VEGETATIVE CHARACTERISTICS

leaves: alternate, compound, trifoliate; terminal leaflets obovate (2–9 cm long, 2–8 cm wide), lateral leaflets obovate (3.5–5 cm long, 2–4 cm wide), crenate or lobed to serrate toward apices, entire at base; adaxially dull, glabrous or pubescent; abaxially pale, densely to lightly pubescent; leaflets sessile to subsessile, apices and bases acute to rounded; petioles to 1.5 cm long; unpleasantly aromatic

stems: twigs grayish- to reddish-brown, slender, glabrous or pubescent; **leaf scars elongate, raised, U-shaped**; bark dark brown, smooth to fissured; fragrant when bruised

other: highly variable species consisting of many varieties

HISTORICAL, FOOD, AND MEDICINAL USES: some tribes of Native Americans made use of the drupes as food and medicine and made drinks with a flavor similar to lemonade; slender shoots were used for basket-weaving

LIVESTOCK LOSSES: none

FORAGE VALUES: poor to fair for cattle, horses, and sheep; good browse for wildlife and goats; birds and small mammals eat the fruits

HABITATS: prairies, ravines, thickets, fence lines, open woodlands, and roadsides; adapted to a wide range of soils; planted for wildlife habitat and erosion control

Poison hemlock
Conium maculatum L.

SYN = *Coriandrum maculatum* (L.) Roth

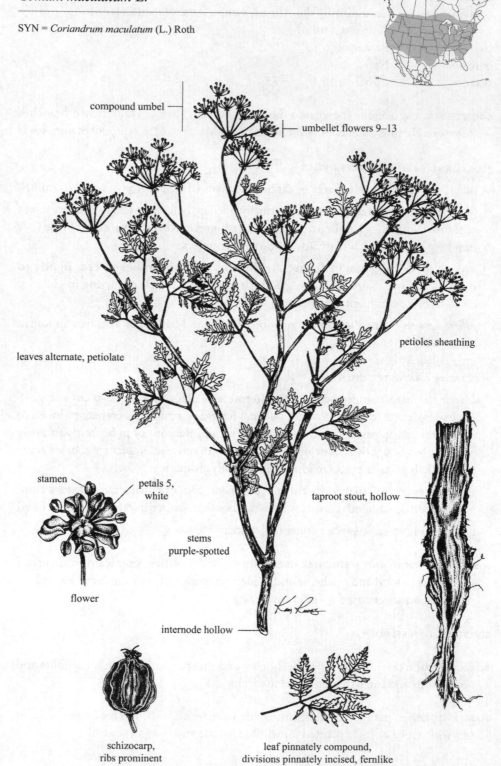

compound umbel

umbellet flowers 9–13

petioles sheathing

leaves alternate, petiolate

stamen

petals 5,
white

taproot stout, hollow

stems
purple-spotted

flower

internode hollow

schizocarp,
ribs prominent

leaf pinnately compound,
divisions pinnately incised, fernlike

FAMILY:	APIACEAE
SPECIES:	*Conium maculatum* L.
COMMON NAME:	Poison hemlock (cicuta major, hemlock, spotted hemlock)
LIFE SPAN:	Perennial (biennial)
ORIGIN:	Introduced (from Europe)
SEASON:	Cool

GROWTH FORM: forbs (0.5–3 m tall); erect, stout, highly branched; flowers May to August; reproduces by seed

FLORAL AND FRUIT CHARACTERISTICS

inflorescences: **umbels compound** (4–7 cm wide), convex; involucre bracts short, ovate, acuminate; rays 8–17 (1.5–2.5 cm long), unequal, spreading to ascending; involucres of numerous bractlets similar to bracts, midveins conspicuous, shorter than pedicels; umbellet flowers 9–13; pedicels spreading (4–6 mm long)

flowers: perfect, regular; sepals absent; petals 5 (1–1.5 mm long), notched, clawed, white

fruits: schizocarps; broadly ovoid (2–3 mm long), laterally flattened, glabrous, grayish-brown; **ribs prominent**, undulate, crenate; seeds 2; seeds compressed to concave, deeply and narrowly sulcate

VEGETATIVE CHARACTERISTICS

leaves: alternate, pinnately compound; divisions pinnately incised, broadly ovate to triangular (15–30 cm long, 5–30 cm wide), pinnately compound 2–4 times, fernlike; lobes oblong to lanceolate; glabrous; petioles sheathing

stems: erect from stout and hollow taproots, highly branched, glabrous, glaucous, **purple-spotted; internodes hollow**

other: plants are biennials producing a rosette of leaves the summer or autumn of the first year and flowering the second year; plants have a sour, musty or mousy odor; may be fatal if eaten by humans

HISTORICAL, FOOD, AND MEDICINAL USES: a folktale says that an extract from this species was used to put Socrates to death, his symptoms would suggest that this could be true; all parts of the plant are poisonous

LIVESTOCK LOSSES: highly poisonous to livestock, contains several alkaloids; alkaloid poisons in the roots, leaves, and fruits make them the most toxic

FORAGE VALUES: usually unpalatable

HABITATS: moist soils of pastures, meadows, roadsides, floodplains, stream banks, and disturbed sites; shaded or open ground

Yarrow
Achillea millefolium L.

SYN = *Achillea lanulosa* Nutt., *Chamaemelum millefolium* (L.) E.H.L. Krause

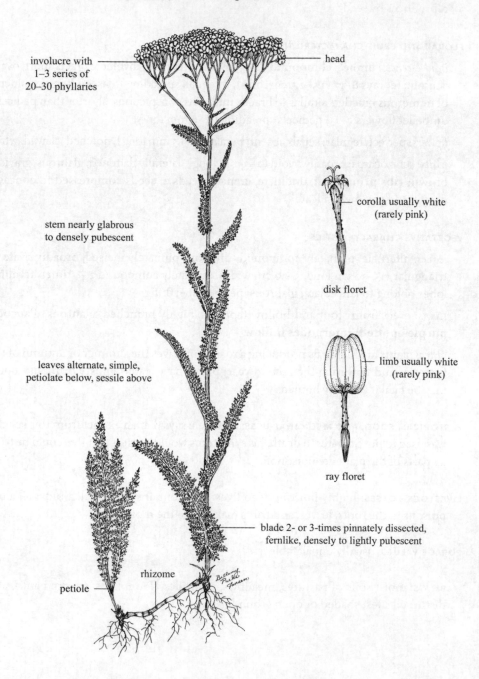

heads numerous in compound corymbiform arrangements

involucre with 1–3 series of 20–30 phyllaries

head

corolla usually white (rarely pink)

disk floret

stem nearly glabrous to densely pubescent

limb usually white (rarely pink)

leaves alternate, simple, petiolate below, sessile above

ray floret

blade 2- or 3-times pinnately dissected, fernlike, densely to lightly pubescent

petiole

rhizome

FAMILY:	ASTERACEAE
TRIBE:	ANTHEMIDEAE
SPECIES:	*Achillea millefolium* L.
COMMON NAME:	Yarrow (ramo de novia, hierba del oro, western yarrow, milfoil, milenrama)
LIFE SPAN:	Perennial
ORIGIN:	Native
SEASON:	Cool

GROWTH FORM: forbs (0.2–1 m tall), rhizomatous; erect, arising singly or as a loose cluster of stems; flowers April to October, reproduces from seeds and rhizomes; colonial

FLORAL AND FRUIT CHARACTERISTICS

inflorescences: heads numerous (5–7 mm long, to 8 mm in diameter) in simple or **compound corymbiform arrangements, flat**- to round-topped; phyllaries imbricate, 20–30, greenish

flowers: perfect, heads radiate; involucres with 1–3 series of 30–40 phyllaries; phyllaries imbricate, green with brown midveins, margins scarious; ray florets 5–8, typically about 5; limbs ovate (2–5 mm long), white (rarely pink, pinkish-white, or purple), teeth 5; disk florets 10–20, corolla tubes somewhat compressed, white (rarely pink or pinkish-white)

fruits: cypselae; oblong, compressed, glabrous; pappi absent

VEGETATIVE CHARACTERISTICS

leaves: alternate, simple, widely spaced; **blades pinnately dissected 2–3 times** (3–15 cm long, 5–30 mm wide), reduced upwards, fernlike, lanceolate in outline, densely to lightly white-villous; basal blades petiolate; cauline leaves sessile; aromatic

stems: erect, simple to sparingly branched above, nearly glabrous to densely pubescent; aromatic

HISTORICAL, FOOD, AND MEDICINAL USES: a decoction of leaves and flowers was used by some Blackfoot as an eyewash; some Winnebago steeped whole plants and poured the liquid into aching ears; green leaves were used to relieve itching, chewed for toothaches, and used as a mild laxative; leaves boiled for tea were used as a cold remedy; during the Civil War, powdered leaves were applied to wounds to stop bleeding, and its common name was soldiers' woundwort

LIVESTOCK LOSSES: not generally considered to be poisonous, but may contain toxic alkaloids, glycosides, and volatile oils

FORAGE VALUES: poor to fair for cattle and fair to good for sheep; usually grazed only when green; heads may be eaten by pronghorn, deer, and sheep; palatability and forage quality vary greatly and depend on locality and seasonal development

HABITATS: prairies, meadows, plains, pastures, open woodlands, roadsides, and disturbed sites; adapted to a wide range of soils; frequently planted in flower gardens

Silver sagebrush
Artemisia cana Pursh

SYN = *Seriphidium canum* (Pursh) W.A. Weber

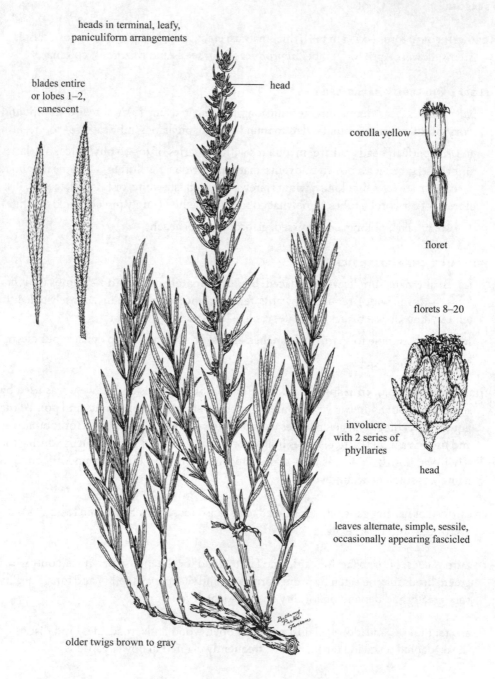

heads in terminal, leafy,
paniculiform arrangements

head

blades entire
or lobes 1–2,
canescent

corolla yellow

floret

florets 8–20

involucre
with 2 series of
phyllaries

head

leaves alternate, simple, sessile,
occasionally appearing fascicled

older twigs brown to gray

FAMILY:	ASTERACEAE
TRIBE:	ANTHEMIDEAE
SPECIES:	*Artemisia cana* Pursh
COMMON NAME:	Silver sagebrush (istafiate, estafiate, white sagebrush, dwarf sagebrush, hoary sagebrush)
LIFE SPAN:	Perennial
ORIGIN:	Native
SEASON:	Evergreen

GROWTH FORM: shrubs (to 1.5 m tall), rhizomatous; densely branched from the base, crowns rounded; often forming colonies from extensive rhizomes; flowers August to September, reproduces from seeds and rhizomes

FLORAL AND FRUIT CHARACTERISTICS

inflorescences: heads in paniculiform arrangements (15–30 cm long, 2–6 cm wide), terminal, leafy, dense; heads in groups of 3–10, subtended by leaflike phyllaries often surpassing the heads

flowers: perfect, heads discoid; involucres campanulate (3.5–5 mm tall, 2.3–4.5 mm in diameter) with 2 series of phyllaries; outer series phyllaries ovate to lanceolate, acute, densely canescent; inner series phyllaries elliptical, canescent to glabrous, margins scarious; florets 8–20, corollas yellowish and tubular (2.5 mm long)

fruits: cypselae; cylindrical (2.5 mm long, 1 mm in diameter), angular, ribs 5–6, light brown, resinous; pappi absent

VEGETATIVE CHARACTERISTICS

leaves: alternate, occasionally appearing fascicled, simple, silvery or grayish **canescent**, sessile; **blades linear** or narrowly elliptical, or lanceolate (2–9 cm long, 1–7 mm wide), acute; margins entire to irregularly lobes 1–2

stems: twigs ascending, green to stramineous, rigid, finely canescent; older twigs brown to gray, glabrous; trunk bark tan to grayish-brown, exfoliating into long fibrous strips

other: plants often dry to a goldish-hue with yellowish stems, mildly aromatic

HISTORICAL, FOOD, AND MEDICINAL USES: decoction was used by some Native Americans to stop coughing; an extract was thought to restore hair

LIVESTOCK LOSSES: none

FORAGE VALUES: good to excellent in autumn and winter for cattle and sheep, increases under browsing by cattle and decreases under browsing by sheep; some varieties have low palatablty

HABITATS: river valleys, terraces, floodplains, and grasslands in moist to moderately dry soils; most abundant in deep loamy to sandy soils, alkali tolerant

Sand sagebrush
Artemisia filifolia Torr.

SYN = *Artemisia plattensis* Nutt., *Oligosporus filifolius* (Torr.) Poljakov

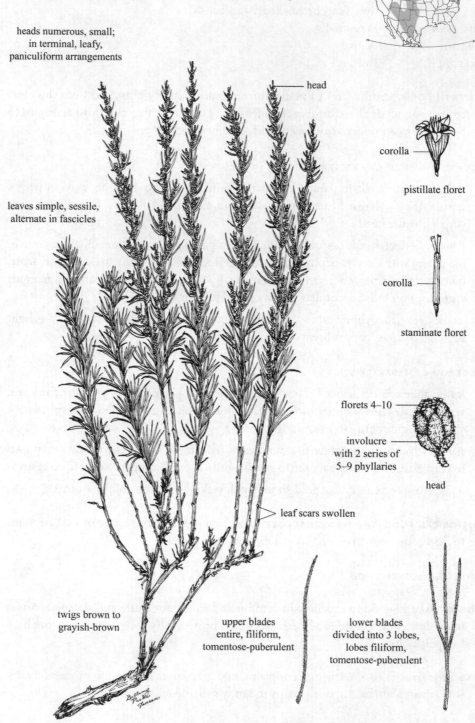

heads numerous, small;
in terminal, leafy,
paniculiform arrangements

head

corolla

pistillate floret

leaves simple, sessile,
alternate in fascicles

corolla

staminate floret

florets 4–10

involucre
with 2 series of
5–9 phyllaries

head

leaf scars swollen

twigs brown to
grayish-brown

upper blades
entire, filiform,
tomentose-puberulent

lower blades
divided into 3 lobes,
lobes filiform,
tomentose-puberulent

FAMILY:	ASTERACEAE
TRIBE:	ANTHEMIDEAE
SPECIES:	*Artemisia filifolia* Torr.
COMMON NAME:	Sand sagebrush (istafiate de arena, artemisa, sandsage, sandhill sage)
LIFE SPAN:	Perennial
ORIGIN:	Native
SEASON:	Warm

GROWTH FORM: shrubs (to 1.8 m tall); freely branching, crowns rounded; flowers August to October, reproduces from seeds

FLORAL AND FRUIT CHARACTERISTICS

inflorescences: heads in **leafy paniculiform arrangements** (15–20 cm long, 5–20 mm wide), terminal, plumelike, subtended by leaflike bracts often surpassing the heads; heads numerous; peduncles short and stiff

flowers: unisexual, heads discoid; involucres globose with 2 series of 5–9 phyllaries (1.8 mm tall), outer series short and indurate; inner series elliptical and thin, obtuse, woolly; pistillate florets 1–4 (0.6–0.7 mm long); staminate florets 3–6 (1.5 mm long); corollas pale yellow

fruits: cypselae; obovoid (0.2–1 mm long, 0.1–0.5 mm in diameter), lobes 5, brown; ribs 4–5, obscure; glabrous; pappi absent

VEGETATIVE CHARACTERISTICS

leaves: alternate in fascicles, simple (3.5–7.5 cm long), **tomentose-puberulent**, grayish-green, sessile; upper blades entire, filiform; lower blades divided into **3 lobes, lobes filiform**, apices acute; margins entire

stems: twigs brown to grayish-brown, erect, slender, freely branching, erect to spreading, striate, pubescent to glabrous, eventually exfoliating into thin shreds; leaf scars swollen

other: entire plant is aromatic

HISTORICAL, FOOD, AND MEDICINAL USES: a decoction of leaves was taken for intestinal worms and stomach problems in Mexico

LIVESTOCK LOSSES: may cause sage sickness in horses

FORAGE VALUES: poor to worthless for cattle, poor to fair for horses and sheep; furnishes fair forage for pronghorn and deer; rarely grazed if other forages are present; cypselae are an important food for grouse and other birds

HABITATS: plains, pastures, and dunes in well-drained sandy soil, generally considered an indicator of sandy soil; often abundant on sand

Fringed sagebrush
Artemisia frigida Willd.

SYN = *Absinthium frigidum* (Willd.) Besser

heads in paniculiform or racemiform arrangements, exserted well above the vegetative mat

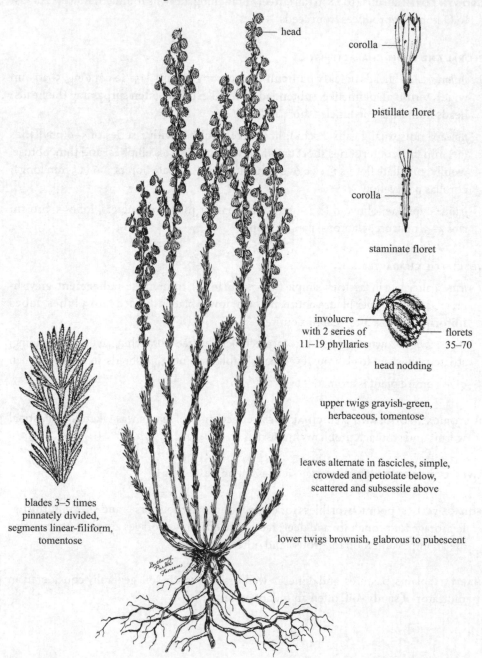

head

corolla

pistillate floret

corolla

staminate floret

involucre with 2 series of 11–19 phyllaries

florets 35–70

head nodding

upper twigs grayish-green, herbaceous, tomentose

leaves alternate in fascicles, simple, crowded and petiolate below, scattered and subsessile above

blades 3–5 times pinnately divided, segments linear-filiform, tomentose

lower twigs brownish, glabrous to pubescent

Bellamy
Piehl
Jansen

FAMILY:	ASTERACEAE
TRIBE:	ANTHEMIDEAE
SPECIES:	*Artemisia frigida* Willd.
COMMON NAME:	Fringed sagebrush (artemisa, prairie sagewort, wormwood)
LIFE SPAN:	Perennial
ORIGIN:	Native
SEASON:	Cool

GROWTH FORM: subshrubs (to 50 cm tall); ascending to decumbent, forming mats; flowers August to September, reproduces from seeds

FLORAL AND FRUIT CHARACTERISTICS

inflorescences: heads in paniculiform or racemiform arrangements (5–30 cm long, 1–10 cm wide), **exerted well above the vegetative mat; leafy**; heads sessile, **nodding**

flowers: pistillate and perfect, heads discoid; involucres loosely tomentose (2–3 mm tall); with 2 series of 11–19 phyllaries, lanceolate to ovate; pistillate florets 10–20; perfect florets 25–50, tubular; corollas 1.5–2 mm long

fruits: cypselae; subcylindrical (1–1.5 mm long), narrowing to bases, ribbed, glabrous; pappi absent

VEGETATIVE CHARACTERISTICS

leaves: alternate in fascicles, simple, persistent; crowded below and scattered above, **tomentose**, silvery-gray; blade outlines ovate (5–12 mm long), **pinnately divided** 3–5 times; segments linear-filiform; petiolate below, subsessile above

stems: twigs branching at the bases; upper stems herbaceous, grayish-green, tomentose; lower twigs brownish, glabrous to pubescent

other: entire plant tomentose, strongly aromatic

HISTORICAL, FOOD, AND MEDICINAL USES: some Native Americans called it "woman sage" and used it to eliminate the greasy smell from dried meat, to bandage cuts after it was chewed, to make mats, fans, menstrual pads, and as toilet paper

LIVESTOCK LOSSES: none

FORAGE VALUES: fair to good for sheep, goats, and pronghorn; poor to fair for cattle; important winter feed for pronghorn, elk, and deer; increases rapidly under heavy grazing; drought tolerant

HABITATS: prairies, plains, abandoned fields, and foothills; most abundant in well-drained sandy or rocky soils; occasionally planted as an ornamental

Cudweed sagewort
Artemisia ludoviciana Nutt.

SYN = *Artemisia albula* Wooton, *Artemisia ghiesbreghtii* Rydb., *Artemisia mexicana* Willd. *ex* Spreng., *Artemisia muelleri* Rydb., *Artemisia redolens* A. Gray, *Artemisia revoluta* Rydb., *Artemisia sulcata* Rydb., *Artemisia vulgaris* C.B. Clarke

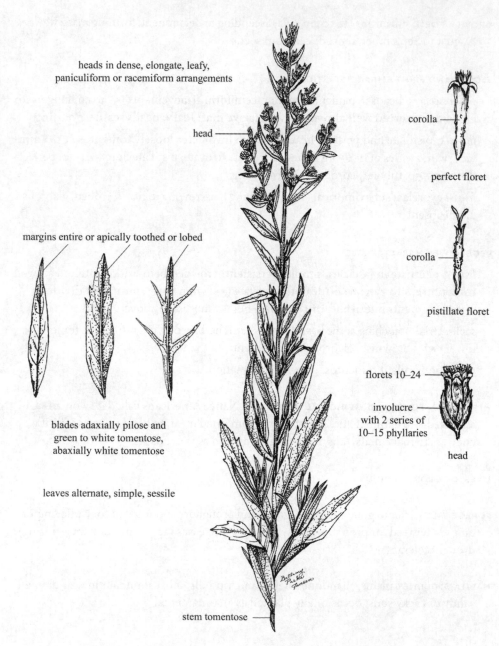

heads in dense, elongate, leafy, paniculiform or racemiform arrangements

head

corolla

perfect floret

corolla

pistillate floret

margins entire or apically toothed or lobed

blades adaxially pilose and green to white tomentose, abaxially white tomentose

leaves alternate, simple, sessile

florets 10–24

involucre with 2 series of 10–15 phyllaries

head

stem tomentose

234

FAMILY:	ASTERACEAE
TRIBE:	ANTHEMIDEAE
SPECIES:	*Artemisia ludoviciana* Nutt.
COMMON NAME:	Cudweed sagewort (rosábari, iztafiate, Louisiana wormwood, white sage, western mugwort)
LIFE SPAN:	Perennial
ORIGIN:	Native
SEASON:	Warm

GROWTH FORM: forbs (0.3–1.2 m tall), rhizomatous; erect, 1 to few stems; flowers August to September, reproduces from seeds and rhizomes, often forming colonies

FLORAL AND FRUIT CHARACTERISTICS

inflorescences: heads in leafy paniculiform or racemiform arrangements (20–50 cm long), dense, elongate; heads fascicled or on spicate branches, sessile or subsessile; erect or nodding

flowers: pistillate and perfect, heads discoid; involucres campanulate or turbinate (2–4 mm tall, 2–5 mm in diameter) with 2 series of 10–15 phyllaries; outer series small, tomentose; inner series elliptical and obtuse, tomentose; pistillate florets 6–12 (1 mm long); perfect florets 5–12 (1.4–2.5 mm long); corollas yellowish, sometimes reddish-tinged

fruits: cypselae; elliptical (about 1 mm long), brownish, smooth, glabrous, obscurely veined; pappi absent

other: inflorescences vary from compact and dense to loose and relatively open

VEGETATIVE CHARACTERISTICS

leaves: alternate, simple, mostly cauline, adaxially pilose and green to white tomentose, **abaxially white tomentose**; blades linear to lanceolate or elliptical (2–12 cm long, to 1 cm wide, excluding the lobes), reduced above; **margins entire or apically toothed or lobed; sessile**

stems: erect, few to numerous, simple or shortly branched above, slender, tomentose

other: variable in size, leaves, and pubescence; all plant parts are aromatic

HISTORICAL, FOOD, AND MEDICINAL USES: some Native Americans called it "man sage" and used it for ceremonial and purification purposes; to deodorize feet; cure headaches, treat coughs, hemorrhoids, stomach troubles, and wounds on horses; made into pillows and saddle pads; burned to drive mosquitos away

LIVESTOCK LOSSES: none

FORAGE VALUES: fair to poor for cattle and sheep; somewhat palatable to elk, deer, and pronghorn; relative forage value decreases from south to north in North America

HABITATS: prairies, plains, foothills, open woodlands, disturbed sites, and roadsides

Black sagebrush
Artemisia nova A. Nelson

SYN = *Artemisia arbuscula* Nutt. subsp. *nova* (A. Nelson) G.H. Ward,
Seriphidium novum (A. Nelson) W.A. Weber

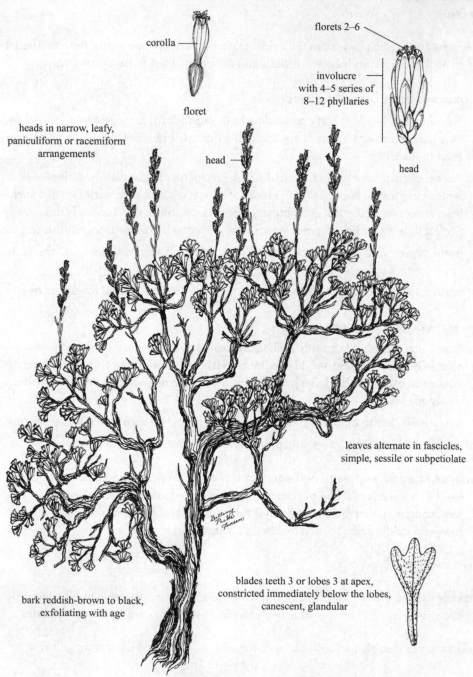

corolla

floret

florets 2–6

involucre
with 4–5 series of
8–12 phyllaries

head

heads in narrow, leafy,
paniculiform or racemiform
arrangements

head

leaves alternate in fascicles,
simple, sessile or subpetiolate

bark reddish-brown to black,
exfoliating with age

blades teeth 3 or lobes 3 at apex,
constricted immediately below the lobes,
canescent, glandular

FAMILY:	ASTERACEAE
TRIBE:	ANTHEMIDEAE
SPECIES:	*Artemisia nova* A. Nelson
COMMON NAME:	Black sagebrush (black sage)
LIFE SPAN:	Perennial
ORIGIN:	Native
SEASON:	Evergreen

GROWTH FORM: shrubs (to 70 cm tall); branches numerous, decumbent, spreading, crowns rounded; flowers August to September, reproduces from seeds

FLORAL AND FRUIT CHARACTERISTICS

inflorescences: heads in narrow paniculiform or racemiform arrangements (5–15 cm long, 1–2.5 cm wide), narrow, leafy, **elevated above the foliage**; heads few, small, subsessile, erect to drooping, occasionally 2–3 in a cluster, **rust- to gold-colored when mature**

flowers: perfect, heads discoid; involucres narrowly turbinate (2–5 mm tall, 1 mm in diameter) with 4–5 series of 8–12 phyllaries; outer phyllaries short, ovate to elliptical, stramineous to light green, canescent; inner phyllaries elliptical, lightly canescent to glabrous; florets 2–6, tubular (2.5–3.5 mm long)

fruits: cypselae; obovoid (1.5–2 mm long), brownish, compressed, glabrous, usually resinous; pappi absent

other: reddish-brown flower stalks, persistent through winter

VEGETATIVE CHARACTERISTICS

leaves: alternate in fascicles, simple, persistent, dark to pale green, often canescent, **glandular**; blades obdeltoid (3–20 mm long, 2–8 mm wide), spatulate to linear above; apical teeth 3 or lobes 3; lobes to one-third blade lengths; **constricted immediately below the lobes**; blades may be reduced and entire above; sessile or subpetiolate

stems: twigs round, short, rigid, light to dark reddish-brown becoming black with age, occasionally canescent; trunk bark dark, reddish-brown to black; exfoliating with age

other: appearing darker in color than the other species of *Artemisia*; entire plant aromatic, many leaves are persistent throughout the winter

HISTORICAL, FOOD, AND MEDICINAL USES: some Native Americans drank a decoction of the boiled stems, leaves, and twigs for bronchitis; leaves were crushed and the vapor inhaled to relieve nasal congestion

LIVESTOCK LOSSES: none

FORAGE VALUES: good for livestock but infrequently consumed in summer; good winter browse for livestock and wildlife

HABITATS: rocky slopes and windswept ridges in dry, shallow soils at elevations between 1500 and 2900 m

Budsage
Artemisia spinescens D.C. Eaton

SYN = *Picrothamnus desertorum* Nutt.

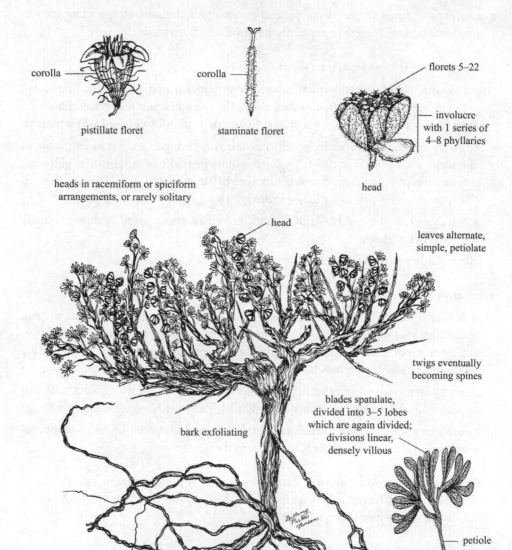

corolla

pistillate floret

corolla

staminate floret

florets 5–22

involucre
with 1 series of
4–8 phyllaries

heads in racemiform or spiciform
arrangements, or rarely solitary

head

head

leaves alternate,
simple, petiolate

twigs eventually
becoming spines

blades spatulate,
divided into 3–5 lobes
which are again divided;
divisions linear,
densely villous

bark exfoliating

petiole

238

FAMILY:	ASTERACEAE
TRIBE:	ANTHEMIDEAE
SPECIES:	*Artemisia spinescens* D.C. Eaton
COMMON NAME:	Budsage (bud sagewort, spiny sagebrush, bud sagebrush)
LIFE SPAN:	Perennial
ORIGIN:	Native
SEASON:	Cool

GROWTH FORM: shrubs (to 60 cm tall); much-branched, crowns rounded, rigid, spinescent; flowers April to June, reproduces from seeds

FLORAL AND FRUIT CHARACTERISTICS

inflorescences: heads in racemiform or spiciform arrangements (1–5 cm long), rarely solitary; axillary and/or terminal, leafy; heads small, nodding, subsessile

flowers: unisexual, heads inconspicuously radiate; involucres obconic (2–3.5 mm tall, 3–4.5 mm in diameter) with 1 series of 4–8 phyllaries; phyllaries ovate, obtuse, densely villous; pistillate florets 2–6, corollas hairy; staminate florets 3–16; corollas pale yellow, villous

fruits: cypselae; oblong or ellipsoid (1–1.5 mm long), ribs obscure, densely villous; pappi absent

VEGETATIVE CHARACTERISTICS

leaves: alternate, simple; petiolate or rarely sessile; blades spatulate (5–15 mm long, less than 10 mm wide), **lobes 3–5, which are again divided**, divisions linear, glandular; densely villous

stems: older twigs rigid, thickened, bark gray to dark brown, **exfoliating**; newer twigs ascending, tomentose to hairy, eventually becoming spines; trunk bark gray to brown, exfoliating; **leaves usually absent** most of the year, leaving tawny-colored stems

other: leaves usually produced in early spring, deciduous by midsummer, aromatic

HISTORICAL, FOOD, AND MEDICINAL USES: pollen commonly causes hay fever

LIVESTOCK LOSSES: reported to be poisonous to cattle, calves, and lambs when consumed alone; not poisonous to mature sheep

FORAGE VALUES: good for sheep in Utah, Nevada, and California where it is especially valuable for early browse; poor to worthless for cattle and horses; good to fair for wildlife

HABITATS: desert mesas, hills, and grasslands; adapted to a broad range of dry or well-drained soils, an indicator of alkaline soils

Big sagebrush
Artemisia tridentata Nutt.

SYN = *Seriphidium tridentatum* (Nutt.) W.A. Weber

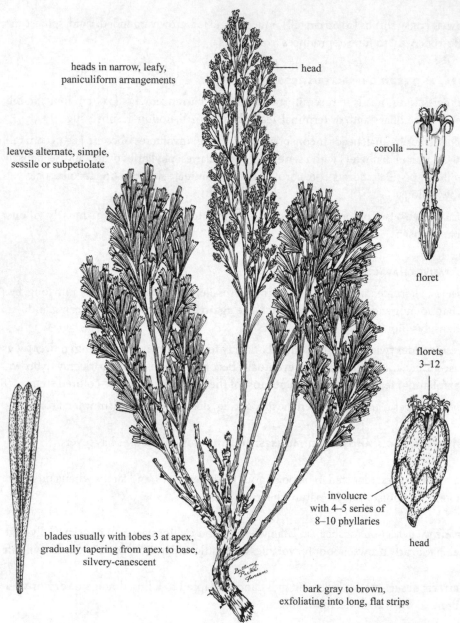

heads in narrow, leafy, paniculiform arrangements

head

corolla

floret

leaves alternate, simple, sessile or subpetiolate

florets 3–12

blades usually with lobes 3 at apex, gradually tapering from apex to base, silvery-canescent

involucre with 4–5 series of 8–10 phyllaries

head

bark gray to brown, exfoliating into long, flat strips

FAMILY:	ASTERACEAE
TRIBE:	ANTHEMIDEAE
SPECIES:	*Artemisia tridentata* Nutt.
COMMON NAME:	Big sagebrush (chamiso hediondo, sagebrush)
LIFE SPAN:	Perennial
ORIGIN:	Native
SEASON:	Evergreen

GROWTH FORM: shrubs (to 5 m tall); erect to occasionally decumbent, highly branched upward, crowns rounded or spreading, trunks short; flowers August to September, reproduces from seeds

FLORAL AND FRUIT CHARACTERISTICS

inflorescences: heads in paniculiform arrangements (5–15 cm long, 1.5–15 cm wide), usually above the foliage, narrow, leafy; heads numerous, erect to drooping, subsessile

flowers: perfect, heads discoid; involucres (2–4 mm tall, 2–2.8 mm in diameter) with 4–5 series of 8–10 phyllaries; outer phyllaries short, ovate, canescent; inner phyllaries oblong, canescent to almost glabrous; florets 3–12; corollas 2.5–3.5 mm long

fruits: cypselae; obovoid (1–3 mm long, 1 mm in diameter), brownish, compressed, pubescent or glabrous, resinous; pappi absent

VEGETATIVE CHARACTERISTICS

leaves: alternate, simple, persistent, crowded, silvery-canescent; blades cuneate to narrowly obdeltoid (1–5 cm long, 3–14 mm wide), **apical lobes 3**, sometimes entire; gradually tapering from the apices to the base; sessile or subpetiolate; elongate

stems: twigs round, rigid, green above, brown below, new growth pubescent; bark gray to brown, exfoliating into long, flat strips; central trunk crooked

other: entire plant aromatic, highly variable with several varieties; leaf size variable between seasons, usually larger in spring and summer than in autumn and winter

HISTORICAL, FOOD, AND MEDICINAL USES: once used for thatch and firewood; decoctions made from leaves and inflorescences were used by some Native Americans as a laxative; pollen causes hay fever

LIVESTOCK LOSSES: volatile oils may cause rumen stasis

FORAGE VALUES: good for sheep and wildlife in winter, fair for cattle; high in protein but also high in volatile oils; food source and cover for sage grouse, other birds, and small mammals

HABITATS: valleys, plains, basins, and mountain slopes; most abundant in dry, well-drained, gravelly or rocky soils

Annual broomweed
Amphiachyris dracunculoides (DC.) Nutt. *ex* Rydb.

SYN = *Brachyris dracunculoides* DC., *Gutierrezia dracunculoides* (DC.)
O. Hoffm., *Gutierrezia lindheimeriana* Scheele, *Xanthocephalum
dracunculoides* (DC.) Shinners

heads numerous in loose, terminal,
corymbiform arrangements

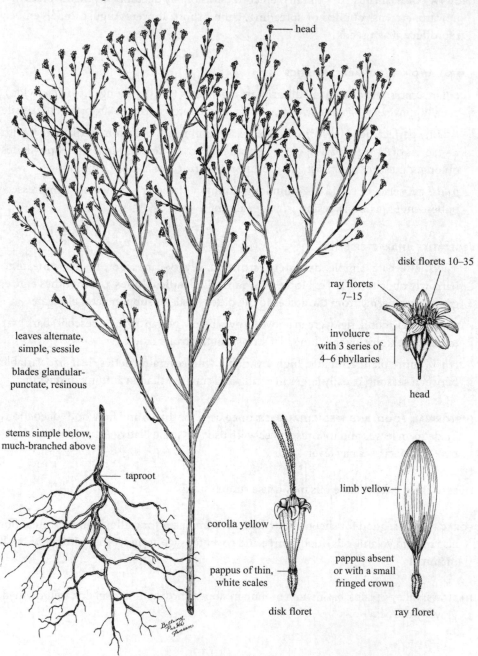

head

leaves alternate,
simple, sessile

blades glandular-
punctate, resinous

stems simple below,
much-branched above

taproot

disk florets 10–35

ray florets
7–15

involucre
with 3 series of
4–6 phyllaries

head

corolla yellow

pappus of thin,
white scales

disk floret

limb yellow

pappus absent
or with a small
fringed crown

ray floret

FAMILY:	ASTERACEAE
TRIBE:	ASTEREAE
SPECIES:	*Amphiachyris dracunculoides* (DC.) Nutt. *ex* Rydb.
COMMON NAME:	Annual broomweed (escobilla anual, common broomweed, prairie broomweed)
LIFE SPAN:	Annual
ORIGIN:	Native
SEASON:	Warm

GROWTH FORM: forbs (1–2 m tall); erect, single-stemmed at base from a taproot, freely branching above, appearing bushy; flowers July to November, reproduces from seeds

FLORAL AND FRUIT CHARACTERISTICS

inflorescences: heads in numerous corymbiform arrangements, loose, terminal

flowers: unisexual, heads radiate; involucres narrowly campanulate to turbinate (3–5 mm tall, 2–4 mm in diameter), with 3 (sometimes 1 or 2) series of 4–6 phyllaries; phyllaries imbricate, veins 1, indurate, **stramineous, with green apices**, shiny; ray florets 7–15, pistillate, limbs yellow (4–6 mm long); disk florets 10–35, staminate, corollas yellow

fruits: cypselae (1.2–2.2 mm long); ribs 7–9, purplish-black, pubescent; ray floret cypselae with pappi absent or with a small fringed crown; disk floret cypselae with pappi of 5 or more basally united scales; scales thin, white, awnlike

VEGETATIVE CHARACTERISTICS

leaves: alternate, simple; blades lanceolate to linear or filiform (1–5 cm long, 0.6–3 mm wide); margins entire; **glandular-punctate**, resinous, glabrous; sessile; terpenoid odor when crushed

stems: **simple below, much-branched above** (compare to *Gutierrezia sarothrae*), branches erect or ascending, slender, glabrous or slightly resinous; upper stems notably glandular-punctate

HISTORICAL, FOOD, AND MEDICINAL USES: none

LIVESTOCK LOSSES: herbage and pollen may cause dermatitis in livestock (and humans) and a condition similar to pinkeye

FORAGE VALUES: worthless to cattle, will be consumed if no other forage is present; increases in abundance under heavy grazing pressure and during dry periods

HABITATS: upland prairies, limestone barrens, roadsides, railroad rights-of-way, and other disturbed sites

Little rabbitbrush
Chrysothamnus viscidiflorus (Hook.) Nutt.

SYN = *Aster viscidiflorus* (Hook.) Kuntze, *Bigelowia viscidiflora* (Hook.) DC.,
Crinitaria viscidiflora Hook., *Ericameria viscidiflora* (Hook.)
L.C. Anderson, *Linosyris viscidiflora* (Hook.) Torr. & A. Gray

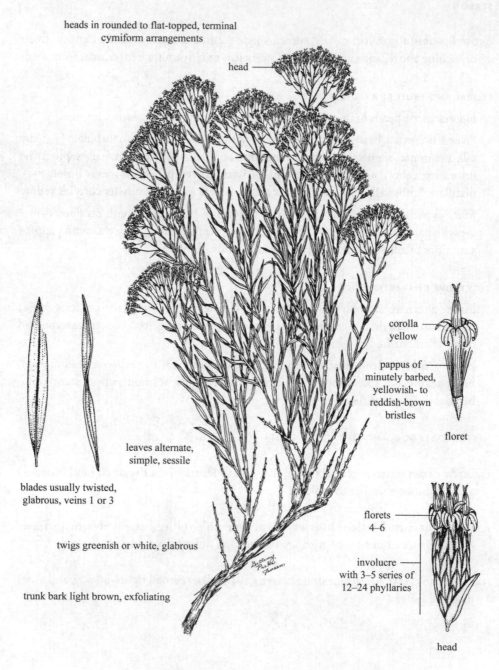

heads in rounded to flat-topped, terminal
cymiform arrangements

head

corolla
yellow

pappus of
minutely barbed,
yellowish- to
reddish-brown
bristles

floret

leaves alternate,
simple, sessile

blades usually twisted,
glabrous, veins 1 or 3

florets
4–6

twigs greenish or white, glabrous

involucre
with 3–5 series of
12–24 phyllaries

trunk bark light brown, exfoliating

head

FAMILY:	ASTERACEAE
TRIBE:	ASTEREAE
SPECIES:	*Chrysothamnus viscidiflorus* (Hook.) Nutt.
COMMON NAME:	Little rabbitbrush (chamisa, Douglas rabbitbrush, yellow rabbitbrush, sticky rabbitbrush)
LIFE SPAN:	Perennial
ORIGIN:	Native
SEASON:	Warm

GROWTH FORM: shrubs (about 1 m tall); erect, much-branched from near the base, branches ascending; crowns rounded; flowers July to September, reproduces from seeds

FLORAL AND FRUIT CHARACTERISTICS

inflorescences: heads in cymiform arrangements (5–6 cm long and wide), terminal, sessile to short-pedunculate, often crowded; collective inflorescences moderately rounded to flat-topped

flowers: perfect, heads discoid; involucres cylindrical to obconic to campanulate (6–8 mm tall) with 3–5 series of phyllaries; phyllaries 12–24, linear, obtuse or acute, glabrous or pubescent, yellowish, midveins usually evident; florets 4–6, corollas yellow (4–7 mm long)

fruits: cypselae; turbinate (3–4 mm long), compressed, light brown, ribs 5, densely villous; **pappi of yellowish- to reddish-brown bristles** (5 mm long), minutely barbed

VEGETATIVE CHARACTERISTICS

leaves: alternate, simple; blades linear to lanceolate (1–5 cm long, 1–4 mm wide), often involute, **usually twisted**, apices acute or obtuse; margins entire, scabrous; **glabrous**, veins 1 or 3; sessile

stems: twigs mostly ascending to erect, stiff, brittle, striate, **greenish or white, glabrous** or with scattered pubescence, often resinous, semiglossy; trunk bark light brown becoming grayish, exfoliating

other: a highly variable species with numerous varieties, leaves especially variable

HISTORICAL, FOOD, AND MEDICINAL USES: roots were chewed as gum in the Southwest; contains rubber, especially when growing in alkali soils

LIVESTOCK LOSSES: none

FORAGE VALUES: poor, occasionally browsed by sheep and cattle when other forage is not available, browsed lightly by deer and pronghorn in the summer and winter, utilized by elk in winter; small mammals eat the heads

HABITATS: dry prairies, valleys, and hillsides in sagebrush, ponderosa pine, lodgepole pine, or aspen belts

Rubber rabbitbrush
Ericameria nauseosa (Pall. *ex* Pursh) G.L. Nesom & G.I. Baird

SYN = *Chondrophora nauseosa* (Pall. *ex* Pursh) Britton, *Chrysocoma nauseosa* Pall. *ex* Pursh, *Chrysothamnus nauseosus* (Pall. *ex* Pursh) Britton

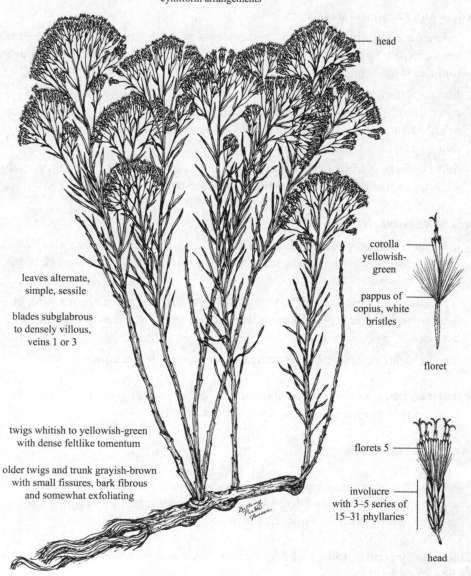

heads in rounded, terminal, cymiform arrangements

head

leaves alternate, simple, sessile

blades subglabrous to densely villous, veins 1 or 3

twigs whitish to yellowish-green with dense feltlike tomentum

older twigs and trunk grayish-brown with small fissures, bark fibrous and somewhat exfoliating

corolla yellowish-green

pappus of copius, white bristles

floret

florets 5

involucre with 3–5 series of 15–31 phyllaries

head

FAMILY:	ASTERACEAE
TRIBE:	ASTEREAE
SPECIES:	*Ericameria nauseosa* (Pall. *ex* Pursh) G.L. Nesom & G.I. Baird
COMMON NAME:	Rubber rabbitbrush (chamiso blanco, gray rabbitbrush, golden rabbitbrush)
LIFE SPAN:	Perennial
ORIGIN:	Native
SEASON:	Warm

GROWTH FORM: shrubs (to 2 m tall); crowns rounded from several erect stems from the base, much-branched; flowers June to September, reproduces from seeds and basal sprouts

FLORAL AND FRUIT CHARACTERISTICS

inflorescences: heads in cymiform arrangements, terminal; collective inflorescences rounded

flowers: perfect, heads discoid; involucres obconic to subcylindrical (7–14 mm tall, 2–4 mm in diameter) with 3–5 series of phyllaries; phyllaries 15–31, lanceolate to linear (2–8 mm long, 0.5–1 mm wide), imbricate in vertical rows, acute to obtuse, glabrous, tan, margins ciliate; florets usually 5, corollas yellowish-green (7–11 mm long), glabrous

fruits: cypselae; turbinate to cylindrical to oblanceoloid (4–5 mm long), angles 5, pubescent or less often glabrous; pappi of copious, dull white capillary bristles

other: inflorescence and phyllaries often persisting well into the next year

VEGETATIVE CHARACTERISTICS

leaves: alternate, simple; blades linear to narrowly oblanceolate (2–8 cm long, 0.5–3 mm wide), apices acute; subglabrous to densely villous, veins 1 or 3 (rarely 5), lateral veins often obscure; adaxial surfaces sometimes concave to sulcate, often glandular; margins entire; sessile

stems: twigs erect or ascending to spreading, flexible, whitish- to yellowish-green with a **dense feltlike tomentum**; trunks and older twigs grayish-brown with small fissures; bark fibrous and somewhat exfoliating

other: highly variable species with many subspecies, entire plant has a nauseous odor

HISTORICAL, FOOD, AND MEDICINAL USES: some Native Americans made chewing gum from pulverized wood and bark, also used as tea, cough syrup, yellow dye, and for chest pains; the Hopi stripped bark from the branches and used the branches for basket-making; small commercial source for rubber

LIVESTOCK LOSSES: occasionally reported to be toxic to livestock

FORAGE VALUES: worthless to poor for livestock, fair for pronghorn and deer on rangelands in winter; dense stands may indicate inappropriate range management

HABITATS: dry plains, hillsides, mesas, arroyos, roadsides, and ravine walls in sandy or clayey soils

Curlycup gumweed
Grindelia squarrosa (Pursh) Dunal

SYN = *Donia squarrosa* Pursh, *Grindelia aphanactis* Rydb., *Grindelia nuda*
    Alph. Wood, *Grindelia serrulata* Rydb.

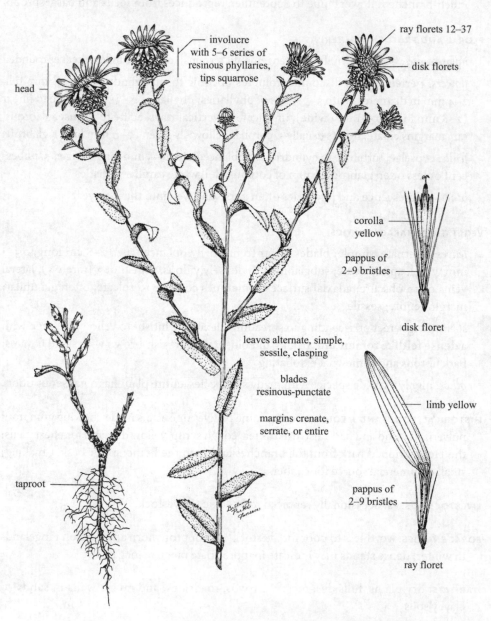

heads solitary or several to numerous
in loose corymbiform arrangements

ray florets 12–37

involucre
with 5–6 series of
resinous phyllaries,
tips squarrose

disk florets

head

corolla
yellow

pappus of
2–9 bristles

disk floret

leaves alternate, simple,
sessile, clasping

blades
resinous-punctate

margins crenate,
serrate, or entire

limb yellow

taproot

pappus of
2–9 bristles

ray floret

248

FAMILY:	ASTERACEAE
TRIBE:	ASTEREAE
SPECIES:	*Grindelia squarrosa* (Pursh) Dunal
COMMON NAME:	Curlycup gumweed (gumweed, curlytop gumweed)
LIFE SPAN:	Perennial (biennial)
ORIGIN:	Native
SEASON:	Warm

GROWTH FORM: forbs (0.2–1 m tall); erect, 1 to several stems from a taproot, branching above; starts growth in early spring, flowers July to August, reproduces from seeds

FLORAL AND FRUIT CHARACTERISTICS

inflorescences: heads solitary or several to numerous in loose corymbiform arrangements; heads (0.5–2.5 cm long, 0.7–3 cm in diameter) resinous-sticky, darkened with drying

flowers: perfect or some unisexual, heads radiate or discoid; involucres hemispheric or globose (7–9 mm tall) with 5–6 series of phyllaries; phyllaries imbricate, **tips squarrose and resinous** (especially the lower series); ray florets (if present) 12–37, pistillate, limbs yellow (7–15 mm long); disk florets perfect or staminate, numerous, corollas tubular, yellow

fruits: cypselae; oblong (1.5–4.5 mm long), angles 4, glabrous; pappi of 2–9 bristles, shorter than the disk florets

VEGETATIVE CHARACTERISTICS

leaves: alternate, simple; blades ovate to oblong to oblanceolate (1.5–7 cm long, 4–15 mm wide), thick; apices obtuse to acute; margins crenate, serrate or entire; **resinous-punctate; sessile and clasping**

stems: erect, 1 to several, glabrous, whitish or stramineous

other: aromatic, variable flowers, leaf serrations, and leaf sizes; several varieties

HISTORICAL, FOOD, AND MEDICINAL USES: some Native Americans used the resinous secretions to relieve asthma, bronchitis, and colic; Pawnee boiled leaves and flowering tops to treat saddle sores and raw skin; flower extract is used in modern medicine to treat whooping cough and asthma

LIVESTOCK LOSSES: may accumulate selenium, but the resinous covering of the herbage usually discourages consumption

FORAGE VALUES: worthless to livestock and most wildlife, although sheep and pronghorn occasionally eat the heads

HABITATS: waste places, improperly grazed rangelands, roadsides, and alluvial deposits; increases under drought conditions

Broom snakeweed
Gutierrezia sarothrae (Pursh) Britton & Rusby

SYN = *Brachyachyris euthamiae* Spreng., *Brachyris divaricata* Nutt., *Galinsoga linearifolia* (Lag.) Spreng., *Gutierrezia corymbosa* A. Nelson, *Gutierrezia fasciculata* Greene, *Gutierrezia scoparia* Rydb., *Solidago sarothrae* Pursh, *Xanthocephalum sarothrae* (Pursh) Shinners

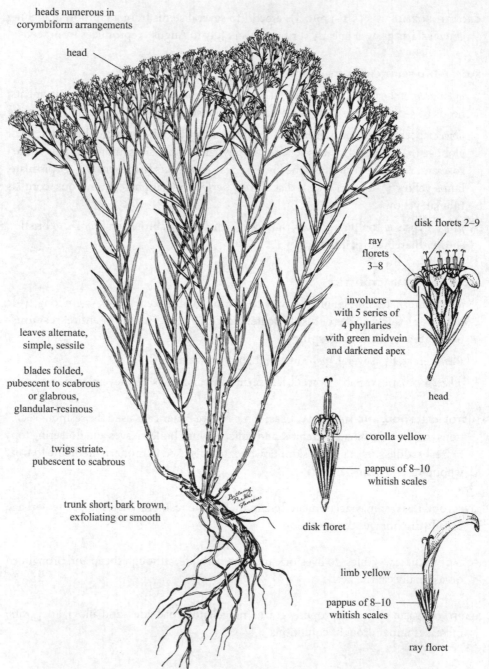

heads numerous in corymbiform arrangements

head

leaves alternate, simple, sessile

blades folded, pubescent to scabrous or glabrous, glandular-resinous

twigs striate, pubescent to scabrous

trunk short; bark brown, exfoliating or smooth

disk florets 2–9

ray florets 3–8

involucre with 5 series of 4 phyllaries with green midvein and darkened apex

head

corolla yellow

pappus of 8–10 whitish scales

disk floret

limb yellow

pappus of 8–10 whitish scales

ray floret

FAMILY:	ASTERACEAE
TRIBE:	ASTEREAE
SPECIES:	*Gutierrezia sarothrae* (Pursh) Britton & Rusby
COMMON NAME:	Broom snakeweed (escobilla común, yerba de víbora, perennial broomweed, turpentine weed, hierba resinosa)
LIFE SPAN:	Perennial
ORIGIN:	Native
SEASON:	Warm

GROWTH FORM: subshrubs (to 70 cm tall), from taproots; stems bushy, highly branched, variable in size and posture; flowers July to October, reproduces from seeds

FLORAL AND FRUIT CHARACTERISTICS

inflorescences: heads numerous in corymbiform arrangements, rounded to flat-topped, loose or dense

flowers: pistillate and perfect, heads inconspicuously radiate; involucres cylindrical to cuneate-campanulate (3–6 mm tall, 2 mm in diameter) with 5 series of 4 phyllaries; phyllaries linear (1.2–3.5 mm long), acute, imbricate, with **green midveins and darkened apices**; ray florets 3–8, pistillate, limbs yellow (1–3 mm long); disk florets 2–9, perfect, corollas yellow

fruits: cypselae; cylindrical (1.7–2 mm long, 0.5 mm in diameter), brown, densely pubescent; pappi of 8–10 scales (0.5–1 mm long), acute, whitish

VEGETATIVE CHARACTERISTICS

leaves: alternate, simple; blades linear to filiform (5–70 mm long, 1–3 mm wide), numerous, folded, veins 1 or 3; margins entire; pubescent to scabrous or glabrous, **glandular-resinous**; produces turpentine-like odor when bruised; sessile

stems: twigs erect, thin, flexible, green to brown, striate, pubescent to scabrous, glandular-resinous; branching from the bases (compare to *Amphiachyris dracunculoides*); trunks short; bark brown, exfoliating or smooth

HISTORICAL, FOOD, AND MEDICINAL USES: crafted by southwestern Native Americans and Mexicans into brooms; decoctions were used for indigestion; pieces of the plant were chewed and placed on bee and wasp stings and snakebites

LIVESTOCK LOSSES: poisonous to sheep and cattle, causing death or abortion; poisonous principle is a saponin that is most toxic during active plant growth; will also accumulate selenium

FORAGE VALUES: fair for sheep and poor for cattle and horses on rangelands in winter; otherwise worthless, indicator of improper management

HABITATS: open plains, upland sites, dry hillsides, improperly grazed rangelands; adapted to a broad range of soils

Hairy goldaster
Heterotheca villosa (Pursh) Shinners

SYN = *Amellus villosus* Pursh, *Chrysopsis angustifolia* Rydb., *Chrysopsis mollis*
Nutt., *Chrysopsis nitidula* Wooton & Standl., *Chrysopsis villosa* (Pursh)
Nutt. *ex* DC., *Diplogon villosum* (Pursh) Kuntze, *Diplopappus villosus*
(Pursh) Hook., *Inula villosa* (Pursh) Nutt.

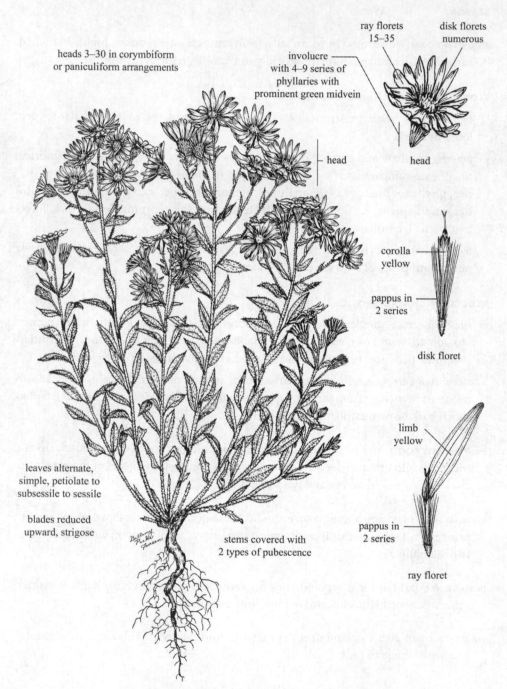

ray florets
15–35

disk florets
numerous

heads 3–30 in corymbiform
or paniculiform arrangements

involucre
with 4–9 series of
phyllaries with
prominent green midvein

head

head

corolla
yellow

pappus in
2 series

disk floret

limb
yellow

leaves alternate,
simple, petiolate to
subsessile to sessile

blades reduced
upward, strigose

stems covered with
2 types of pubescence

pappus in
2 series

ray floret

FAMILY:	ASTERACEAE
TRIBE:	ASTEREAE
SPECIES:	*Heterotheca villosa* (Pursh) Shinners
COMMON NAME:	Hairy goldaster (hierba velluda, telegraph plant, false goldenaster, golden aster)
LIFE SPAN:	Perennial
ORIGIN:	Native
SEASON:	Warm

GROWTH FORM: forbs (20–70 cm tall); stems arising singularly or in a small group from **woody caudices** surmounting taproots, erect or ascending; flowers July to September, reproduces from seeds and rarely from rhizomes

FLORAL AND FRUIT CHARACTERISTICS

inflorescences: heads in corymbiform or rarely paniculiform arrangements; heads 3–30 per branch, often with reduced leaves below the involucres

flowers: pistillate and perfect, heads radiate; involucres campanulate to narrowly cylindrical (7–12 mm tall, 7–15 mm in diameter) with 4–9 series of phyllaries; phyllaries imbricate, narrow, linear, inner series longest, midveins prominent, green, pubescent; ray florets 15–35, pistillate, yellow, revolute, limbs yellow (8–14 mm long), apical teeth 5; disk florets (5–8 mm long) numerous, perfect, corollas yellow

fruits: cypselae; oblong, small, villous, somewhat flattened, ribs 4–8; pappi in 2 series, outer pappi of uneven scales, inner pappi of uneven bristles

VEGETATIVE CHARACTERISTICS

leaves: alternate, simple; blades (1–5 cm long, 3–10 mm wide), usually reduced upward; lower blades oblanceolate and petiolate, midcauline blades oblanceolate and subsessile, upper blades linear to oblanceolate and sessile; apices acute to obtuse; bases cuneate; margins entire, flat, rarely with apical teeth 1–2; **strigose**, hairs appressed to ascending

stems: simple or branched above, erect to occasionally decumbent; covered with 2 types of pubescence; first pustulate-hispid, elongate, persistent; second divergent to appressed; reddish-brown to brown

other: a variable species with several varieties

HISTORICAL, FOOD, AND MEDICINAL USES: some Native Americans consumed a decoction from the tops and stems as a soothing, quieting medicine to aid sleep; human allergen

LIVESTOCK LOSSES: none

FORAGE VALUES: fair for sheep in semidesert areas of the West, otherwise considered worthless

HABITATS: prairies, plains, and semidesert areas; most abundant in dry open areas in sandy and rocky or calcareous soils

Prairie goldenrod
Solidago missouriensis Nutt.

SYN = *Aster missouriensis* (Nutt.) Kuntze, *Doria concinna* (A. Nelson) Lunell, *Solidago concinna* A. Nelson, *Solidago duriuscula* Greene, *Solidago glaberrima* M. Martens, *Solidago glaucophylla* Rydb., *Solidago tenuissima* Wooton & Standl.

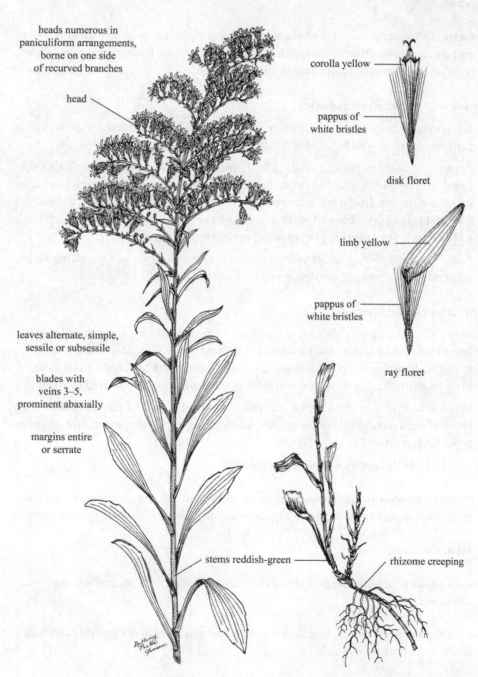

heads numerous in paniculiform arrangements, borne on one side of recurved branches

head

corolla yellow

pappus of white bristles

disk floret

limb yellow

pappus of white bristles

ray floret

leaves alternate, simple, sessile or subsessile

blades with veins 3–5, prominent abaxially

margins entire or serrate

stems reddish-green

rhizome creeping

FAMILY:	ASTERACEAE
TRIBE:	ASTEREAE
SPECIES:	*Solidago missouriensis* Nutt.
COMMON NAME:	Prairie goldenrod (vara de oro, Missouri goldenrod)
LIFE SPAN:	Perennial
ORIGIN:	Native
SEASON:	Warm

GROWTH FORM: forbs (0.2–1 m tall); erect or ascending, arising singularly or as a group from creeping rhizomes or woody caudices; flowers June to October, reproduces from seeds and rhizomes

FLORAL AND FRUIT CHARACTERISTICS

inflorescences: heads numerous in **paniculiform arrangements** (3–20 cm long, 3–12 cm wide) usually broader than long, heads spreading and arching to ascending, **borne on one side** of the recurved branches

flowers: pistillate and perfect, heads radiate; involucres (3–5 mm tall) with 3–4 series of 4–6 phyllaries; phyllaries unequal, lanceolate, obtuse to acute; ray florets 5–13, pistillate, limbs yellow; disk florets 8–20, perfect, shorter than ray florets, corollas yellow

fruits: cypselae; obconic to cylindrical (1–2.2 mm long), glabrous to sparingly pubescent; pappi of numerous white bristles (2.5–3 mm long)

VEGETATIVE CHARACTERISTICS

leaves: alternate, simple; blades thick, firm; **lowermost blades largest, blades reduced upward; blades oblanceolate to elliptical to linear above** (4–6 cm long, 0.5–3 cm wide); apices acute, bases long-tapered; margins entire or serrate, glabrous, with **veins 3–5, prominent abaxially**; blades of small plants may have a single vein; sessile or subsessile or on winged petioles

stems: simple or rarely branched, erect, stiff, glabrous to sparingly strigose, **reddish-green**

HISTORICAL, FOOD, AND MEDICINAL USES: some Native Americans chewed leaves and flowers to relieve sore throats, roots were chewed to relieve toothache; pollen highly desirable to several species of bees for honey production; human allergen

LIVESTOCK LOSSES: may be toxic to sheep

FORAGE VALUES: poor, but will be grazed by cattle and sheep in spring and early summer; deer and pronghorn will graze the lower leaves; generally becomes more abundant in poorly managed areas

HABITATS: upland prairies, plains, meadows, open woodlands, and roadsides

False dandelion
Agoseris glauca (Pursh) Raf.

SYN = *Macrorhynchus glaucus* (Pursh) D.C. Eaton, *Troximon glaucum* Pursh

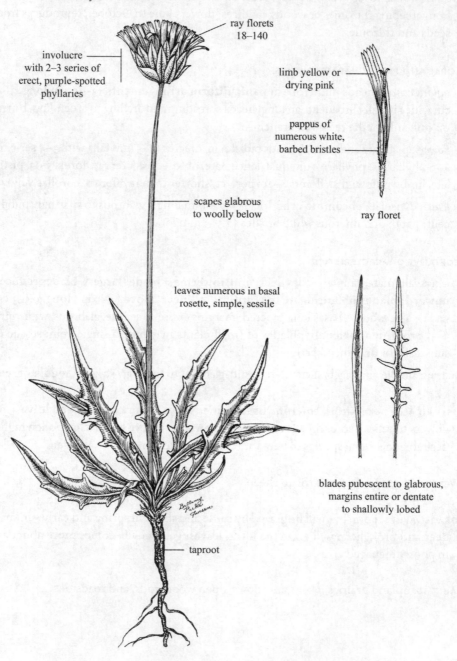

heads solitary

ray florets
18–140

involucre
with 2–3 series of
erect, purple-spotted
phyllaries

limb yellow or
drying pink

pappus of
numerous white,
barbed bristles

scapes glabrous
to woolly below

ray floret

leaves numerous in basal
rosette, simple, sessile

blades pubescent to glabrous,
margins entire or dentate
to shallowly lobed

taproot

FAMILY:	ASTERACEAE
TRIBE:	CICHORIEAE
SPECIES:	*Agoseris glauca* (Pursh) Raf.
COMMON NAME:	False dandelion (falso diente de león, mountain dandelion, pale agoseris)
LIFE SPAN:	Perennial
ORIGIN:	Native
SEASON:	Cool

GROWTH FORM: forbs (5–60 cm tall); scapose, from taproots; flowers May to September, reproduces from seeds

FLORAL AND FRUIT CHARACTERISTICS

inflorescences: heads solitary (1–6 cm in diameter), terminal, erect, stout; scapes slender (to 55 cm tall), glabrous to woolly below

flowers: perfect, heads ligulate; involucres obconic to hemispheric (1–3 cm tall), with **2–3 series of phyllaries**; phyllaries lanceolate, subequal, **erect**, tapering from bases, imbricate, acute, glabrous; inner phyllaries **frequently purple-spotted**; outer series of phyllaries the shortest and broadest; ray florets 18–140, limbs yellow or drying pink (1.2–1.5 cm long), apices toothed, exceeding phyllaries; receptacles flattened, naked

fruits: cypselae, fusiform to narrowly conic (5–12 mm long), stout, ribs 10, slightly hirsute, not flattened, tapering to a beak, one-half the length of the body; pappi of numerous white, barbed capillary bristles (to 2 cm long)

VEGETATIVE CHARACTERISTICS

leaves: numerous in basal rosettes, simple; blades (5–40 cm long, 2–40 mm wide) **highly variable in outline**, linear to lanceolate or oblanceolate; margins entire or dentate to shallowly lobed; pubescent to glabrous, glaucous; midveins distinct, white; sessile

stems: scapes glabrous to woolly below, somewhat glaucous

other: entire plant contains a milky latex; variable species with several varieties

HISTORICAL, FOOD, AND MEDICINAL USES: sap was chewed by some Native Americans to clean teeth

LIVESTOCK LOSSES: none

FORAGE VALUES: fair to good for sheep, pronghorn, and deer; lightly grazed by cattle and horses; tends to decrease under grazing by sheep and increase under grazing by cattle; abundance usually indicates rangeland deterioration

HABITATS: moist meadows, open woodlands, and prairie swales, common on disturbed or eroded rangelands, adapted to a wide range of soils

Tapertip hawksbeard
Crepis acuminata Nutt.

SYN = *Crepis angustata* Rydb., *Crepis seselifolia* Rydb., *Hieraciodes acuminatum* (Nutt.) Kuntze, *Psilochenia acuminata* (Nutt.) W.A. Weber

heads 15–100 in corymbiform arrangements

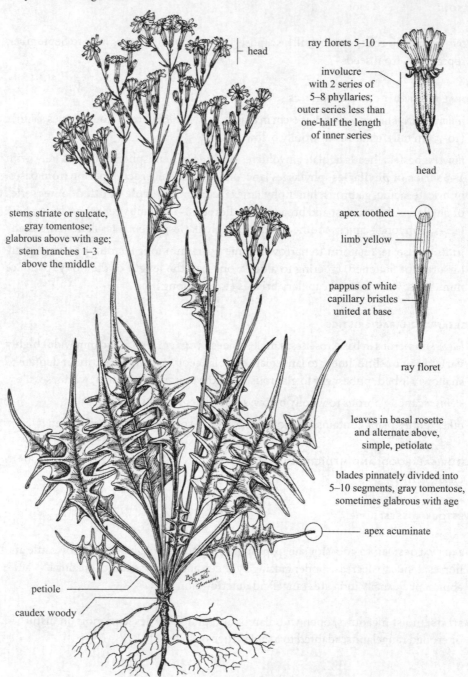

head

ray florets 5–10

involucre with 2 series of 5–8 phyllaries; outer series less than one-half the length of inner series

head

stems striate or sulcate, gray tomentose; glabrous above with age; stem branches 1–3 above the middle

apex toothed

limb yellow

pappus of white capillary bristles united at base

ray floret

leaves in basal rosette and alternate above, simple, petiolate

blades pinnately divided into 5–10 segments, gray tomentose, sometimes glabrous with age

apex acuminate

petiole

caudex woody

FAMILY:	ASTERACEAE
TRIBE:	CICHORIEAE
SPECIES:	*Crepis acuminata* Nutt.
COMMON NAME:	Tapertip hawksbeard (longleaf hawksbeard)
LIFE SPAN:	Perennial
ORIGIN:	Native
SEASON:	Cool

GROWTH FORM: forbs (20–70 cm tall); arising as 1–3 stems from woody caudices surmounting taproots, branched above; flowers May to August, reproduces from seeds

FLORAL AND FRUIT CHARACTERISTICS

inflorescences: heads 15–100 in corymbiform arrangements; heads small, slender; peduncles short

flowers: perfect, heads of ray florets 5–10; apices toothed apices toothed; involucres cylindrical (9–16 mm tall, 2.5–4 mm in diameter), **with 2 series of 5–8 phyllaries; outer phyllaries less than one-half the length of inner phyllaries,** lanceolate to deltoid, glabrous or tomentose; inner phyllaries lanceolate, glabrous or tomentose; ray floret limbs yellow (9–20 mm long), apices toothed

fruits: cypselae; linear (5.5–9 mm long), yellowish-brown, ribs 12; pappi of white capillary bristles (6–9 mm long), bristles united at base

VEGETATIVE CHARACTERISTICS

leaves: basal rosette, alternate above, simple; basal and lower blades lanceolate to elliptical (12–40 cm long, 0.5–11 cm wide); petioles winged; pinnately divided into segments 5–10; segments entire to dentate, **apices acuminate**; middle cauline leaves few and similar to basal leaves; upper cauline leaves reduced to bracts; blades taper to indistinct petioles; gray tomentose, sometimes glabrous with age

stems: erect, stout, branches 1–3 above the middle, striate or sulcate; gray tomentose especially when young, glabrous above with age but remaining tomentose below

other: entire plant contains a milky latex

HISTORICAL, FOOD, AND MEDICINAL USES: some Native Americans ate the leaves and shoots raw or boiled

LIVESTOCK LOSSES: none

FORAGE VALUES: good for deer, pronghorn, cattle, sheep, and horses; preferred forage of sheep; most palatable in late spring and early summer

HABITATS: prairies, meadows, hillsides, and broken slopes; most abundant in open areas with dry, well-drained or shallow soils

Dandelion
Taraxacum officinale F.H. Wigg.

SYN = *Leontodon taraxacum* L., *Leontodon vulgare* Lam., *Taraxacum dens-leonis* Desf., *Taraxacum mexicanum* DC., *Taraxacum retroflexum* H. Lindb., *Taraxacum subspathulatum* A.J. Richards, *Taraxacum sylvanicum* R. Doll, *Taraxacum vulgare* Schrank

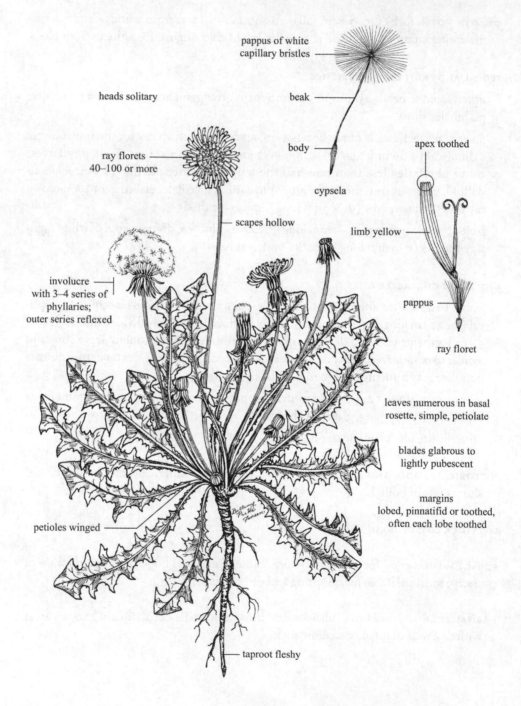

pappus of white capillary bristles

heads solitary

beak

body

apex toothed

ray florets 40–100 or more

cypsela

scapes hollow

limb yellow

involucre with 3–4 series of phyllaries; outer series reflexed

pappus

ray floret

leaves numerous in basal rosette, simple, petiolate

blades glabrous to lightly pubescent

margins lobed, pinnatifid or toothed, often each lobe toothed

petioles winged

taproot fleshy

FAMILY:	ASTERACEAE
TRIBE:	CICHORIEAE
SPECIES:	*Taraxacum officinale* F.H. Wigg.
COMMON NAME:	Dandelion (diente de león, common dandelion)
LIFE SPAN:	Perennial
ORIGIN:	Introduced (from Europe)
SEASON:	Cool

GROWTH FORM: forbs (3–60 cm tall); scapose, from fleshy taproots; flowers March to November, reproduces from seeds

FLORAL AND FRUIT CHARACTERISTICS

inflorescences: heads solitary (1–5 cm in diameter); terminating erect scapes; scapes slender (to 50 cm tall), hollow, greenish-white to reddish

flowers: perfect, heads ligulate; involucres obconic to campanulate (1–2.5 cm tall) with 3–4 series of 13–18 phyllaries; **outer phyllaries shorter**, linear, graduated in length, **loosely reflexed**, sometimes reddish tipped; inner phyllaries in 2 series, linear, erect, spreading with age, imbricate; ray florets 40–100 or more, limbs yellow, apical teeth 4–5; outer limbs (10–15 mm long), about twice as long as the inner limbs

fruits: cypselae; flattened (3–4 mm long), rust- to grayish-brown, ribs 4–12, spinose, tipped with a **long filiform beak** that is longer than cypselae body; **pappi of white capillary bristles** (4–5 mm long); bristles abundant

other: heads form at ground level, then scapes elongate in about 48 hours

VEGETATIVE CHARACTERISTICS

leaves: numerous in basal rosettes, simple; **blades oblanceolate** (5–40 cm long, 2–15 cm wide); apices acute to acuminate; margins variously lobed, pinnatifid, or toothed, often with each lobe toothed; **terminal lobe usually longest**; glabrous to lightly pubescent; petioles narrowly winged, may be indistinct

stems: scapes erect to ascending, glabrous to puberulent, greenish-white to purplish, hollow

other: entire plant contains a milky latex

HISTORICAL, FOOD, AND MEDICINAL USES: young leaves can be eaten as spring greens; roots can be ground and used as a coffee substitute, mild laxative, or to treat heartburn; good honey plant; tea and wine can be made from the flowers, heads can be fried in batter and eaten

LIVESTOCK LOSSES: none

FORAGE VALUES: fair to good for livestock and wildlife, readily eaten since it is relatively succulent; generally most abundant on improperly managed rangelands but can also occur on well-managed rangelands; production of forage is relatively low

HABITATS: meadows, pastures, open stream banks, disturbed sites, and lawns; common on a wide variety of soils

Dotted gayfeather
Liatris punctata Hook.

SYN = *Lacinaria punctata* (Hook.) Kuntze, *Liatris mucronata* DC.

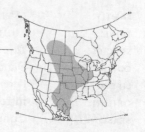

heads few to numerous,
sessile or short-peduncled in
a spiciform arrangement

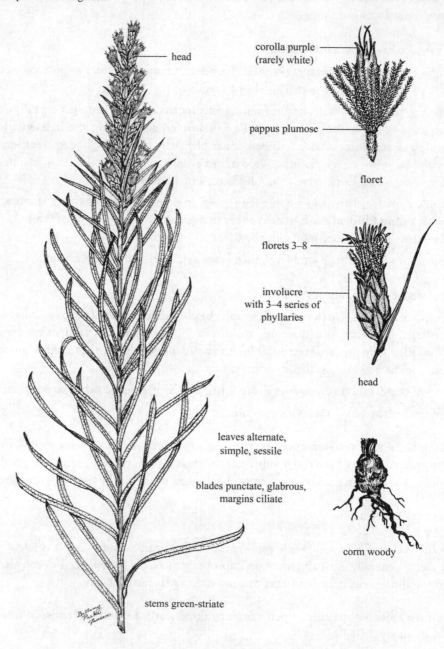

head

corolla purple
(rarely white)

pappus plumose

floret

florets 3–8

involucre
with 3–4 series of
phyllaries

head

leaves alternate,
simple, sessile

blades punctate, glabrous,
margins ciliate

corm woody

stems green-striate

FAMILY:	ASTERACEAE
TRIBE:	EUPATORIEAE
SPECIES:	*Liatris punctata* Hook.
COMMON NAME:	Dotted gayfeather (liatris, blazing star)
LIFE SPAN:	Perennial
ORIGIN:	Native
SEASON:	Warm

GROWTH FORM: forbs (10–80 cm tall); arising singularly or clustered from thick, **woody corms**; flowers July to October, reproduces from seeds and corms

FLORAL AND FRUIT CHARACTERISTICS

inflorescences: heads few to numerous, sessile or short-peduncled in **spiciform arrangements** (6–30 cm long); upper stems usually obscured by the heads

flowers: perfect; heads discoid; involucres cylindrical to campanulate (1.5–2 cm tall, 8–10 mm in diameter), loose with 3–4 series of phyllaries; **phyllaries thick, punctate**, appressed, apices acute to rounded, margins ciliate; disk florets 3–8; **corollas purple** (rarely white), pilose inside; stamens and styles short-exserted in late flowering

fruits: cypselae; prismatic (6–7.5 mm long), ribs 10, pubescent; **pappi plumose** (9–11 mm long), exceeding corollas, numerous

VEGETATIVE CHARACTERISTICS

leaves: alternate, simple; blades linear to lanceolate (8–15 cm long, 1.5–6 mm wide), reduced above, numerous, imbricate, rigidly ascending or arching; veins 1; margins thickened, whitish, ciliate; **punctate, glabrous**; sessile

stems: erect, simple or rarely branched, green-striate, glabrous

HISTORICAL, FOOD, AND MEDICINAL USES: corms reportedly used by some Native Americans for food; plants of this genus were eaten in New England during the 19th century for treatment of gonorrhea

LIVESTOCK LOSSES: none

FORAGE VALUES: fair for cattle; good for sheep, deer, and pronghorn; most valuable when plants are young; greatly reduced or eliminated with continuous overuse

HABITATS: prairies, plains, hills, gravelly and rocky slopes, roadsides, and uplands; most abundant on sandy soils

Triangleleaf bursage
Ambrosia deltoidea (Torr.) W.W. Payne

SYN = *Franseria deltoidea* Torr., *Gaertnera deltoidea* (Torr.) Kuntze

staminate heads in terminal
spiciform or racemiform
arrangements; pistillate heads in
axillary clusters below

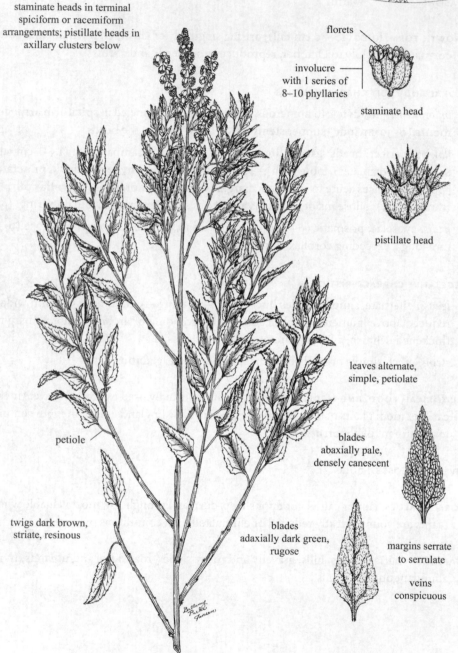

florets

involucre
with 1 series of
8–10 phyllaries

staminate head

pistillate head

leaves alternate,
simple, petiolate

petiole

blades
abaxially pale,
densely canescent

twigs dark brown,
striate, resinous

blades
adaxially dark green,
rugose

margins serrate
to serrulate

veins
conspicuous

FAMILY:	ASTERACEAE
TRIBE:	HELIANTHEAE
SPECIES:	*Ambrosia deltoidea* (Torr.) W.W. Payne
COMMON NAME:	Triangleleaf bursage (chicurilla, chamizo forrajero, triangle burroweed, triangle bursage, canyon ragweed)
LIFE SPAN:	Perennial
ORIGIN:	Native
SEASON:	Cool

GROWTH FORM: monoecious **shrubs** (to 1 m tall); crowns rounded to flat-topped, branches erect; flowers March to April, reproduces from seeds

FLORAL AND FRUIT CHARACTERISTICS

inflorescences: staminate heads (6–7 mm in diameter) in spiciform or racemiform arrangements, terminal; pistillate heads (5–7 mm long, 5–7 mm in diameter) in axillary clusters below

flowers: unisexual, heads discoid; staminate involucres saucer-shaped (3–5 mm in diameter) with 1 series of 8–10 triangular phyllaries; phyllaries tomentose when young, then glabrous; staminate heads on peduncles (0.5–3 mm long); disk floret corollas yellow (2–2.5 mm long), puberulent; pistillate heads sessile

fruits: cypselae; globose to fusiform (3–6 mm long), glandular, tomentose, with 2–3 compressed beaks; **bearing 15–30 flattened spines in 2–3 series**; spines puberulent, straight, not hooked

VEGETATIVE CHARACTERISTICS

leaves: alternate, simple; blades **narrowly deltoid to ovate** (2–3 cm long, 1–1.5 cm wide); apices acute or attenuate; bases truncate to cuneate; margins serrate to serrulate; adaxially dark green and rugose, abaxially pale and densely canescent, veins conspicuous, thick; petioles (up to 1.5 cm long) slightly winged above

stems: twigs erect, dark brown, striate, resinous, rigid, glabrous

other: principal nurse plants for saguaro cactus [*Carnegiea gigantea* (Engelm.) Britton & Rose]

HISTORICAL, FOOD, AND MEDICINAL USES: bitter tea made from its leaves has anti-inflammatory properties and can be taken internally or used as a wash; southwestern Native Americans used it in sweat baths to relieve arthritic pain; root preparations were used for stomach and intestinal cramps

LIVESTOCK LOSSES: fruits may contaminate fleece and reduce its value, considered to be one of the worst hay fever plants of the Southwest

FORAGE VALUES: worthless to livestock and wildlife

HABITATS: alluvial plains, hillsides, mesas, plains, and gullies; most abundant in rocky or gravelly soils; grows in almost pure stands

White bursage
Ambrosia dumosa (A. Gray) W.W. Payne

SYN = *Franseria dumosa* A. Gray, *Gaertnera dumosa* (A. Gray) Kuntze

heads in terminal and lateral
spiciform or racemiform arrangements;
staminate and pistillate heads intermingled

disk florets

involucre
with 1 series of
5–9 united phyllaries

staminate head

pistillate head

leaves alternate
or fascicled,
simple, petiolate

blades
pinnately divided
1–3 times

twigs white, striate-fissured

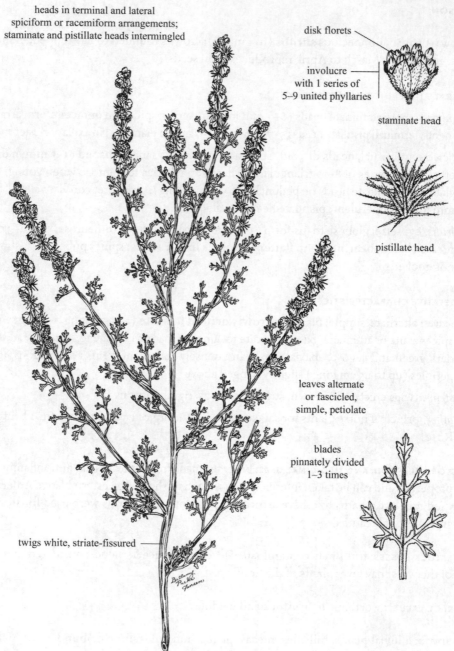

FAMILY:	ASTERACEAE
TRIBE:	HELIANTHEAE
SPECIES:	*Ambrosia dumosa* (A. Gray) W.W. Payne
COMMON NAME:	White bursage (hierba del burro, burroweed, burrobush)
LIFE SPAN:	Perennial
ORIGIN:	Native
SEASON:	Warm

GROWTH FORM: monoecious **shrubs** (to 80 cm tall), bushy; crowns rounded, compact, much-branched, stiff, spinescent; flowers March to May (to November), reproduces from seeds

FLORAL AND FRUIT CHARACTERISTICS

inflorescences: heads in terminal and lateral spiciform or racemiform arrangements; staminate and pistillate heads intermingled within the inflorescence

flowers: unisexual, heads discoid; staminate involucres saucer-shaped (2–2.5 mm tall, 3–5 mm in diameter), with 1 series of 5–9 united phyllaries; phyllaries ovate to triangular, often unequal, canescent, acute; disk floret corollas funnelform, yellow, puberulent; staminate heads numerous (3–5 mm in diameter), peduncles 0.2–3 mm long; pistillate head florets 1–2, sessile

fruits: cypselae; ovoid to subglobose (4–9 mm long), glandular-puberulent, beaks 2, tips with **15–40 straight spines in 2–3 series** (1.5–4 mm long); spines flattened at base, sharply pointed, not hooked, striate above

VEGETATIVE CHARACTERISTICS

leaves: alternate or fascicled, simple; blades ovate to elliptical (1.5–5 cm long, 8–25 mm wide), **pinnately divided** 1–3 times; divisions ovate or obovate (0.5–3 mm long, 0.5–2 mm wide); **grayish-green canescent** on both surfaces; petiolate (2–20 mm long)

stems: twigs erect to ascending, rigid, white, **somewhat striate-fissured**, lightly canescent; inflorescence branches from the previous year forming spines during the current year; trunks tan to light brown, striate

HISTORICAL, FOOD, AND MEDICINAL USES: plant extracts are being evaluated for pharmaceutical properties, wind-blown pollen often causes hay fever, sometimes used for specimen plantings in landscapes

LIVESTOCK LOSSES: may accumulate nitrates

FORAGE VALUES: fair for cattle and horses, fair to good for goats, often preferred by horses and donkeys

HABITATS: dry plains, mesas, alluvial slopes, and rocky or sandy gullies, may become locally abundant

Western ragweed
Ambrosia psilostachya DC.

SYN = *Ambrosia coronopifolia* Torr. & A. Gray, *Ambrosia rugelii* Rydb.

staminate heads in racemiform arrangements;
pistillate heads in axillary clusters or
spiciform arrangements below staminate heads

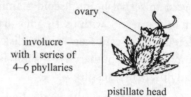

florets

involucre
with 1 series of
5–12 phyllaries,
united to partially united

staminate head

ovary

involucre
with 1 series of
4–6 phyllaries

pistillate head

cypsela

leaves alternate above and opposite below,
simple, sessile or petiolate

blades once-pinnatifid with appressed
papilla-based hairs, resinous

stems hirsute to pubescent, striate

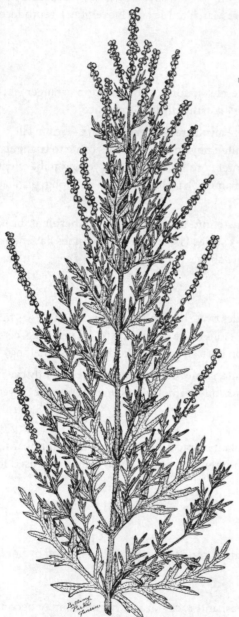

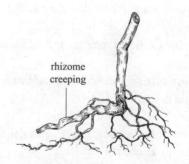

rhizome
creeping

FAMILY:	ASTERACEAE
TRIBE:	HELIANTHEAE
SPECIES:	*Ambrosia psilostachya* DC.
COMMON NAME:	Western ragweed (amargosa, Cuman ragweed)
LIFE SPAN:	Perennial
ORIGIN:	Native
SEASON:	Warm

GROWTH FORM: monoecious **forbs** (0.3–1 m tall); erect, branching above; flowers July to October, reproduces from deep, slender, creeping rhizomes and seeds; forms extensive colonies

FLORAL AND FRUIT CHARACTERISTICS

inflorescences: staminate heads in racemiform arrangements, nodding; pistillate heads in axillary clusters or spiciform arrangements below staminate heads

flowers: unisexual, heads discoid; staminate involucres cupulate to turbinate (2–3 mm tall, 1–5 mm in diameter), with 1 series of 5–12 united to partially united phyllaries; phyllaries hirsute, truncate to shallowly crenate-lobed; staminate disk florets several, corollas absent, anthers yellow; pistillate involucres obovoid, hirsute (2–4 mm tall), with 1 series of 4–6 phyllaries; phyllaries short with blunt spines; florets 1–2, corollas reduced

fruits: cypselae; ovoid to fusiform (1–2 mm long); enclosed in hard spiny "burs" (2–3 mm long) formed by the indurate involucres; bodies obpyramidal to globose, hirsutulous; spines or tubercles 1–6 (0.1–1 mm long); pappi absent

VEGETATIVE CHARACTERISTICS

leaves: alternate above and opposite below, simple; blades lanceolate to narrowly deltate (2–15 cm long, 1–8 cm wide), grayish-green, usually once-pinnatifid, divisions linear to lanceolate; apices acute; margins entire to toothed; covered with appressed, papilla-based hairs, resinous; sessile to petiolate with a winged petiole (to 2.5 cm long)

stems: erect to decumbent, simple below, branched above, hirsute to pubescent with ascending hairs, striate

HISTORICAL, FOOD, AND MEDICINAL USES: leaves were steeped by some Native Americans and used as a treatment for sore eyes; the pollen is one of the most important causes of hay fever in North America

LIVESTOCK LOSSES: may accumulate nitrates, milk produced from cows grazing this forb has a bitter taste, resin may induce dermatitis

FORAGE VALUES: worthless for livestock and wildlife, generally unpalatable, but cattle may graze it in early spring and late summer

HABITATS: prairies, barrens, hills, pastures, swales, fields, waste places, and roadsides

Desert marigold
Baileya multiradiata Harv. & A. Gray

SYN = *Baileya australis* Rydb., *Baileya pleniradiata* Harv. & A. Gray,
Baileya thurberi Rydb.

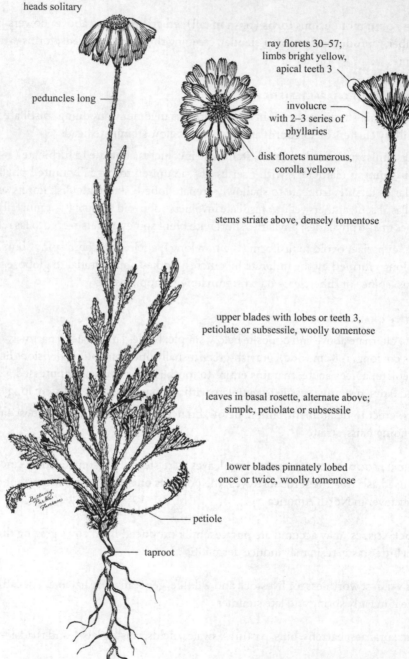

heads solitary

peduncles long

ray florets 30–57;
limbs bright yellow,
apical teeth 3

involucre
with 2–3 series of
phyllaries

disk florets numerous,
corolla yellow

stems striate above, densely tomentose

upper blades with lobes or teeth 3,
petiolate or subsessile, woolly tomentose

leaves in basal rosette, alternate above;
simple, petiolate or subsessile

lower blades pinnately lobed
once or twice, woolly tomentose

petiole

taproot

FAMILY:	ASTERACEAE
TRIBE:	HELIANTHEAE
SPECIES:	*Baileya multiradiata* Harv. & A. Gray
COMMON NAME:	Desert marigold (telempacate, hierba amarilla, paperdaisy, desert baileya)
LIFE SPAN:	Perennial (biennial)
ORIGIN:	Native
SEASON:	Warm

GROWTH FORM: forbs (10–60 cm tall); erect or ascending, stems 1 to several from woody branched caudices surmounting taproots, sparingly branched; short-lived; flowers March to November, reproduces from seeds

FLORAL AND FRUIT CHARACTERISTICS

inflorescences: **heads solitary** (3.5–5 cm in diameter), showy; terminating long peduncles (10–30 cm long)

flowers: pistillate and perfect, heads radiate; involucres hemispheric to campanulate (6–8 mm tall, 1.5–2 cm in diameter) with 2–3 series of phyllaries; phyllaries 20–34, subequal, elliptical to lanceolate, floccose-tomentose, acute; ray florets 30–57, showy, pistillate, limbs linear (1–2 cm long, 4–7 mm wide), bright yellow, **apical teeth 3, shallow**; disk florets numerous, perfect, corollas tubular, yellow, teeth 5, **pubescent**

fruits: cypselae; clavate (2.5–4 mm long), greenish-white, truncate, ribs 15–20, glabrous to resinous; pappi absent

VEGETATIVE CHARACTERISTICS

leaves: basal rosettes, alternate above, simple; basal and lower blades ovate (2–15 cm long, 1–5 cm wide), lower blades longest, **pinnately lobed once or twice**; divisions ovate or linear, entire; upper blades linear, lobes or teeth 3; woolly tomentose; petiolate (2–6 cm long) or subsessile

stems: erect, much-branched after second year, branched only near the bases, somewhat striate above, densely tomentose; first year a rosette

HISTORICAL, FOOD, AND MEDICINAL USES: cultivated as an ornamental; contains some substances that show promise as cancer treatments

LIVESTOCK LOSSES: sheep and goats may die after consuming large quantities over a relatively long period of time; cattle and horses seem to be unaffected

FORAGE VALUES: poor to worthless for livestock and wildlife, generally will not be grazed if other forage plants are present

HABITATS: dry plains, brushy hills, mesas, floodplains, roadsides, and disturbed sites; most abundant in sandy and gravelly soils

Arrowleaf balsamroot
Balsamorhiza sagittata (Pursh) Nutt.

SYN = *Balsamorhiza helianthoides* (Nutt.) Nutt., *Buphthalmum sagittatum* Pursh, *Espeletia helianthoides* Nutt., *Espeletia sagittata* (Pursh) Nutt.

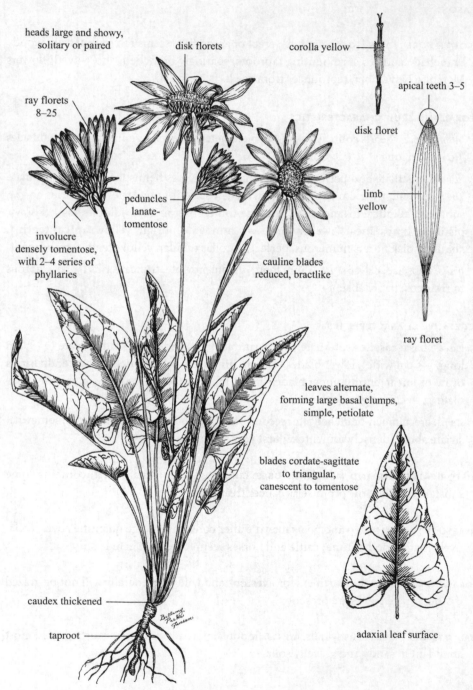

heads large and showy, solitary or paired

disk florets

corolla yellow

apical teeth 3–5

disk floret

ray florets 8–25

peduncles lanate-tomentose

involucre densely tomentose, with 2–4 series of phyllaries

cauline blades reduced, bractlike

limb yellow

ray floret

leaves alternate, forming large basal clumps, simple, petiolate

blades cordate-sagittate to triangular, canescent to tomentose

caudex thickened

taproot

adaxial leaf surface

FAMILY:	ASTERACEAE
TRIBE:	HELIANTHEAE
SPECIES:	*Balsamorhiza sagittata* (Pursh) Nutt.
COMMON NAME:	Arrowleaf balsamroot (gray dock, breadroot)
LIFE SPAN:	Perennial
ORIGIN:	Native
SEASON:	Cool

GROWTH FORM: forbs (20–80 cm tall), scapiform, from branched **thickened caudices** and taproots; **caudices often with an exfoliating appearance** due to the many old, persisting leaf petioles; flowers May to July, reproduces from seeds

FLORAL AND FRUIT CHARACTERISTICS

inflorescences: heads large and showy, solitary or paired; long, peduncles lanate-tomentose

flowers: heads radiate; involucres (1.5–3 cm in diameter) with 2–4 series of phyllaries; phyllaries subequal, ovate to lanceolate, acuminate, **densely tomentose**, outer phyllaries congested (2.5 cm long); ray florets 8–25, pistillate, limbs yellow (2.5–4 cm long), apical teeth 3–5; disk florets perfect, corollas tubular, yellow

fruits: cypselae; oblong (7–8 mm long), glabrous; pappi absent

VEGETATIVE CHARACTERISTICS

leaves: alternate, simple, forming large basal clumps; **basal blades cordate-sagittate to triangular** (15–50 cm long, 5–15 cm wide), apices acute, margins entire, silvery canescent to tomentose, usually glandular; cauline blades few, reduced, narrow, bract-like; basal petioles (20–55 cm long), canescent to tomentose

stems: scapiform, but usually with several reduced leaves; tomentose, arising from a large basal cluster of leaves

HISTORICAL, FOOD, AND MEDICINAL USES: some Cheyenne boiled roots, stems, and leaves and drank the decoction for stomach pains and headaches; they also steamed the plant and inhaled the vapors for the same purposes; ripe fruits were pounded into flour; roots were commonly eaten raw or boiled

LIVESTOCK LOSSES: none

FORAGE VALUES: good for sheep and big game and fair for cattle when green, heads are most palatable

HABITATS: plains, valleys, hillsides, and open woodlands; most abundant in deep, well-drained soils at moderate elevations (to 3000 m)

Tarbush
Flourensia cernua DC.

SYN = *Helianthus cernuus* (DC.) Benth. & Hook. f.

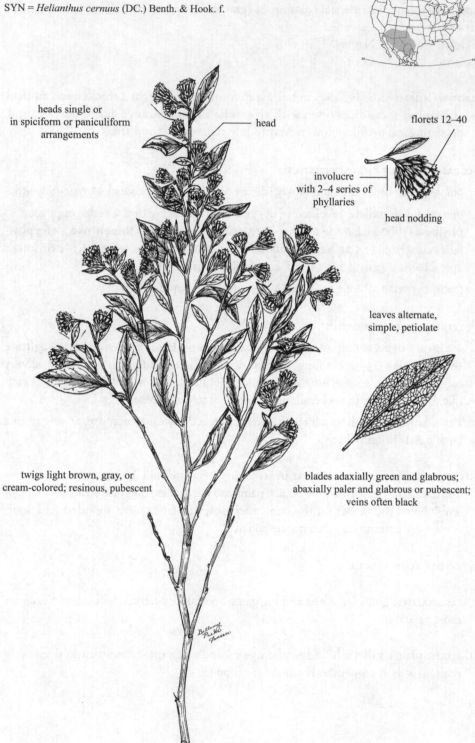

heads single or
in spiciform or paniculiform
arrangements

head

florets 12–40

involucre
with 2–4 series of
phyllaries

head nodding

leaves alternate,
simple, petiolate

twigs light brown, gray, or
cream-colored; resinous, pubescent

blades adaxially green and glabrous;
abaxially paler and glabrous or pubescent;
veins often black

FAMILY:	ASTERACEAE
TRIBE:	HELIANTHEAE
SPECIES:	*Flourensia cernua* DC.
COMMON NAME:	Tarbush (hojasén, blackbrush, varnishbush, Mexican tarwort, American tarwort)
LIFE SPAN:	Perennial
ORIGIN:	Native
SEASON:	Evergreen

GROWTH FORM: shrubs (to 2 m tall), erect to procumbent, densely leafy, highly branched throughout, resinous; taproots; flowers July to December, reproduces from seeds

FLORAL AND FRUIT CHARACTERISTICS

inflorescences: heads single or in spiciform or paniculiform arrangements, eventually nodding; peduncles short and curved, or heads sessile

flowers: perfect, heads discoid; involucres campanulate to hemispheric (1 cm tall and in diameter) with 2–4 series of 12–36 unequal phyllaries; phyllaries linear, with spreading tips, **sticky, often turning blackish with drying**; disk florets 12–40, corollas dark yellow (3–4 mm long), resinous

fruits: cypselae; oblong (4–7 mm long), flattened, villous; pappi of 2–4 unequal bristles (2.5–3.2 mm long)

VEGETATIVE CHARACTERISTICS

leaves: alternate, simple; blades elliptical to oval (1.6–2.5 cm long, 6–12 mm wide), acute, thick; margins entire, rarely undulate; adaxially green and glabrous, abaxially paler and glabrous or pubescent, **veins often black**, somewhat sunken; petioles puberulent (1–2.5 mm long)

stems: twigs erect, light brown or gray or cream-colored, slender, resinous, pubescent; trunk gray to dark gray, glabrous, sparingly striate, resinous

other: entire plant has a tarlike odor

HISTORICAL, FOOD, AND MEDICINAL USES: a decoction was made in Mexico from the leaves and flowers to treat indigestion

LIVESTOCK LOSSES: mature fruits may cause losses of sheep and goats in January to March with the ingestion of 1 percent of their body weight before the fruits fall from the plants; foliage may be toxic in moderate quantities

FORAGE VALUES: worthless to livestock, generally not utilized, increases with improper grazing; may be utilized by jackrabbits, small mammals, and other wildlife

HABITATS: deserts and dry soils of valleys, mesas, flats, and foothills, especially on limestone, alkaline, or clay soils

Orange sneezeweed
Hymenoxys hoopesii (A. Gray) Bierner

SYN = *Dugaldia hoopesii* (A. Gray) Rydb., *Helenium hoopesii* A. Gray

heads solitary
or in corymbiform or paniculiform
arrangements

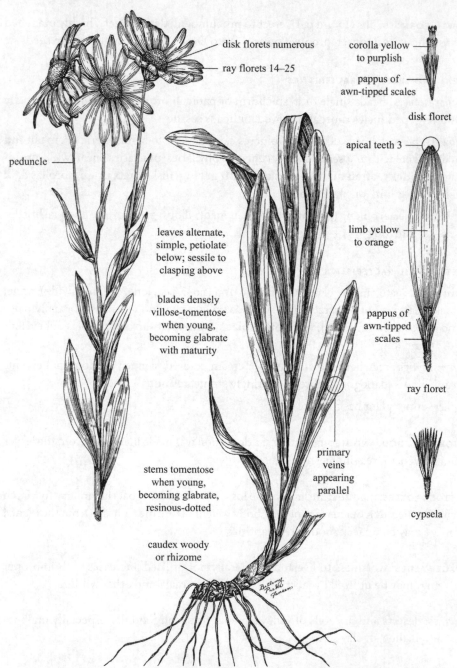

disk florets numerous

ray florets 14–25

corolla yellow
to purplish

pappus of
awn-tipped scales

disk floret

apical teeth 3

limb yellow
to orange

pappus of
awn-tipped
scales

ray floret

cypsela

peduncle

leaves alternate,
simple, petiolate
below; sessile to
clasping above

blades densely
villose-tomentose
when young,
becoming glabrate
with maturity

primary
veins
appearing
parallel

stems tomentose
when young,
becoming glabrate,
resinous-dotted

caudex woody
or rhizome

FAMILY:	ASTERACEAE
TRIBE:	HELIANTHEAE
SPECIES:	*Hymenoxys hoopesii* (A. Gray) Bierner
COMMON NAME:	Orange sneezeweed (owlsclaws, western sneezeweed)
LIFE SPAN:	Perennial
ORIGIN:	Native
SEASON:	Warm

GROWTH FORM: forbs (0.2–1 m tall), stems 1 to several from rhizomes or woody caudices; flowers May to September, reproduces from seeds and rhizomes; rhizomes stout, blackish

FLORAL AND FRUIT CHARACTERISTICS

inflorescences: heads solitary or in corymbiform or paniculiform arrangements with up to 12 heads, large, showy; peduncles erect (4–20 cm long); tomentose near involucres

flowers: heads radiate; involucres (6–10 mm tall) with 2–3 series of phyllaries; phyllaries linear to lanceolate (8–11 mm long), acute, pubescent, thin, loosely imbricate, reflexed with age; ray florets 14–25, pistillate, limbs narrow (1.5–3 cm long), **yellow to orange, apical teeth 3**; disk florets numerous, perfect, corollas tubular (about 5 mm long), **curved, yellow to purplish**

fruits: cypselae; obpyramidal (2.5–4.5 mm long), veins 4–10, pubescent; pappi of awn-tipped scales (2.5–4 mm long), white

VEGETATIVE CHARACTERISTICS

leaves: alternate, simple; basal leaves longest (10–30 cm long, 1–5 cm wide), reduced above, oblanceolate to lanceolate, thick; apices attenuate; **primary veins appearing parallel**, especially prominent near the base; margins entire; densely villose-tomentose when young, becoming glabrate with maturity, glandular; upper leaves sessile to clasping; lower leaves petiolate, petioles winged

stems: erect, 1 to several, simple to little-branched, usually yellowish-green, may be reddish-purple tinted, tomentose when young, becoming glabrate, resinous-dotted

HISTORICAL, FOOD, AND MEDICINAL USES: some Native Americans used the flowers to make a yellow textile dye; chewing gum was made from the roots

LIVESTOCK LOSSES: poisonous to sheep, contains a glycoside, causes "spewing sickness"; considered poisonous to cattle but it is unpalatable and seldom grazed

FORAGE VALUES: poor to worthless for livestock and wildlife, most commonly eaten by sheep; will only be consumed when other forage is not readily available

HABITATS: moist slopes, well-drained meadows, stream banks, forest edges, open woodlands, and valleys

Bitterweed
Hymenoxys odorata DC.

SYN = *Actinella odorata* (DC.) A. Gray, *Picradenia odorata* (DC.) Britton,
Ptilepedia odorata (DC.) Britton

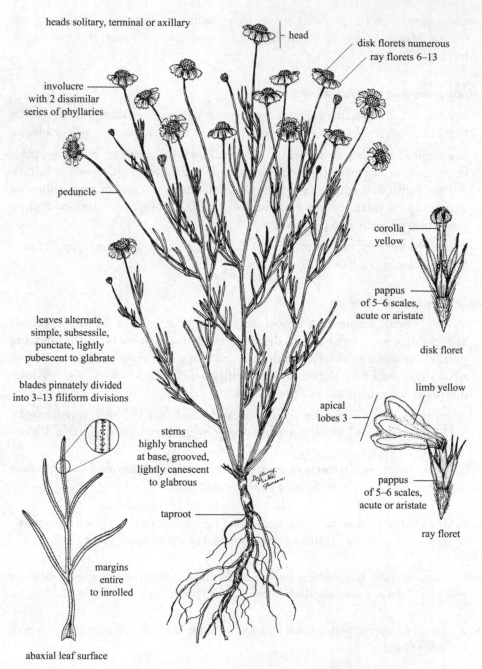

heads solitary, terminal or axillary

head

disk florets numerous
ray florets 6–13

involucre
with 2 dissimilar
series of phyllaries

peduncle

corolla
yellow

pappus
of 5–6 scales,
acute or aristate

disk floret

leaves alternate,
simple, subsessile,
punctate, lightly
pubescent to glabrate

blades pinnately divided
into 3–13 filiform divisions

limb yellow

apical
lobes 3

pappus
of 5–6 scales,
acute or aristate

ray floret

stems
highly branched
at base, grooved,
lightly canescent
to glabrous

taproot

margins
entire
to inrolled

abaxial leaf surface

FAMILY:	ASTERACEAE
TRIBE:	HELIANTHEAE
SPECIES:	*Hymenoxys odorata* DC.
COMMON NAME:	Bitterweed (girasolillo dentado, hierba amarga, hierba apestosa, limoncillo, bitter rubberweed)
LIFE SPAN:	Annual
ORIGIN:	Native
SEASON:	Cool

GROWTH FORM: forbs (20–50 cm tall, rarely to 80 cm tall); highly branched producing a bushy appearance, arising from a taproot; flowers March to August, reproduces from seeds

FLORAL AND FRUIT CHARACTERISTICS

inflorescences: heads usually solitary; terminal or axillary, numerous, small (3–12 mm wide), elevated on peduncles (3–15 cm long); peduncles sparsely hairy

flowers: heads radiate; involucres campanulate to subhemispheric (4–6.5 mm tall), with 2 dissimilar series of phyllaries; outer 8–13 phyllaries united and thickened at the bases, obovate to lanceolate, acute; inner phyllaries long convergent, exceeding the outer phyllaries, obovate, acuminate to acute; ray florets 6–13, pistillate, limbs yellow (5–11 mm long), apical lobes 3, reflexed, persistent; disk florets perfect, numerous, corollas yellow

fruits: cypselae; turbinate to narrowly obpyramidal (1.5–2 mm long), angles 4, pubescent; pappi of 5–6 scales (1.5–2.5 mm long), acute or aristate

VEGETATIVE CHARACTERISTICS

leaves: alternate, simple or lobed; initially in a basal rosette but the basal leaves caducous; cauline leaves numerous (2–10 cm long), **pinnately divided into 3–13 filiform divisions**; margins entire to inrolled; **punctate**, lightly pubescent to glabrate; subsessile

stems: erect, highly branched at the bases, granulate, grooved, lightly canescent to glabrous, often reddish-tinged

other: entire plant aromatic and somewhat resinous

HISTORICAL, FOOD, AND MEDICINAL USES: plant extracts are being evaluated for pharmaceutical properties

LIVESTOCK LOSSES: poisonous to sheep especially in winter, toxicity increases with water stress, poison is cumulative in animals; toxin is water soluble, identity is unknown; plants retain toxicity when dry but not after leaching by rain

FORAGE VALUES: poor to worthless for all classes of livestock and wildlife

HABITATS: disturbed sites, ditches, stream banks, deteriorated pastures and rangelands, and roadsides; most abundant in limestone soils

Prairie coneflower
Ratibida columnifera (Nutt.) Wooton & Standl.

SYN = *Lepachys columnifera* (Nutt.) J.F. Macbr., *Ratibida columnaris*
(Pursh) D. Don, *Rudbeckia columnifera* Nutt.

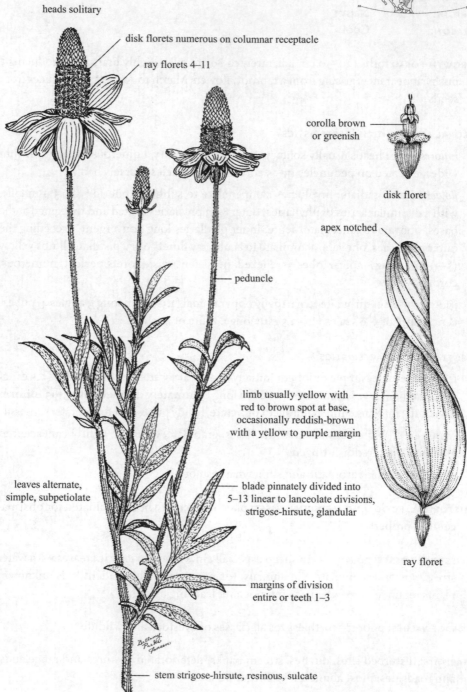

heads solitary

disk florets numerous on columnar receptacle

ray florets 4–11

corolla brown
or greenish

disk floret

apex notched

peduncle

limb usually yellow with
red to brown spot at base,
occasionally reddish-brown
with a yellow to purple margin

leaves alternate,
simple, subpetiolate

blade pinnately divided into
5–13 linear to lanceolate divisions,
strigose-hirsute, glandular

ray floret

margins of division
entire or teeth 1–3

stem strigose-hirsute, resinous, sulcate

FAMILY:	ASTERACEAE
TRIBE:	HELIANTHEAE
SPECIES:	*Ratibida columnifera* (Nutt.) Wooton & Standl.
COMMON NAME:	Prairie coneflower (sombrero de Zapata, upright prairieconeflower, Mexican hat)
LIFE SPAN:	Perennial
ORIGIN:	Native
SEASON:	Warm

GROWTH FORM: forbs (0.2–1 m tall); solitary or several stems from woody caudices surmounting taproots; flowers April to October, reproduces from seeds

FLORAL AND FRUIT CHARACTERISTICS

inflorescences: heads solitary; several to many per stem, cylindrical (1–5.5 cm long, 2–3 cm in diameter), rounded at the top, terminal, on naked peduncles (6–45 cm long), exserted well above the leaves

flowers: heads radiate; involucres short, with 2 series of 5–14 phyllaries; phyllaries linear to lanceolate, acuminate, hirsute; outer phyllaries longer (4–12 mm long) than the inner (about 3 mm long); ray florets 4–11, sterile, limbs oblong (1–3.5 cm long), usually **yellow with a red to brown spot at the bases**, occasionally reddish-brown with a yellow to purple margin, drooping, **apices notched; disk florets on a columnar receptacle** (1.5–5 cm long, 8–12 mm in diameter), numerous, perfect, corollas brown or greenish to purplish (1.5–2.5 mm long)

fruits: cypselae; oblong (1.5–3 mm long), compressed, gray, ciliate on inner edge; pappi of 1–2 toothlike awns, sometimes absent

VEGETATIVE CHARACTERISTICS

leaves: alternate, simple; blades oblong (2.5–15 cm long, 2–6 cm wide), pinnately divided into 5–13 linear to lanceolate divisions (most 1.5 cm or more long), often very unequal, divided nearly to the midveins; margins of divisions entire or teeth 1–3; **strigose-hirsute**, glandular; subpetiolate

stems: erect, branched, leafy, **strigose-hirsute, resinous, sulcate**

HISTORICAL, FOOD, AND MEDICINAL USES: some Cheyenne boiled leaves and stems to make a yellow solution applied externally to draw poison out of snakebites and for relief from rash caused by poison ivy [*Toxicodendron radicans* (L.) Kuntze]; members of some other Native American tribes made tea from the flowers and leaves; planted as an ornamental

LIVESTOCK LOSSES: none

FORAGE VALUES: fair to good for sheep and wildlife, fair for cattle

HABITATS: prairies, plains, pastures, roadsides, and disturbed sites; adapted to a broad range of soils

Mulesears
Wyethia amplexicaulis (Nutt.) Nutt.

SYN = *Espeletia amplexicaulis* Nutt., *Wyethia lanceolata* Howell

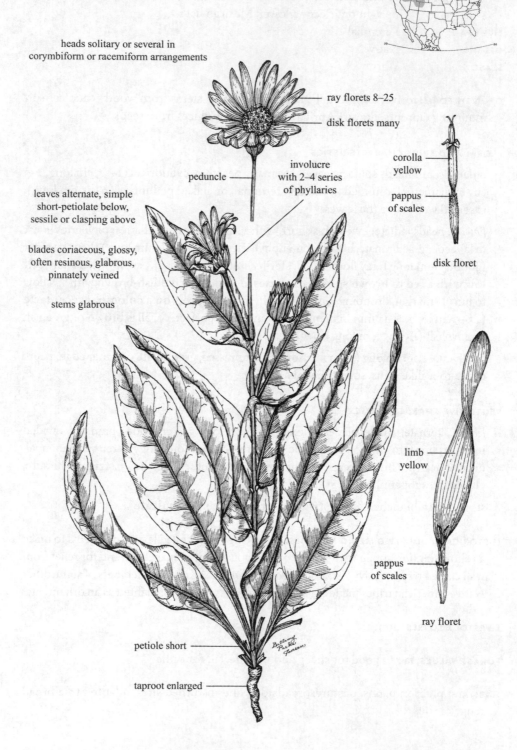

heads solitary or several in corymbiform or racemiform arrangements

ray florets 8–25

disk florets many

peduncle

involucre with 2–4 series of phyllaries

corolla yellow

pappus of scales

disk floret

leaves alternate, simple short-petiolate below, sessile or clasping above

blades coriaceous, glossy, often resinous, glabrous, pinnately veined

stems glabrous

limb yellow

pappus of scales

ray floret

petiole short

taproot enlarged

FAMILY:	ASTERACEAE
TRIBE:	HELIANTHEAE
SPECIES:	*Wyethia amplexicaulis* (Nutt.) Nutt.
COMMON NAME:	Mulesears (pe-ik, dock)
LIFE SPAN:	Perennial
ORIGIN:	Native
SEASON:	Cool

GROWTH FORM: forbs (30–80 cm tall); erect or ascending, solitary to few stems from **simple or branching caudices** surmounting **enlarged taproots**; flowers May to July, reproduces from seeds and branching caudices

FLORAL AND FRUIT CHARACTERISTICS

inflorescences: heads solitary or several in corymbiform or racemiform arrangements; terminal and axillary, large and showy (2–4 cm tall, 4–10 cm in diameter), terminal heads larger than axillary heads; both peduncled

flowers: heads radiate; involucres hemispheric to turbinate (1.8–4 cm tall, 1.5–3 cm in diameter) with 2–4 series of phyllaries; phyllaries 18–36, outer phyllaries (1.8–3 cm long), exceeding inner phyllaries; phyllaries lanceolate to obovate, acute or obtuse, **broad, glabrous**; ray florets 8–25, pistillate, limbs yellow (2–6 cm long); disk florets perfect, corollas yellow

fruits: cypselae; oblong (6–15 mm long), angles 4, glabrous or puberulent; pappi of scales (about 2 mm long), 1–2 awns (1 mm long)

VEGETATIVE CHARACTERISTICS

leaves: alternate, simple, mostly basal; blades oblong-lanceolate (15–60 cm long, 5–15 cm wide), **coriaceous**, reduced above; apices acute or lance-elliptical; margins entire or denticulate; green, glossy, often resinous, **glabrous**, pinnately veined; short-petiolate below (compare to *Wyethia mollis*), sessile or clasping above

stems: simple or rarely branched, leafy, glabrous

other: entire plant aromatic

HISTORICAL, FOOD, AND MEDICINAL USES: some Native Americans fermented roots for 2 days in a pit heated with hot stones to make a sweet-flavored food

LIVESTOCK LOSSES: none

FORAGE VALUES: poor for cattle, horses, deer, and elk; poor to fair for sheep; immature foliage is most frequently consumed; heads are eaten by all classes of livestock and big game

HABITATS: hillsides, open woodlands, dry meadows, and moist draws

Woolly mulesears
Wyethia mollis A. Gray

heads solitary to few
in racemiform or
corymbiform arrangements

ray florets 5–17

disk florets many

peduncle

leaves alternate, simple,
petiolate below,
subsessile above

involucre with 2–4 series of phyllaries

blades densely
tomentose, pubescence
decreasing with maturity

corolla yellowish

pappus a short, erose crown
with 2–5 unequal awns
or absent

disk floret

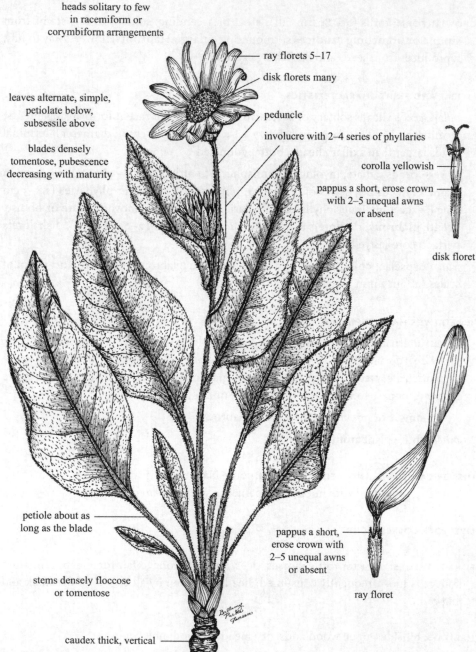

petiole about as
long as the blade

pappus a short,
erose crown with
2–5 unequal awns
or absent

stems densely floccose
or tomentose

ray floret

caudex thick, vertical

FAMILY:	ASTERACEAE
TRIBE:	HELIANTHEAE
SPECIES:	*Wyethia mollis* A. Gray
COMMON NAME:	Woolly mulesears (woolly wyethia)
LIFE SPAN:	Perennial
ORIGIN:	Native
SEASON:	Cool

GROWTH FORM: forbs (0.4–1 m tall); cespitose; solitary to a few stems arising from thick, vertical caudices surmounting taproots; forms dense stands; flowers June to August, reproduces from seeds

FLORAL AND FRUIT CHARACTERISTICS

inflorescences: heads solitary to few (2–3) in racemiform or corymbiform arrangements, broad (4–9 cm wide), showy; terminal and axillary, peduncled

flowers: heads radiate; involucres campanulate to hemispheric (1.3–3.5 cm tall) with 2–4 series of phyllaries; phyllaries 12–22, erect, subequal or unequal, triangular to oblong, acute, **tomentose**; ray florets 5–17, pistillate, limbs yellow (1.5–4.5 cm long, 1–1.5 cm wide), acute; disk florets many, perfect, corollas yellowish

fruits: cypselae; oblong (8–11 mm long), angles 4, glaucous, glandular, pubescent above; pappi short, erose crowns with 2–5 unequal awns (about 7 mm long) or absent

VEGETATIVE CHARACTERISTICS

leaves: alternate, simple, principal leaves basal; blades lanceolate to oblong-ovate (9–50 cm long, 3–15 cm wide), narrow at the bases, reduced above; apices acute; margins entire; **densely tomentose**, especially when young, becoming more green with maturity as pubescence decreases; resinous, pinnately veined; petiolate below (compare to *Wyethia amplexicaulis*), subsessile above; petioles about as long as blades

stems: erect or ascending, simple, leafy, densely floccose or tomentose

HISTORICAL, FOOD, AND MEDICINAL USES: roots and heads were eaten by some Native Americans

LIVESTOCK LOSSES: none

FORAGE VALUES: poor for cattle and fair for sheep and deer; most palatable in early spring immediately following snowmelt; heads readily eaten by livestock and big game

HABITATS: open woodlands, rocky openings, and ridges; most abundant on dry, open sites

Threadleaf groundsel
Senecio flaccidus Less.

SYN = *Senecio douglasii* DC., *Senecio longilobus* Benth.

heads in corymbiform arrangements

head

involucre
with 1 series of
13–30 phyllaries

bracteoles
well developed

leaves simple, alternate,
sometimes fascicled,
subpetiolate to sessile

blades deeply pinnatifid
into linear divisions,
densely lanate-tomentose

stems herbaceous above,
greenish-white, tomentose;
woody below, gray,
somewhat tomentose

limb light yellow

caudex woody

taproot

corolla
yellowish

pappus of
numerous white
capillary bristles

leaf margins
entire, revolute

abaxial leaf surface

disk floret

ray floret

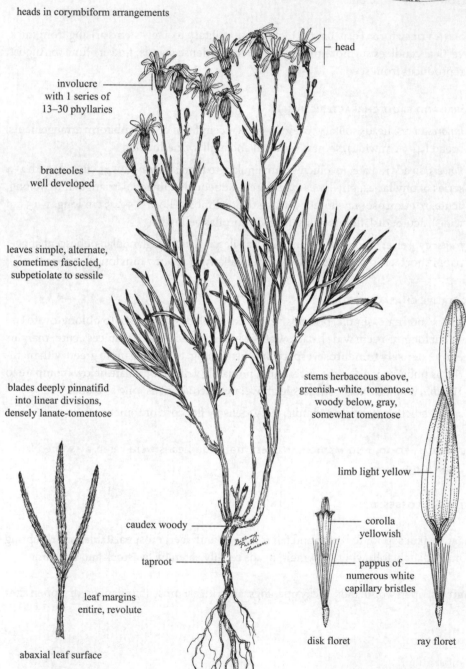

FAMILY:	ASTERACEAE
TRIBE:	SENECIONEAE
SPECIES:	*Senecio flaccidus Less.*
COMMON NAME:	Threadleaf groundsel (senecio, cenicillo, shrubby butterweed, threadleaf ragwort)
LIFE SPAN:	Perennial
ORIGIN:	Native
SEASON:	Warm

GROWTH FORM: shrubs or subshrubs (to 2 m tall); woody stems arising from woody caudices surmounting taproots, erect to ascending, few-branched, usually herbaceous above; flowers April to November, reproduces from seeds

FLORAL AND FRUIT CHARACTERISTICS

inflorescences: heads 3–20 in corymbiform arrangements; clusters of flowering branches **often aggregated to form large and showy inflorescences**

flowers: heads radiate; bracteoles well developed, unequal; involucres campanulate to cylindrical (7–11 mm tall) with **1 series of 13–30 phyllaries**; phyllaries linear (5–8 mm long), narrow, herbaceous; **ray florets 7–16**, pistillate, limbs light yellow (1–1.5 cm long); disk florets perfect, numerous, corollas yellowish

fruits: cypselae; narrowly cylindrical (1.5–2.5 mm long), ribs 5–10, canescent; pappi of numerous white capillary bristles

VEGETATIVE CHARACTERISTICS

leaves: alternate, sometimes with fascicles of small leaves in the axils of larger leaves, simple; **many deeply pinnatifid; blades and divisions linear or linear-filiform**; divisions unequal (4–12 cm long, 0.5–5 mm wide); apices acute or obtuse; margins entire to remotely toothed and revolute; **densely lanate-tomentose** to unevenly glabrescent, sessile to subpetiolate or obscurely petiolate

stems: arching to erect, usually multiple, woody below, gray, somewhat tomentose; herbaceous above, greenish-white, tomentose

HISTORICAL, FOOD, AND MEDICINAL USES: some Navajo drank tea made from the whole plant to ensure a good voice for ceremonial songs, used for numerous medicinal purposes by other Native Americans

LIVESTOCK LOSSES: poisonous to cattle, horses, and sheep (to a lesser extent); contains several alkaloids, poison is cumulative in animals

FORAGE VALUES: worthless to poor for cattle; sheep and goats may lightly browse it; poor for wildlife

HABITATS: plains, mesas, sandy washes, gravelly streambeds, open slopes, and grasslands; most abundant in well-drained sandy or rocky soils

Sawtooth butterweed
Senecio serra Hook.

SYN = *Senecio inornatus* DC.

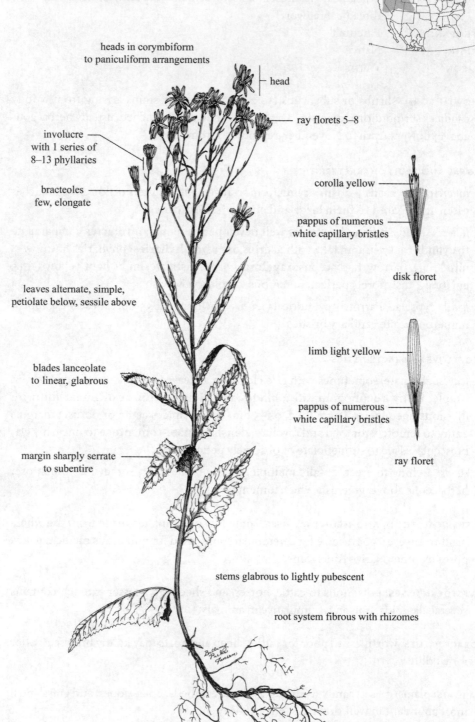

heads in corymbiform
to paniculiform arrangements

head

ray florets 5–8

involucre
with 1 series of
8–13 phyllaries

bracteoles
few, elongate

corolla yellow

pappus of numerous
white capillary bristles

disk floret

leaves alternate, simple,
petiolate below, sessile above

blades lanceolate
to linear, glabrous

limb light yellow

pappus of numerous
white capillary bristles

ray floret

margin sharply serrate
to subentire

stems glabrous to lightly pubescent

root system fibrous with rhizomes

FAMILY:	ASTERACEAE
TRIBE:	SENECIONEAE
SPECIES:	*Senecio serra* Hook.
COMMON NAME:	Sawtooth butterweed (butterweed groundsel, tall groundsel)
LIFE SPAN:	Perennial
ORIGIN:	Native
SEASON:	Warm

GROWTH FORM: forbs (0.6–2 m tall); erect to ascending, stout, several stems from a fibrous root system with rootstocks; flowers June to September, reproduces from seeds and rhizomes

FLORAL AND FRUIT CHARACTERISTICS

inflorescences: heads in corymbiform to paniculiform arrangements of **a few rather large heads** (11–13 mm tall, 5–8 mm in diameter) **and many smaller heads** (6–9 mm tall, 3–5 mm in diameter); numerous, erect, on slender peduncles

flowers: heads radiate; bracteoles few, elongate; involucres cylindrical to campanulate (5–9 mm tall) with **1 series of 8–13 phyllaries**; phyllaries narrow (4–9 mm long), linear; apices acute, usually green, sometimes black; ray florets mostly 5–8, pistillate, limbs light yellow (5–10 mm long); disk florets perfect, numerous, corolla yellow

fruits: cypselae; flattened (0.5–5 mm long), ribs 5–10, glabrous or glabrate; pappi of numerous white capillary bristles

VEGETATIVE CHARACTERISTICS

leaves: alternate, simple; blades lanceolate to linear (7–15 cm long, 1–4 cm wide), numerous, lower leaves caducous; apices long-acute or acuminate; margins sharply serrate to subentire; glabrous, tomentose when immature; petiolate below, sessile above

stems: erect, simple below, branching near inflorescence, glabrous to lightly pubescent

HISTORICAL, FOOD, AND MEDICINAL USES: some Native Americans drank a tea from the foliage to aid digestion

LIVESTOCK LOSSES: none

FORAGE VALUES: good to excellent for sheep, good for elk and deer, fair to poor for cattle; most palatable during spring and summer

HABITATS: meadows, damp ground, open woodlands, and moist stream banks; most abundant in rich, well-drained, sandy or gravelly loams

Gray horsebrush
Tetradymia canescens DC.

SYN = *Tetradymia inermis* Nutt.

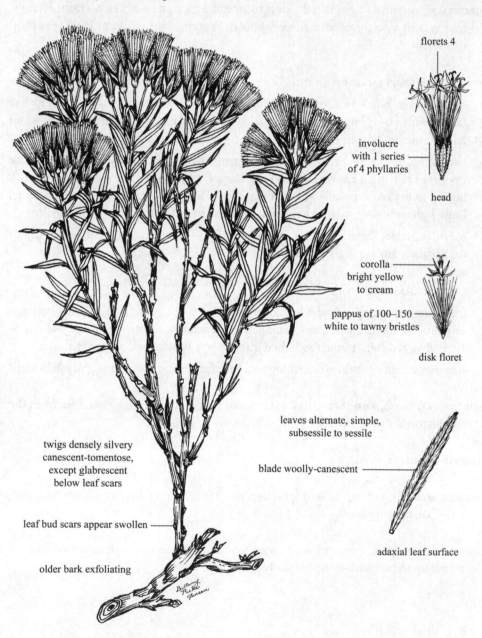

heads in corymbiform arrangements

florets 4

involucre with 1 series of 4 phyllaries

head

corolla bright yellow to cream

pappus of 100–150 white to tawny bristles

disk floret

leaves alternate, simple, subsessile to sessile

blade woolly-canescent

adaxial leaf surface

twigs densely silvery canescent-tomentose, except glabrescent below leaf scars

leaf bud scars appear swollen

older bark exfoliating

FAMILY:	ASTERACEAE
TRIBE:	SENECIONEAE
SPECIES:	*Tetradymia canescens* DC.
COMMON NAME:	Gray horsebrush (spineless horsebrush, common horsebrush)
LIFE SPAN:	Perennial
ORIGIN:	Native
SEASON:	Warm

GROWTH FORM: shrubs (to 80 cm tall); highly branched, spreading to ascending; flowers May to September, reproduces from seeds

FLORAL AND FRUIT CHARACTERISTICS

inflorescences: heads (3–15) in corymbiform arrangements; compact, terminal, peduncles 5–20 mm long

flowers: heads discoid; involucres turbinate to cylindrical with **1 series of 4 phyllaries** (6–15 mm tall); phyllaries linear to oblong, acute or obtuse, imbricate, thickened, tomentose; disk florets 4; corollas funnelform (4–7 mm long), limbs bright yellow to cream, lobes often longer than the tubes

fruits: cypselae; linear to slightly obovoid (2.5–5 mm long), ribs 5, densely silky; pappi of 100–150 white to tawny bristles (6–11 mm long), in 2–3 series

VEGETATIVE CHARACTERISTICS

leaves: alternate, simple; blades narrowly lanceolate to oblanceolate or spatulate (1–4 cm long, 1–6 mm wide); apices acute; margins entire; midveins prominent; **woolly-canescent**; subsessile to sessile

stems: erect, short, stout, highly branched; new growth densely **silvery canescent-tomentose**, except **glabrescent below leaf scars**; older bark gray, glabrescent, exfoliating; leaf bud scars appear swollen

HISTORICAL, FOOD, AND MEDICINAL USES: some Hopi made a tonic from leaves and roots for internal disorders

LIVESTOCK LOSSES: causes photosensitization in sheep, symptoms are called "big head" or "swell head" from swelling of the head and facial features; alkaloids may also cause liver damage in sheep followed by death

FORAGE VALUES: poor to worthless for cattle, sheep, goats, and big game; consumed only when other forage is unavailable

HABITATS: barren plains, foothills, and deserts; most abundant on sandy or rocky soils

Mountain alder
Alnus incana (L.) Moench

SYN = *Alnus februaria* Kuntze, *Alnus rugosa* (Du Roi) Spreng., *Alnus tenuifolia*
     Nutt., *Betula incana* (L.) L. f.

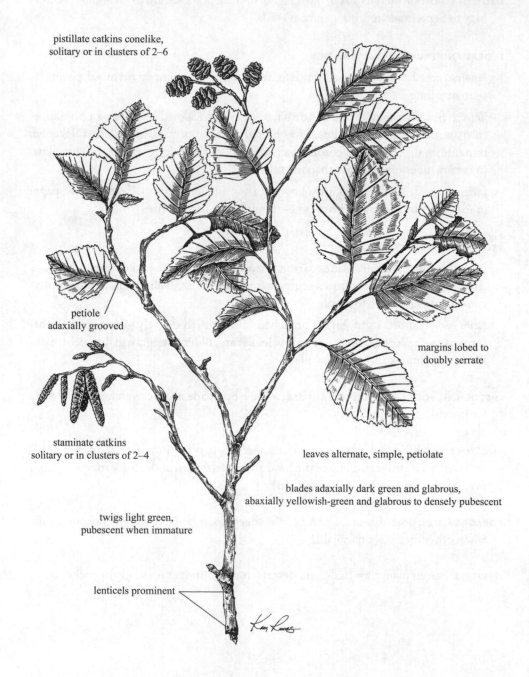

pistillate catkins conelike,
solitary or in clusters of 2–6

petiole
adaxially grooved

margins lobed to
doubly serrate

staminate catkins
solitary or in clusters of 2–4

leaves alternate, simple, petiolate

blades adaxially dark green and glabrous,
abaxially yellowish-green and glabrous to densely pubescent

twigs light green,
pubescent when immature

lenticels prominent

FAMILY:	BETULACEAE
SPECIES:	*Alnus incana* (L.) Moench
COMMON NAME:	Mountain alder (aliso gris, thinleaf alder, speckled alder, gray alder)
LIFE SPAN:	Perennial
ORIGIN:	Native
SEASON:	Cool

GROWTH FORM: monoecious shrubs to small trees (to 10 m tall), often with clustered trunks; flowers April to June; sometimes forming dense thickets; fruits mature in September; reproduces from seeds and basal sprouts

FLORAL AND FRUIT CHARACTERISTICS

inflorescences: catkins; staminate catkins solitary or in clusters of 2–4 (4–9 cm long, 2–9 mm wide), pendulous; **pistillate catkins conelike** (1–1.7 cm long, 8–12 mm wide), ovoid, solitary or in clusters of 2–6, brown, comprised of woody scales

flowers: unisexual, apetalous; staminate flowers in clusters of 3–4, each cluster subtended by a purplish, peltate bracteole on a stalk (about 2 mm long); perianth lobes 4, lobes obovate (1 mm long), cupped; stamens 4, attached to each calyx lobe; pistillate flowers paired, subtended by a fleshy obovate bract (1 mm long), persistent, becoming woody

fruits: samaras; elliptical to obovate (2–3 mm long and wide), brown, flat

VEGETATIVE CHARACTERISTICS

leaves: alternate, simple; blades highly variable, ovate to broadly elliptical (5–8 cm long, 2.5–5 cm wide), apices acute to obtuse; bases cuneate to narrowly rounded; **margins lobed to doubly serrate**, teeth unequal; adaxially dark green, glabrous; abaxially yellowish-green, glabrous to densely pubescent; often more pubescent on the veins; petioles (1–3 cm long) adaxially grooved; stipules oblong (6–10 mm long)

stems: twigs light green, pubescent when immature, becoming reddish-brown; lenticels prominent, horizontal, orangish; trunk bark grayish- to reddish-brown with whitish lenticels, thin, smooth or with age broken into irregular plates

HISTORICAL, FOOD, AND MEDICINAL USES: some Native Americans made a red dye from the bark

LIVESTOCK LOSSES: none

FORAGE VALUES: limited value for cattle and horses; fair to good for deer, elk, and moose

HABITATS: wet to moist soils, usually sandy or gravelly soils along streams, rivers, ponds, or in swamps, or moist woodlands; intolerant of shade

Tansymustard
Descurainia pinnata (Walter) Britton

SYN = *Descurainia canescens* (Nutt.) Prantl, *Erysimum pinnatum* Walter, *Sophia pinnata* (Walter) Howell

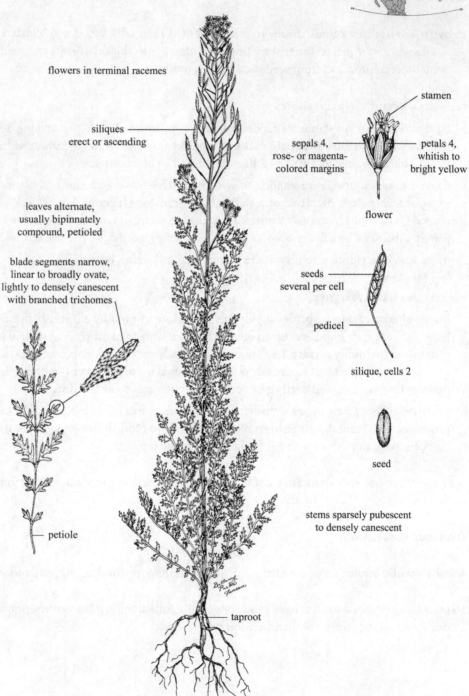

flowers in terminal racemes

siliques erect or ascending

leaves alternate, usually bipinnately compound, petioled

blade segments narrow, linear to broadly ovate, lightly to densely canescent with branched trichomes

petiole

stamen

sepals 4, rose- or magenta-colored margins

petals 4, whitish to bright yellow

flower

seeds several per cell

pedicel

silique, cells 2

seed

stems sparsely pubescent to densely canescent

taproot

FAMILY:	BRASSICACEAE
SPECIES:	*Descurainia pinnata* (Walter) Britton
COMMON NAME:	Tansymustard (mostácilla, western tansymustard, green tansymustard)
LIFE SPAN:	Annual
ORIGIN:	Native
SEASON:	Cool

GROWTH FORM: forbs (5–80 cm tall); erect to ascending, simple or branched, often branched above; stems 1 to several from a taproot; flowers March to August, reproduces from seeds

FLORAL AND FRUIT CHARACTERISTICS

inflorescences: racemes; terminal, elongating with maturity

flowers: perfect, regular; sepals 4 (1–2.5 mm long), oblong to ovate, petallike, margins membranous and rose- or magenta-colored; petals 4, obovate to spatulate (1–3 mm long), clawed, whitish to bright yellow

fruits: **siliques; clavate** (4–20 mm long, 1–1.5 mm wide at maturity), cells 2 with seeds several per cell; pedicels (to 2 cm long) in fruit, erect or ascending

VEGETATIVE CHARACTERISTICS

leaves: alternate, simple, usually **bipinnately compound** (1.5–10 cm long, including petiole); blades variable, reduced in size and usually simply pinnate above; segments narrow, linear to broadly ovate; **lightly to densely canescent, surfaces with branched trichomes**; petiolate

stems: sparsely pubescent to densely canescent with **usually branched trichomes**, sometimes glandular

other: a variable species with several forms

HISTORICAL, FOOD, AND MEDICINAL USES: seeds were used by some Native Americans to make pinole flour, young plants were used as potherbs

LIVESTOCK LOSSES: known to be poisonous to cattle in the Southwest, symptoms in cattle include "paralyzed tongue" or an inability to swallow food or water; generally results from consuming large quantities of plant material, which may not be easily avoided since it is green earlier in the spring than most forage species

FORAGE VALUES: poor for cattle, fair to good for sheep, goats, and pronghorn; unpalatable to horses; palatability declines with maturity

HABITATS: waste places, prairies, open woodlands, fields, disturbed sites, and roadsides; adapted to a broad range of soils; most abundant in dry or sandy soils

Desert princesplume
Stanleya pinnata (Pursh) Britton

SYN = *Cleome pinnata* Pursh, *Stanleya arcuata* Rydb., *Stanleya canescens* Rydb.,
Stanleya fruticosa Nutt., *Stanleya glauca* Rydb., *Stanleya heterophylla* Nutt.,
Stanleya pinnatifida Nutt.

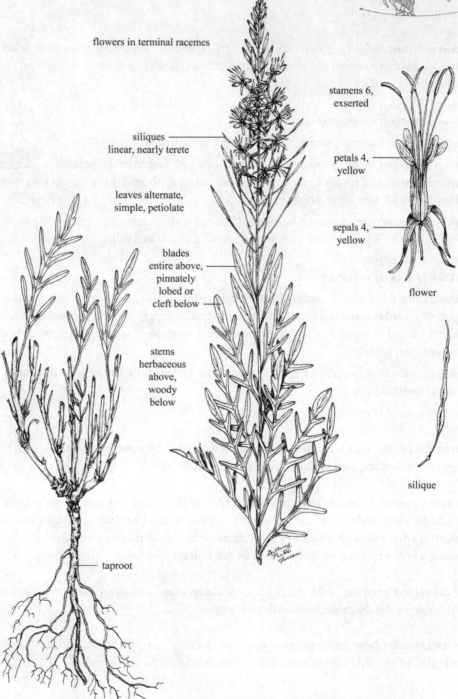

flowers in terminal racemes

stamens 6, exserted

siliques
linear, nearly terete

petals 4, yellow

leaves alternate, simple, petiolate

sepals 4, yellow

blades
entire above,
pinnately
lobed or
cleft below

flower

stems
herbaceous
above,
woody
below

silique

taproot

FAMILY:	BRASSICACEAE
SPECIES:	*Stanleya pinnata* (Pursh) Britton
COMMON NAME:	Desert princesplume
LIFE SPAN:	Perennial
ORIGIN:	Native
SEASON:	Cool

GROWTH FORM: subshrubs (0.6–1.5 m tall), suffrutescent; 1 to several stems from woody bases surmounting taproots, simple to branched above; flowers April to August, reproduces from seeds

FLORAL AND FRUIT CHARACTERISTICS

inflorescences: **racemes (10–35 cm long); terminal, flowers many**, crowded, showy, elongated in fruit (to 50 cm)

flowers: perfect, regular, pedicellate (about 1 cm long); sepals 4 (7–17 mm long), linear-oblong, spreading, yellow; petals 4 (9–17 mm long), pilose on inner surface, yellow; claws narrowing upward, brownish; stamens 6, exserted, nearly equal

fruits: **siliques; linear** (2–8 cm long), nearly terete, straight or curved, seeds in 1 row, seeds many; pedicels elongate (1–3 cm long)

VEGETATIVE CHARACTERISTICS

leaves: alternate, simple; blades lanceolate to elliptical or obovate (5–12 cm long, 3.5–5 cm wide), thick; **margins pinnately lobed or cleft below**, entire above; glabrous or glaucous or sparsely pubescent; cauline leaves petiolate; basal leaves frequently absent by flowering

stems: erect, glabrous or rarely densely pubescent below, green, glaucous; **woody below**, herbaceous above

other: a variable species with several forms

HISTORICAL, FOOD, AND MEDICINAL USES: some pioneers and Native Americans cooked and ate the stems and leaves

LIVESTOCK LOSSES: poisonous throughout the growing season, accumulates selenium and is an indicator of selenium in the soil; rarely consumed when other forage is available in early spring; poisoning on rangelands is rare, but force-feeding animals has caused poisoning

FORAGE VALUES: worthless to livestock and wildlife

HABITATS: dry hills, plains, valleys, and desert washes; reliable indicator of seleniferous soils; adapted to a broad range of soil textures; most abundant in dry soils on abused rangelands

Spiny hackberry
Celtis pallida Torr.

SYN = *Celtis spinosa* Spreng., *Celtis tala* Gillies *ex* Planch., *Momisia pallida* (Torr.) Planch.

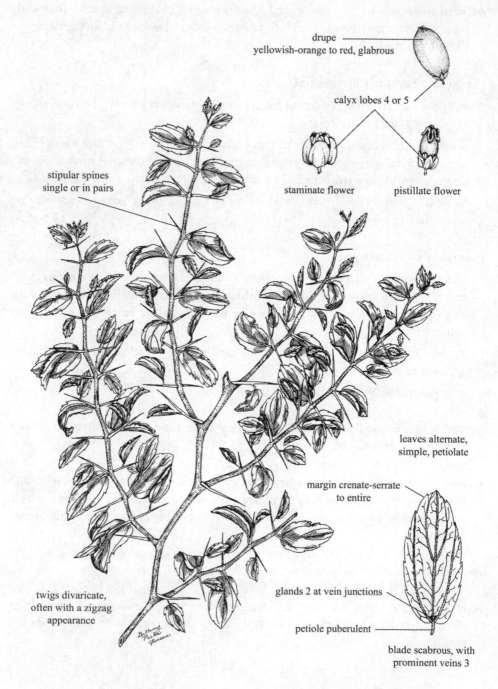

drupe
yellowish-orange to red, glabrous

calyx lobes 4 or 5

staminate flower

pistillate flower

stipular spines
single or in pairs

leaves alternate,
simple, petiolate

margin crenate-serrate
to entire

glands 2 at vein junctions

petiole puberulent

twigs divaricate,
often with a zigzag
appearance

blade scabrous, with
prominent veins 3

FAMILY:	CANNABACEAE
SPECIES:	*Celtis pallida* Torr.
COMMON NAME:	Spiny hackberry (granjeno, desert hackberry, acebuche, siempreverde)
LIFE SPAN:	Perennial
ORIGIN:	Native
SEASON:	Warm (semi-evergreen)

GROWTH FORM: polygamous or monoecious shrubs (to 6 m tall); densely branched, forming thickets, crowns rounded; flowers March to May, reproduces from seeds

FLORAL AND FRUIT CHARACTERISTICS

inflorescences: cymes, polygamous; branches 2; flowers 3–5; axillary; 1 flower usually staminate

flowers: unisexual and perfect, apetalous; calyx lobes 4 or 5 (small), greenish-white; stamens as many as the calyx lobes; styles absent; stigmas 2, each 2-cleft

fruits: drupes; subglobose or ovoid (6–7 mm long, 5–7 mm in diameter), yellowish-orange to red, glabrous; thin-fleshed, mealy, acidulous, edible; pedicels 1–2 mm long

VEGETATIVE CHARACTERISTICS

leaves: alternate, simple; blades ovate to ovate-oblong to elliptical (2–5.5 cm long, 1–2.5 cm wide), thick; apices rounded to acute; **bases with veins 3 (prominent), glands 2 at vein junctions abaxially; margins crenate-serrate to entire**; dark green, puberulent, slightly scabrous; petiole puberulent (2–5 mm long)

stems: twigs divaricate, flexuous, spreading, whitish-gray, often with a **zigzag appearance; stipular spines** (4–25 mm long), straight, stout, single or in pairs, sometimes forming a "V," puberulent; bark smooth, gray

HISTORICAL, FOOD, AND MEDICINAL USES: wood used for fence posts and firewood; sweet, insipid fruits were ground and eaten by some Native Americans with parched corn or fat; good honey plant

LIVESTOCK LOSSES: spines may cause minor injuries to soft tissues of browsing animals

FORAGE VALUES: worthless for livestock, rarely browsed; poor to fair browse for wildlife; fruits eaten by deer, small mammals, rabbits, quail, and other birds; browsed only when other plants are not available; valuable for erosion control and cover for wildlife

HABITATS: deserts, canyons, mesas, washes, foothills, thickets, brushlands, and grasslands; most abundant in gravelly or well-drained sandy soils

Snowberry
Symphoricarpos albus (L.) S.F. Blake

SYN = *Vaccinium album* L.

calyx lobes 5,
light green

corolla
lobes 5,
white to pinkish

flower

persistent calyx

drupe white

flowers solitary or in
short racemes of 2–5

margin entire,
occasionally sinuate

leaves opposite, simple, petiolate,
averaging less than 3 cm wide

blades adaxially dark green and glabrate,
abaxially pale and densely pubescent

stems generally hollow

petiole usually
pubescent

adaxial leaf surface

FAMILY:	CAPRIFOLIACEAE
SPECIES:	*Symphoricarpos albus* (L.) S.F. Blake
COMMON NAME:	Snowberry (common snowberry, white coralberry)
LIFE SPAN:	Perennial
ORIGIN:	Native
SEASON:	Cool

GROWTH FORM: shrubs (to 1 m tall); stems few to several from woody bases, occasionally from creeping rootstocks, forming thickets; flowers May to July, reproduces from seeds and rhizomes

FLORAL AND FRUIT CHARACTERISTICS

inflorescences: **flowers solitary or in short racemes; terminal** and in upper axils; terminal racemes with flowers 2–5; axillary racemes with flowers 1–2

flowers: perfect, regular; calyx lobes 5 (0.4–0.9 mm long), not spreading, triangular to lanceolate, light green, glabrous to ciliate; corollas campanulate; lobes 5 (5–7 mm long, 3–5 mm wide), white to pinkish, rounded, slightly spreading, outer surface of lobes glabrous, inner surface villous; stamens 5; stamens and style not exceeding corolla; pedicellate

fruits: drupes; ovoid (7–9 mm long, 6–8 mm in diameter), fleshy, white; calyces persistent; solitary or few in upper leaf axils; 2 nutlets per drupe; seeds 1 per nutlet

VEGETATIVE CHARACTERISTICS

leaves: **opposite**, simple; blades oval or ovate (1–4 cm long, 8–30 mm wide, averaging less than 3 cm wide); apices acute to obtuse; bases cuneate to rounded; margins entire or occasionally sinuate, ciliate when immature; adaxially dark green and glabrate; **abaxially pale, densely pubescent especially along the veins;** relatively thin (compare to *Symphoricarpos occidentalis*); petioles (1–3 mm long) usually pubescent

stems: twigs erect, slender, internodes hollow, yellowish-brown, pubescent when young, especially at nodes, glabrous with age; bark of trunk thin, grayish- or reddish-brown, exfoliating; winter buds with pubescent scales (2 mm long)

HISTORICAL, FOOD, AND MEDICINAL USES: some Native Americans made a tonic from the roots and an eyewash from the bark, and all parts were crushed and applied to wounds; consumption of relatively large quantities of fruits may cause vomiting and diarrhea

LIVESTOCK LOSSES: leaves contain saponins which are cathartic

FORAGE VALUES: fair for sheep and goats in winter, otherwise worthless for livestock; important food and cover for song and game birds; occasionally browsed by pronghorn and mule deer; planted as an ornamental

HABITATS: wooded hillsides, prairies, and open woodlands, present in both moist and dry soils

Western snowberry
Symphoricarpos occidentalis Hook.

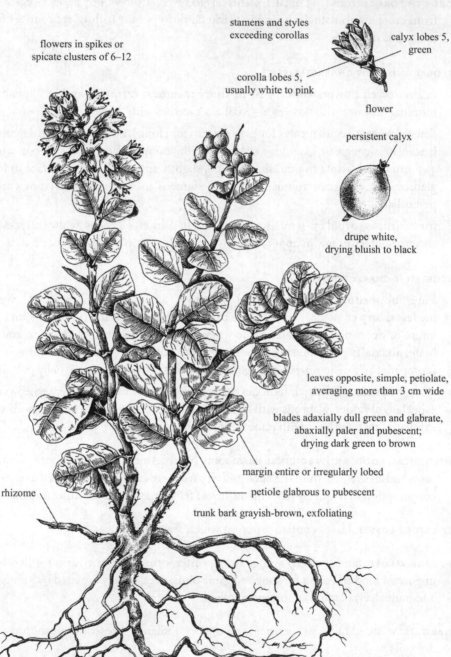

stamens and styles
exceeding corollas

calyx lobes 5,
green

flowers in spikes or
spicate clusters of 6–12

corolla lobes 5,
usually white to pink

flower

persistent calyx

drupe white,
drying bluish to black

leaves opposite, simple, petiolate,
averaging more than 3 cm wide

blades adaxially dull green and glabrate,
abaxially paler and pubescent;
drying dark green to brown

margin entire or irregularly lobed

petiole glabrous to pubescent

rhizome

trunk bark grayish-brown, exfoliating

FAMILY:	CAPRIFOLIACEAE
SPECIES:	*Symphoricarpos occidentalis* Hook.
COMMON NAME:	Western snowberry (wolfberry)
LIFE SPAN:	Perennial
ORIGIN:	Native
SEASON:	Cool

GROWTH FORM: shrubs (to 1 m tall); few to several stems from rhizomes, branches numerous; flowers June to August, reproduces from seeds and rhizomes; forming large colonies

FLORAL AND FRUIT CHARACTERISTICS

inflorescences: **spikes or spicate clusters of 6–12 flowers**, terminal and in upper leaf axils, dense, sometimes globular

flowers: perfect, regular, sessile; calyx lobes 5, spreading, triangular (less than 1 mm long), acute, green; corollas campanulate (5–8 mm long), wider than long; lobes 5, rounded, usually white to pink, less frequently greenish-white to purple, outer surface of lobes glabrous, inner surface hirsute; stamens 5; stamens and styles exceeding the corollas

fruits: drupes; globose (6–9 mm in diameter), numerous, fleshy, white but drying bluish to black, smooth, calyces persistent; nutlets 2 per drupe; each nutlet with 1 seed

VEGETATIVE CHARACTERISTICS

leaves: opposite, simple; blades ovate to elliptical or suborbicular (2–6 cm long, 1–4 cm wide, averaging more than 3 cm wide); apices acute to obtuse; bases cuneate, rounded or truncate; margins entire or irregularly lobed, typically ciliate; adaxially dull green and glabrate, **abaxially paler and pubescent; veins indented on adaxial surfaces**; thicker than *Symphoricarpos albus*; **drying dark green to brown**; petioles (2–7 mm long) glabrous to pubescent

stems: twigs erect, slender, brown, pubescent to glabrate, internodes hollow; trunk bark grayish-brown, exfoliating; buds small (0.5 mm long)

other: juvenile shoots sometimes produce large leaves (to 10 cm long, 8 cm wide); flowers are sweet-scented

HISTORICAL, FOOD, AND MEDICINAL USES: leaves were steeped by some Blackfoot to make a wash for sore eyes, fruits were used as famine food, and boiled fruits were given to horses as a diuretic; Lakota children made lightweight arrows from the stems to use in play

LIVESTOCK LOSSES: leaves contain saponins which are cathartic

FORAGE VALUES: poor for cattle, fair for sheep and goats; good for deer, pronghorn, and occasionally important for other big game; provides food and cover for small mammals and several song and game birds; planted as an ornamental; often considered to be a weed

HABITATS: prairies, ravines, rocky and gravelly hillsides, open woodlands, and wooded valleys

Fourwing saltbush
Atriplex canescens (Pursh) Nutt.

SYN = *Atriplex nuttallii* S. Watson, *Calligonum canescens* Pursh, *Obione canescens* (Pursh) Moq., *Pterochiton canescens* (Pursh) Nutt.

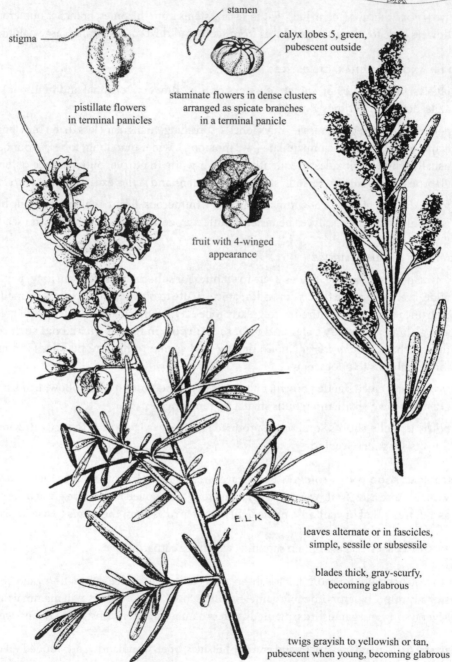

stigma

stamen

calyx lobes 5, green, pubescent outside

pistillate flowers in terminal panicles

staminate flowers in dense clusters arranged as spicate branches in a terminal panicle

fruit with 4-winged appearance

E.L.K

leaves alternate or in fascicles, simple, sessile or subsessile

blades thick, gray-scurfy, becoming glabrous

twigs grayish to yellowish or tan, pubescent when young, becoming glabrous

FAMILY:	CHENOPODIACEAE
SPECIES:	*Atriplex canescens* (Pursh) Nutt.
COMMON NAME:	Fourwing saltbush (costilla de vaca, chamizo, wingscale, chamiza)
LIFE SPAN:	Perennial
ORIGIN:	Native
SEASON:	Evergreen

GROWTH FORM: dioecious or rarely monoecious shrubs (to 3 m tall); erect, stout, much-branched; flowers May to September, reproduces from seeds

FLORAL AND FRUIT CHARACTERISTICS

inflorescences: staminate flowers in dense clusters arranged as spicate branches in a terminal panicle; pistillate flowers in panicles (5–10 cm long), terminal

flowers: unisexual, apetalous; calyx lobes 5; lobes ovate (less than 1 mm long and wide), obtuse, green or yellowish green, pubescent outside; stamens 5; pistillate flowers subtended by 2 bracts (each about 1.6 mm long)

fruits: achenes, enclosed by 2 winged bracteoles (8–10 mm long); wings lacerate, entire or undulate, apices entire or bifid; surfaces smooth, reticulate; sessile or short-pedicelled; pair of bracteoles enclosing achenes gives a **4-winged appearance**

VEGETATIVE CHARACTERISTICS

leaves: alternate or occurring in fascicles, simple, persistent; **blades oblong to obovate or lanceolate** (2.5–4.5 cm long, 3–7 mm wide), thick; apices obtuse to retuse; bases cuneate; margins entire, may be inrolled; **gray-scurfy**, becoming glabrous; sessile or subsessile

stems: twigs rigid, erect, grayish to yellowish or tan, pubescent when young, becoming glabrous; bark thin, tight, gray, lightly sulcate

other: plants highly variable in size, leaf shape, and fruiting bract size and shape

HISTORICAL, FOOD, AND MEDICINAL USES: some Native Americans ground the achenes to make flour for bread or to mix with sugar and water for a drink; some Navajo used an infusion of leaves and stems to make a yellow dye; pollen commonly causes hay fever

LIVESTOCK LOSSES: concentrated feeding causes scours in cattle

FORAGE VALUES: good for cattle, sheep, goats, pronghorn, and deer; furnishes valuable browse in winter; achenes provide good food for wildlife

HABITATS: bluffs, hillsides, deserts, and saline or alkali flats; common in many different soil types

Shadscale saltbush
Atriplex confertifolia (Torr. & Frém.) S. Watson

SYN = *Atriplex collina* Wooton & Standl., *Atriplex subconferta* Rydb., *Obione confertifolia* Torr. & Frém., *Obione rigida* Torr. & Frém.

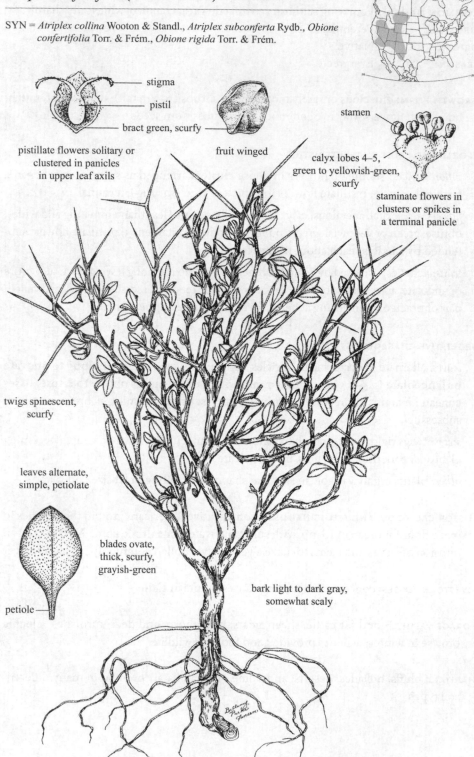

stigma

pistil

bract green, scurfy

pistillate flowers solitary or clustered in panicles in upper leaf axils

fruit winged

stamen

calyx lobes 4–5, green to yellowish-green, scurfy

staminate flowers in clusters or spikes in a terminal panicle

twigs spinescent, scurfy

leaves alternate, simple, petiolate

blades ovate, thick, scurfy, grayish-green

petiole

bark light to dark gray, somewhat scaly

FAMILY:	CHENOPODIACEAE
SPECIES:	*Atriplex confertifolia* (Torr. & Frém.) S. Watson
COMMON NAME:	Shadscale saltbush (saladillo, spiny saltbrush, roundleaf saltbrush)
LIFE SPAN:	Perennial
ORIGIN:	Native
SEASON:	Cool

GROWTH FORM: dioecious shrubs (to 90 cm tall); late deciduous, crowns rounded, densely branched, rigid, spinescent; flowers July to August, reproduces from seeds

FLORAL AND FRUIT CHARACTERISTICS

inflorescences: staminate flowers in clusters or spikes (to 1 cm long) in a terminal panicle; pistillate flowers solitary or clustered in panicles in upper leaf axils

flowers: unisexual, apetalous; staminate calyx lobes 4 or 5; lobes oblong (1–1.5 mm long and wide), green to yellowish-green, scurfy; stamens 5; sessile; pistillate flowers subtended by 2 fleshy bracts, bracts ovate (4–12 mm long, 4–12 mm wide), green, scurfy, united one-third the length

fruits: achenes; pistillate bracteoles 2, persistent; **bracteoles similar to leaves**, sub-orbiculate to elliptical or rhombic (4–10 mm wide); margins herbaceous, entire to undulate or serrulate to denticulate, yellowish-brown; sessile or subsessile; bracteoles enclosing the achenes give the fruit a winged appearance

VEGETATIVE CHARACTERISTICS

leaves: alternate, simple, persistent; **blades ovate**, triangular, or elliptical (9–30 mm long, 5–20 mm wide), thick; apices obtuse or rounded; bases attenuate; margins entire; grayish-green, **scurfy**; petiolate (1–4 mm long)

stems: twigs rigid, erect, **stout, spinescent**, yellowish-brown, smooth, scurfy; trunk irregular, appearing as a cluster of older branches, bark light to dark gray, somewhat scaly

HISTORICAL, FOOD, AND MEDICINAL USES: some Native Americans ground the achenes into flour

LIVESTOCK LOSSES: none

FORAGE VALUES: good to fair for all classes of livestock, pronghorn, and mule deer; provides winter and spring browse; fruits provide food for game and songbirds; resistant to heavy browsing

HABITATS: desert valleys, hills, and bluffs in stony or clayey alkaline soils; forming almost pure stands in some locations; frequently grows in association with *Achnatherum hymenoides*

Saltbush
Atriplex gardneri (Moq.) D. Dietr.

SYN = *Atriplex aptera* A. Nelson, *Obione gardneri* Moq.

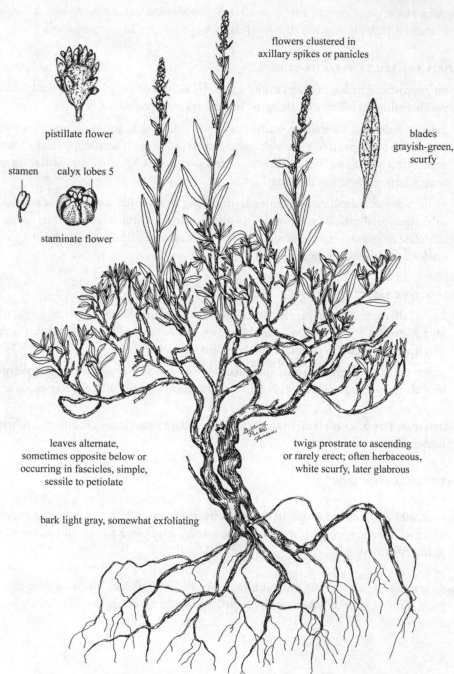

pistillate flower

stamen   calyx lobes 5

staminate flower

flowers clustered in
axillary spikes or panicles

blades
grayish-green,
scurfy

leaves alternate,
sometimes opposite below or
occurring in fascicles, simple,
sessile to petiolate

twigs prostrate to ascending
or rarely erect; often herbaceous,
white scurfy, later glabrous

bark light gray, somewhat exfoliating

FAMILY:	CHENOPODIACEAE
SPECIES:	*Atriplex gardneri* (Moq.) D. Dietr.
COMMON NAME:	Saltbush (saladillo, Gardner saltbush)
LIFE SPAN:	Perennial
ORIGIN:	Native
SEASON:	Evergreen

GROWTH FORM: dioecious or rarely monoecious **subshrubs** (to 80 cm tall); ascending from **decumbent spreading bases**, much-branched; flowers June to August, reproduces from seeds and rootstocks

FLORAL AND FRUIT CHARACTERISTICS

inflorescences: flowers clustered in spikes or panicles, axillary; staminate and pistillate flowers in dense clusters toward the branch tips, giving the appearance of a terminal inflorescence

flowers: unisexual, apetalous; staminate calyx lobes 5 (sometimes 3); lobes ovate (less than 1 mm long and wide), obtuse, yellowish-brown; stamens 5; pistillate flowers subtended by 2 fleshy bracts; bracts ovate to orbicular (3–5 mm long), united to middle, indurate, enclosing the ovary

fruits: achenes, enclosed in pistillate bracteoles (5–8 mm long); bearing 4 deeply lobed lateral wings; margins dentate, surface smooth or with **various linear tubercles**; sessile or short-pedicellate (4–6 mm long)

VEGETATIVE CHARACTERISTICS

leaves: alternate, sometimes opposite below or occurring in fascicles, simple, sometimes persistent; blades narrowly oblong-linear or oblanceolate (2.5–5 cm long, 3–10 mm wide), thick; apices rounded; bases narrowed or cuneate; margins entire, may be inrolled; grayish-green, scurfy; sessile to petiolate

stems: twigs slender, **prostrate to ascending** or rarely erect, erect as seedlings, often herbaceous; white scurfy, later glabrous and dark; trunk bark light gray, smooth, somewhat exfoliating

other: highly variable species

HISTORICAL, FOOD, AND MEDICINAL USES: some Native Americans parched achenes and ground them to make pinole flour

LIVESTOCK LOSSES: none

FORAGE VALUES: good for cattle, sheep, deer, and pronghorn; important winter browse on western rangelands, withstands heavy browsing

HABITATS: plains, valleys, and badlands; usually in saline or alkaline soils

Spiny hopsage
Grayia spinosa (Hook.) Moq.

SYN = *Atriplex grayi* Collotzi *ex* W.A. Weber, *Atriplex spinosa* (Hook.) Collotzi, *Chenopodium spinosum* Hook.

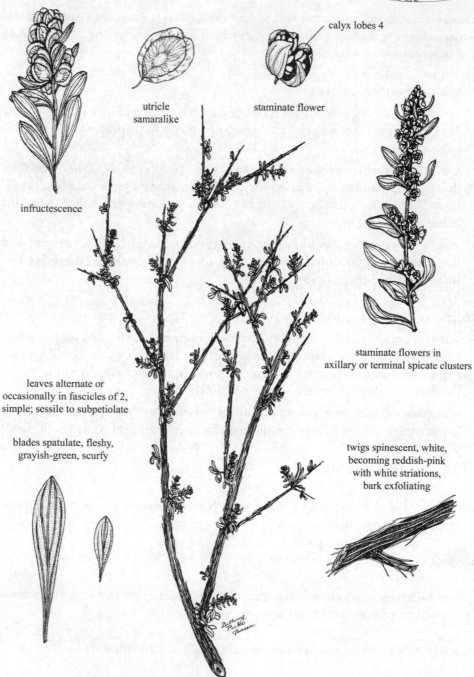

infructescence

utricle samaralike

staminate flower

calyx lobes 4

staminate flowers in axillary or terminal spicate clusters

leaves alternate or occasionally in fascicles of 2, simple; sessile to subpetiolate

blades spatulate, fleshy, grayish-green, scurfy

twigs spinescent, white, becoming reddish-pink with white striations, bark exfoliating

FAMILY:	CHENOPODIACEAE
SPECIES:	*Grayia spinosa* (Hook.) Moq.
COMMON NAME:	Spiny hopsage (Grays saltbush, spiny sage)
LIFE SPAN:	Perennial
ORIGIN:	Native
SEASON:	Warm

GROWTH FORM: dioecious or rarely monoecious shrubs (to 1.5 m tall); erect, highly branched; branches divergent, spinescent; flowers April to July, reproduces from seeds

FLORAL AND FRUIT CHARACTERISTICS

inflorescences: spicate clusters; axillary and terminal

flowers: unisexual, apetalous; staminate calyx lobes 4, stamens 4–5; pistillate flowers subtended by 2 persistent bracts; bracts ovate or orbicular (5–12 mm long, 6 mm wide), enclosing ovary, green to **greenish-pink when mature**

fruits: utricles; singe achenes enclosed in 2 bracteoles; bracteoles thin, glabrous, greenish-white or reddish, entire, sessile, each with a wing on the back or midvein, appearing as a saclike structure covering the ovary; wing somewhat thickened near margins, yellowish-green to pinkish-tinged, glabrous; samaralike

VEGETATIVE CHARACTERISTICS

leaves: alternate or occasionally in fascicles of 2, simple; blades oblanceolate, **spatulate**, or obovate (1–3 cm long), **fleshy**; apices obtuse to subacute, often yellowish-green to whitish; margins entire; grayish-green, scurfy, becoming glabrous with maturity; midveins prominent, sessile to subpetiolate

stems: twigs rigid, erect to ascending, white, becoming **reddish-pink with white striations formed by stringy exfoliating bark**; scurfy when young, turning glabrous, **spinescent**; trunk bark reddish-gray, exfoliating into thin strips, white striations may persist, glabrous

other: bark can be quite variable, from smooth and yellowish-white to red and striate

HISTORICAL, FOOD, AND MEDICINAL USES: some Native Americans parched and ground the achenes to make pinole flour

LIVESTOCK LOSSES: spinescent branches may cause minor injury to soft tissues of browsing animals

FORAGE VALUES: good to fair for sheep, goats, deer, and pronghorn; fair to poor for cattle: poor for horses; browsed in autumn, winter, and spring; fruits valuable for fattening sheep

HABITATS: mesas, flats, and valleys; adapted to alkaline, limestone, gravelly, and dry heavy clay soils

Halogeton
Halogeton glomeratus (M. Bieb.) C.A. Mey.

SYN = *Anabasis glomerata* M. Bieb.

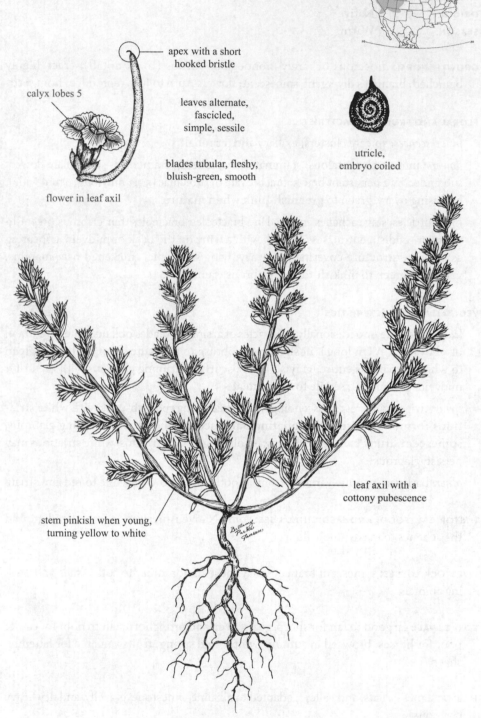

apex with a short
hooked bristle

calyx lobes 5

leaves alternate,
fascicled,
simple, sessile

blades tubular, fleshy,
bluish-green, smooth

utricle,
embryo coiled

flower in leaf axil

leaf axil with a
cottony pubescence

stem pinkish when young,
turning yellow to white

FAMILY:	CHENOPODIACEAE
SPECIES:	*Halogeton glomeratus* (M. Bieb.) C.A. Mey.
COMMON NAME:	Halogeton (saltlover)
LIFE SPAN:	Annual
ORIGIN:	Introduced (from Eurasia)
SEASON:	Warm

GROWTH FORM: forbs (to 30 cm tall); few or highly branched from base, stems spreading first then erect or ascending; flowers July to September, reproduces from seeds

FLORAL AND FRUIT CHARACTERISTICS

inflorescences: axillary glomerules; distributed throughout the aerial portion of the plants

flowers: perfect or pistillate only, apetalous, small and usually inconspicuous; calyx lobes 5; lobes thin (1.5 mm long), rounded, persistent; stamens 3–5

fruits: utricles, dimorphic; lateral ones blackish (0.1–1 mm long); central ones brown (1–2 mm long); seed coiled; calyx lobes enlarging with maturity to form yellowish or **reddish fanlike wings**; pericarps somewhat adherent to the seeds; sessile; seeds orbiculate, flattened, brown or brownish-black; embryo coiled

other: persistent calyx is often misidentified as the fruiting body

VEGETATIVE CHARACTERISTICS

leaves: alternate, fascicled, simple; blades linear (5–14 mm long), tubular, **fleshy**, smooth; rounded at apices with a **short hooked bristle**; margins entire; **bluish-green**; sessile

stems: erect main stem, ascending lateral stems; fleshy, pinkish when young, turning yellow to white; few to many primary branches from the base; numerous secondary branches short, glaucous

other: plants are quite striking when growing with their pinkish stems and bluish-green leaves; leaf axils with a cottony pubescence; halogeton was introduced into North America in the early 1930s and has rapidly spread, becoming a serious weed on some rangelands

HISTORICAL, FOOD, AND MEDICINAL USES: none

LIVESTOCK LOSSES: poisonous, contains toxic amounts of sodium, potassium, and calcium oxalates; sheep are the most susceptible; first signs of poisoning occur 2 to 6 hours after an animal ingests a fatal amount, and death occurs in 9 to 11 hours

FORAGE VALUES: poor to fair for both cattle and sheep; provides usable forage only when mixed in small quantities with other forage plants

HABITATS: dry deserts, barren areas, heavily grazed rangelands, roadsides, and other disturbed sites; especially abundant in alkaline or saline soils

Greenmolly summercypress
Kochia americana S. Watson

SYN = *Bassia americana* (S. Watson) A.J. Scott, *Kochia vestita* (S. Watson) Rydb., *Neokochia americana* (S. Watson) G.L. Chu & S.C. Sand.

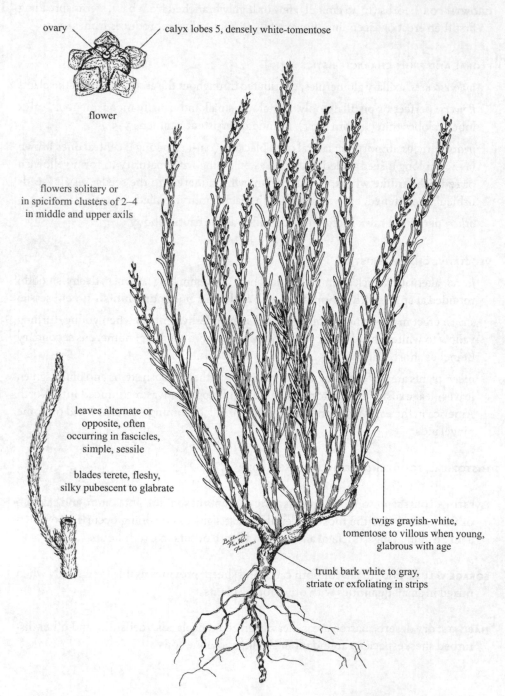

ovary

calyx lobes 5, densely white-tomentose

flower

flowers solitary or
in spiciform clusters of 2–4
in middle and upper axils

leaves alternate or
opposite, often
occurring in fascicles,
simple, sessile

blades terete, fleshy,
silky pubescent to glabrate

twigs grayish-white,
tomentose to villous when young,
glabrous with age

trunk bark white to gray,
striate or exfoliating in strips

FAMILY:	CHENOPODIACEAE
SPECIES:	*Kochia americana* S. Watson
COMMON NAME:	Greenmolly summercypress (perennial summercypress, red sage, green molly)
LIFE SPAN:	Perennial
ORIGIN:	Native
SEASON:	Warm

GROWTH FORM: subshrubs (to 50 cm tall), late deciduous, woody and much-branched below, simple and herbaceous above from the branching crowns of woody roots; flowers July to October, reproduces from seeds

FLORAL AND FRUIT CHARACTERISTICS

inflorescences: flowers solitary or in **spiciform clusters of 2–4 in middle and upper leaf axils**

flowers: perfect or pistillate only, regular (1.2–2.6 mm in diameter); calyx lobes 5; lobes incurved, herbaceous, persistent, densely white-tomentose; stamens 3–5, exserted or aborted in the pistillate flowers

fruits: utricles; depressed globose (2 mm long, 2 mm in diameter); persistent calyces form **5 wedge-shaped, membranous, horizontal wings** (1.5–2 mm long); margins scarious and toothed or erose

other: flowers often dry a dark color and readily fall from the plant upon drying

VEGETATIVE CHARACTERISTICS

leaves: alternate or opposite, often occurring in fascicles, simple; **blades narrow**, linear (6–30 mm long), erect or ascending, **terete, fleshy**, flat when dried; apices acute; margins entire; **dark green, silky pubescent** to glabrate; sessile

stems: twigs simple, **erect or ascending, grayish-white**, numerous, tomentose to **villous when young, glabrous with age**; trunk bark white to gray, striate or exfoliating in strips

other: new growth, especially leaves, often drying black; variable; one variety is glabrate, another is densely and permanently villous

HISTORICAL, FOOD, AND MEDICINAL USES: some Navajo used it to treat venereal diseases

LIVESTOCK LOSSES: may accumulate nitrates

FORAGE VALUES: excellent for sheep and goats, poor for cattle and deer; often used as winter forage for sheep; high in protein during autumn

HABITATS: desert valleys, flats, marshes, roadsides, and foothills of the cold desert region; usually found growing in alkaline soils

Kochia
Kochia scoparia (L.) Schrad.

SYN = *Bassia scoparia* (L.) A.J. Scott, *Chenopodium scoparium* L., *Kochia parodii* Aellen

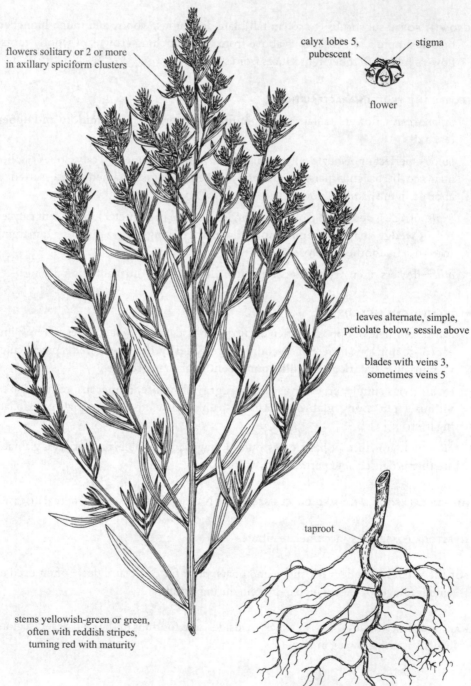

flowers solitary or 2 or more
in axillary spiciform clusters

calyx lobes 5,
pubescent

stigma

flower

leaves alternate, simple,
petiolate below, sessile above

blades with veins 3,
sometimes veins 5

taproot

stems yellowish-green or green,
often with reddish stripes,
turning red with maturity

FAMILY:	CHENOPODIACEAE
SPECIES:	*Kochia scoparia* (L.) Schrad.
COMMON NAME:	Kochia (cochia, coquia, fireweed, Mexican fireweed, summercypress, belvedere)
LIFE SPAN:	Annual
ORIGIN:	Introduced (from Eurasia)
SEASON:	Warm

GROWTH FORM: forbs (0.1–2 m tall); erect and spreading, much-branched from the base; flowers June to August, reproduces from seeds

FLORAL AND FRUIT CHARACTERISTICS

inflorescences: remote to dense; compact cylindrical or oblong-claviform

flowers: perfect or unisexual, apetalous; paired, subtended by pubescent leafy bracts (3–18 mm long); calyces campanulate (0.3–0.6 mm long, 1.5–2 mm in diameter), lobes 5; lobes oblong (2.3–3 mm wide), pubescent, bifid, persistent; stamens 5; sessile

fruits: utricles; depressed-globose; **pericarp wings 5, persistent**

other: some plants floriferous for most of their height

VEGETATIVE CHARACTERISTICS

leaves: alternate, simple; blades linear or lanceolate to narrow-obovate (2–10 cm long, 1–12 mm wide), **flat; veins 3, strong**, sometimes veins 5; apices acute or obtuse; margins entire; green, villous to pilose or glabrate; petiolate below, sessile above

stems: erect or spreading, branches usually many, yellowish-green or green, often with **reddish stripes**, turning red with maturity, glabrous or glabrescent

other: plant size is moisture dependent; plant is often called a tumbleweed, breaking off at ground level and rolling with the wind when it grows in a cylindrical to globose shape

HISTORICAL, FOOD, AND MEDICINAL USES: escaped ornamental, common cause of hay fever and other respiratory problems

LIVESTOCK LOSSES: may cause nitrate poisoning, bloat, and photosensitization in grazing livestock

FORAGE VALUES: good for livestock and wildlife when immature, quality and palatability rapidly decline with maturity; generally considered a weed, but is palatable to livestock and can be highly productive and nutritious

HABITATS: roadsides, pastures, wastelands, disturbed sites, and fields; adapted to a broad range of soils

Winterfat
Krascheninnikovia lanata (Pursh) A. Meeuse & A. Smit

SYN = *Ceratoides lanata* (Pursh) J.T. Howell, *Diotis lanata* Pursh, *Eurotia lanata* (Pursh) Moq.

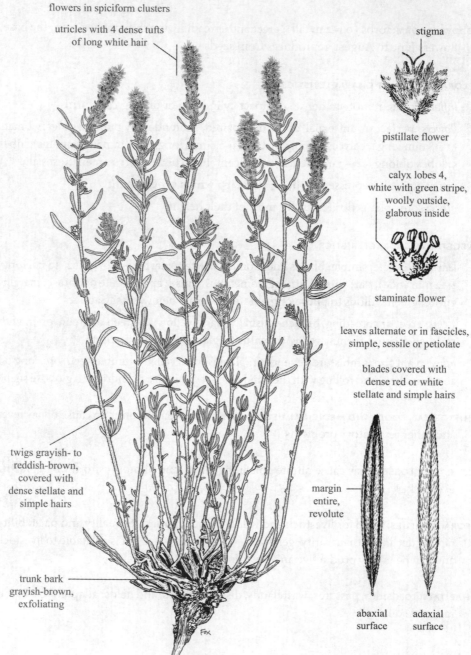

flowers in spiciform clusters

utricles with 4 dense tufts of long white hair

stigma

pistillate flower

calyx lobes 4, white with green stripe, woolly outside, glabrous inside

staminate flower

leaves alternate or in fascicles, simple, sessile or petiolate

blades covered with dense red or white stellate and simple hairs

twigs grayish- to reddish-brown, covered with dense stellate and simple hairs

margin entire, revolute

trunk bark grayish-brown, exfoliating

abaxial surface

adaxial surface

Fox

FAMILY:	CHENOPODIACEAE
SPECIES:	*Krascheninnikovia lanata* (Pursh) A. Meeuse & A. Smit
COMMON NAME:	Winterfat (hierba lanosa, lambstail, winter sage, white sage)
LIFE SPAN:	Perennial
ORIGIN:	Native
SEASON:	Cool

GROWTH FORM: monoecious or rarely dioecious **subshrubs** (to 70 cm tall); highly branched, herbaceous branches ascending; flowers April to September, reproduces from seeds

FLORAL AND FRUIT CHARACTERISTICS

inflorescences: spiciform clusters; staminate flowers terminal and axillary in clusters of 6–8; pistillate flowers axillary, solitary or in clusters of 2–4

flowers: unisexual, apetalous; staminate calyx lobes 4; lobes obovate (1.5–2 mm long, 0.7–1.2 mm wide), white with green stripes, **woolly outside**, glabrous inside; stamens 4; pistillate flowers subtended by 2 bracteoles; bracteoles ovate, green, pubescent

fruits: utricles (2.5–3.5 mm in diameter); bracts 2, lanceolate (5–6 mm long), 2-horned above, **covered with 4 dense tufts of long, white hair**

VEGETATIVE CHARACTERISTICS

leaves: alternate or in fascicles, simple; blades linear to narrowly lanceolate (1–4 cm long, 1.7–2.5 mm wide); apices obtuse (rarely acute); margins entire, **revolute**; covered with **dense red or white stellate and simple hairs**; midveins prominent abaxially, channeled adaxially; sessile or petiolate (1–3 mm long)

stems: twigs grayish- to reddish-brown, stout, ascending to erect, **covered with dense stellate and simple hairs**; trunk bark grayish-brown, exfoliating

HISTORICAL, FOOD, AND MEDICINAL USES: some Blackfoot soaked the leaves in warm water to make a hair wash, others treated fever with a decoction from the leaves

LIVESTOCK LOSSES: none

FORAGE VALUES: good for sheep, pronghorn, elk, and mule deer; fair for cattle; most valuable in winter

HABITATS: hillsides, mesas, foothills, and plains; most abundant in subalkaline or chalky medium to fine soils

Russian thistle
Salsola tragus L.

SYN = *Salsola australis* R. Br., *Salsola dichracantha* Kitag., *Salsola iberica* (Sennen & Pau) Botsch. *ex* Czerep., *Salsola kali* L., *Salsola pestifer* A. Nelson, *Salsola ruthenica* Iljin

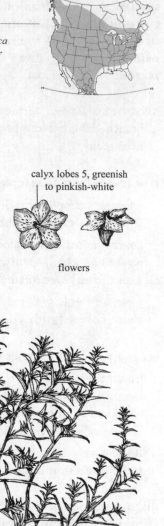

bract apex with a short, straight spine

utricles

calyx lobes 5, greenish to pinkish-white

flowers

flowers solitary in axils of spine-tipped bracts, also subtended by 2 bracteoles

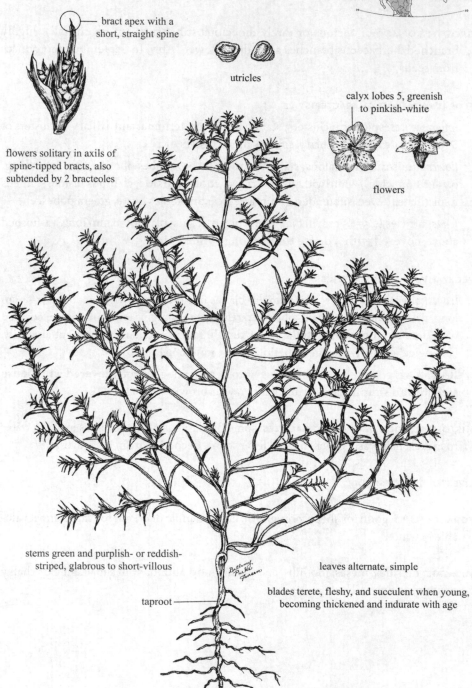

stems green and purplish- or reddish-striped, glabrous to short-villous

taproot

leaves alternate, simple

blades terete, fleshy, and succulent when young, becoming thickened and indurate with age

FAMILY:	CHENOPODIACEAE
SPECIES:	*Salsola tragus* L.
COMMON NAME:	Russian thistle (maroma, rodadora, tumbleweed, cardo ruso, maromera, prickly russianthistle)
LIFE SPAN:	Annual
ORIGIN:	Introduced (from Eurasia)
SEASON:	Warm

GROWTH FORM: forbs (20–80 cm tall); much-branched, ascending or spreading, rounded; flowers July to October, reproduces from seeds

FLORAL AND FRUIT CHARACTERISTICS

inflorescences: flowers solitary (rarely flowers 2–3); **in axils of spine-tipped bracts**, also subtended by 2 bracteoles

flowers: perfect, apetalous; calyx lobes 5 (2.5–3.5 mm long, 3–6 mm wide); lobes ovate or oblong, greenish to pinkish-white, acute, entire, persistent; stamens 3–5, exserted with age

fruits: utricles; ovoid or orbicular (1.5–2.5 mm in diameter), flattened, apices concave or convex; persistent calyces forming 5 wings (3–6 mm wide), white to pink; seeds orbicular, black or brown

VEGETATIVE CHARACTERISTICS

leaves: alternate, simple; blades linear to filiform (2–8 cm long, about 1 mm wide) reduced above, terete, **fleshy** and succulent when young, thickened and indurate with age; **becoming spinose** (> 1.5 mm long) at the apices, spines straight; bases broad; margins entire to denticulate; upper leaves eventually thickening and enclosing the fruits

stems: erect or rarely ascending, freely branching, **green and purplish- or reddish-striped**, glabrous to short-villous

other: forming a rounded plant that breaks off at ground level and is tumbled by the wind; thus, a tumbleweed

HISTORICAL, FOOD, AND MEDICINAL USES: young shoots can be used as a potherb; seeds can be ground into meal; first introduced into Dakota Territory in flax seed in the early 1870s

LIVESTOCK LOSSES: may cause mechanical injury to grazing animals from sharply pointed leaves, may accumulate nitrates, may contain oxalates

FORAGE VALUES: fair for cattle and sheep in early spring, becoming worthless with maturity because of the sharply pointed leaves; young plants can be cut for hay

HABITATS: cultivated fields, heavily grazed pastures, roadsides, waste places, and disturbed sites in nearly all soil types

Greasewood
Sarcobatus vermiculatus (Hook.) Torr.

SYN = *Batis vermiculata* Hook., *Fremontia vermiculata* (Hook.) Torr.,
Sarcobatus baileyi Coville, *Sarcobatus maximilianii* Nees

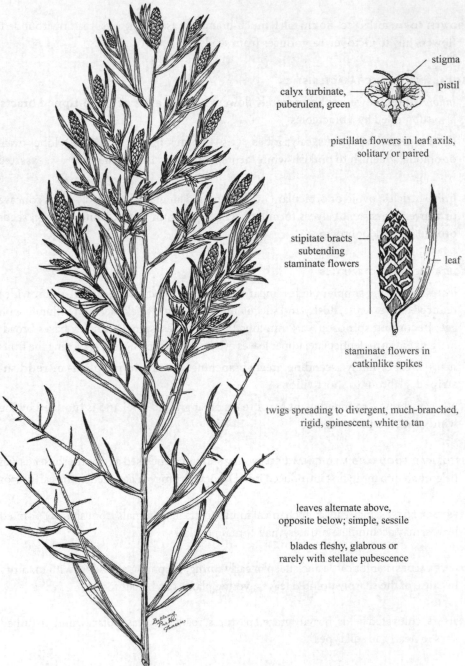

stigma

calyx turbinate,
puberulent, green

pistil

pistillate flowers in leaf axils,
solitary or paired

stipitate bracts
subtending
staminate flowers

leaf

staminate flowers in
catkinlike spikes

twigs spreading to divergent, much-branched,
rigid, spinescent, white to tan

leaves alternate above,
opposite below; simple, sessile

blades fleshy, glabrous or
rarely with stellate pubescence

FAMILY:	CHENOPODIACEAE
SPECIES:	*Sarcobatus vermiculatus* (Hook.) Torr.
COMMON NAME:	Greasewood (chico, black greasewood, chicobush)
LIFE SPAN:	Perennial
ORIGIN:	Native
SEASON:	Warm

GROWTH FORM: monoecious **shrubs** (to 2.5 m tall); late deciduous, erect or spreading, rounded, much-branched, rigid; flowers May to August, reproduces from seeds and basal sprouts

FLORAL AND FRUIT CHARACTERISTICS

inflorescences: staminate flowers in terminal and axillary **catkinlike spikes** (1.2–4 cm long, 4–5 mm wide); pistillate flowers solitary or paired in leaf axils below staminate flowers; sessile or short-peduncled

flowers: unisexual, apetalous; staminate flowers subtended by a stipitate bract; scales rhombic, apices acute, green; stamens 3; calyces of pistillate flowers turbinate (1.5 mm long, 1 mm in diameter), puberulent, green

fruits: achenes; turbinate (3–5 mm long, 2.5–3.5 mm in diameter), enclosed in the persistent perianth that forms a wing (to 1.4 cm wide); wings scarious, veined, green to tan or reddish-pink

VEGETATIVE CHARACTERISTICS

leaves: alternate above, opposite below, simple; **blades linear** to filiform (1–4 cm long, 1.5–2.5 mm wide) elliptical or 4-angled in cross section, **fleshy**; apices obtuse or acute; bases cuneate; glabrous or rarely with stellate pubescence: **hairs bright green** to yellowish-green; sessile

stems: **twigs spreading to divergent**, irregularly much-branched, **rigid, spinescent**, white to tan; trunk bark yellowish-gray to light brown, **fissured and exfoliating**, shiny, with elliptical pits

HISTORICAL, FOOD, AND MEDICINAL USES: wood sometimes used for fuel; some Native Americans used sharpened branches as planting tools

LIVESTOCK LOSSES: soluble oxalates have caused mass mortality in flocks of sheep; cattle and horses are rarely poisoned; young twigs are especially toxic

FORAGE VALUES: good to fair for sheep, goats, and big game in winter; important food and cover for small mammals, jackrabbits, and birds

HABITATS: dry plains, slopes, eroded hills, and flats; especially in alkaline or saline soils

St. Johnswort
Hypericum perforatum L.

SYN = *Hypericum nachitschevanicum* Grossh.

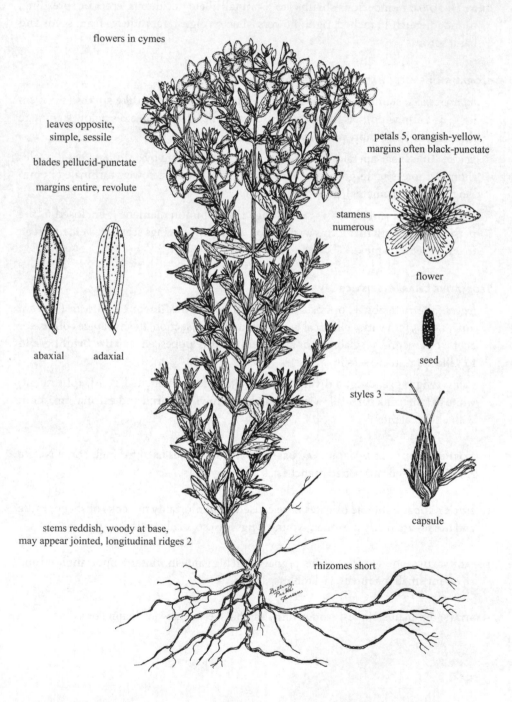

flowers in cymes

leaves opposite,
simple, sessile

blades pellucid-punctate

margins entire, revolute

abaxial   adaxial

petals 5, orangish-yellow,
margins often black-punctate

stamens
numerous

flower

seed

styles 3

capsule

stems reddish, woody at base,
may appear jointed, longitudinal ridges 2

rhizomes short

324

FAMILY:	CLUSIACEAE
SPECIES:	*Hypericum perforatum* L.
COMMON NAME:	St. Johnswort (hierba de San Juan, klamathweed, goatweed)
LIFE SPAN:	Perennial
ORIGIN:	Introduced (from Europe and Africa)
SEASON:	Warm

GROWTH FORM: forbs (0.3–1.5 m tall); erect, somewhat clustered, much-branched above and often from the bases; rhizomatous; often with small, basal, sterile shoots; flowers June to August, reproduces from seeds and short rhizomes

FLORAL AND FRUIT CHARACTERISTICS

inflorescences: cymes; collectively flat-topped or rounded, leafy-bracteate at branch tips, densely flowered

flowers: perfect, regular; sepals 5, linear-lanceolate (4–6 mm long), unequal, acute; petals 5, **orangish-yellow**, obovate to cuneate (8–12 mm long), **twisting when dry, margins often black-punctate**; stamens numerous, about as long as the petals

fruits: capsules; globose to ovoid (4–8 mm diameter), exceeding the sepals, cells 3, tipped by 3 spreading styles, glandular; seeds several; seeds 0.9–1.3 mm long

VEGETATIVE CHARACTERISTICS

leaves: opposite, simple; blades elliptical to linear or oblong (1–4 cm long, 2–8 mm wide), reduced above, diverging from the stem at about 90° angles; apices acute to rounded; bases acuminate; margins entire, revolute; **pellucid-punctate**; sessile

stems: erect, reddish, woody at the bases, **may appear jointed because of opposite leaf scars; longitudinal ridges usually 2**

HISTORICAL, FOOD, AND MEDICINAL USES: some Menominee mixed St. Johnswort with black raspberry root in hot water and drank the tea for tuberculosis; acts as a diuretic, may kill internal worms; yields a drug used as an antidepressant and a red dye used in cosmetics; planted as an ornamental

LIVESTOCK LOSSES: poisonous to livestock; causes a photosensitizing reaction resulting in dermatitis to unpigmented skin of horses, cattle, and sheep; sunlight acts as a catalyst, therefore, symptoms occur only when livestock are exposed to strong sunlight after grazing; frequently fatal

FORAGE VALUES: poor to worthless for cattle, sheep, horses, and wildlife; fair for goats; mourning doves, quail, and other birds eat the seeds

HABITATS: prairies, pastures, waste areas, disturbed fields, and roadsides; most abundant in sandy soils

Redosier dogwood
Cornus sericea L.

SYN = *Cornus alba* L., *Cornus stolonifera* Michx., *Thelycrania sericea* (L.) Dandy

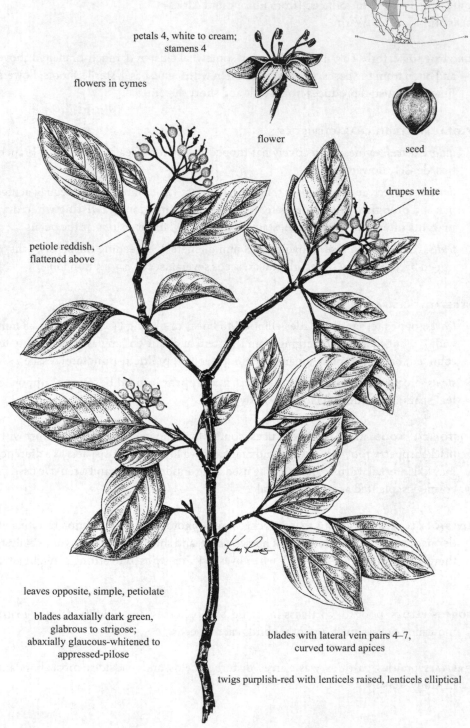

petals 4, white to cream;
stamens 4

flowers in cymes

flower

seed

drupes white

petiole reddish,
flattened above

leaves opposite, simple, petiolate

blades adaxially dark green,
glabrous to strigose;
abaxially glaucous-whitened to
appressed-pilose

blades with lateral vein pairs 4–7,
curved toward apices

twigs purplish-red with lenticels raised, lenticels elliptical

FAMILY:	CORNACEAE
SPECIES:	*Cornus sericea* L.
COMMON NAME:	Redosier dogwood (American dogwood, cornejo, red osier, red dogwood)
LIFE SPAN:	Perennial
ORIGIN:	Native
SEASON:	Cool

GROWTH FORM: shrubs (to 4 m tall); highly branched; erect to procumbent, stoloniferous, forming thickets; flowers May to July; fruits mature August to September, reproduces from seeds and stolons

FLORAL AND FRUIT CHARACTERISTICS

inflorescence: cymes, convex to flat-topped (2–6 cm wide), main branches 4–6, densely flowered; pedicels reddish (4–6 cm long), tomentose, becoming glabrous

flowers: perfect, regular; sepal lobes 4, triangular (to 0.5 mm long), erect to spreading, green, pubescent; corollas ovate (1.5–2.5 mm long); **petals 4**, triangular (2–4 mm long), **white to cream**, erect to spreading at anthesis, margins inrolled, outer surfaces pubescent, inner surfaces papillose; stamens 4, opposite the sepals

fruits: drupes; globose (6–9 mm diameter), white, few appressed hairs, depressed at the pedicel, styles persistent; seeds 1 per drupe; seeds pale to dark brown (5 mm long, 4–6 mm wide), smooth, with 8 white to yellow stripes

VEGETATIVE CHARACTERISTICS

leaves: **opposite**, simple; blades oblong-lanceolate to ovate (5–11 cm long, 2.5–5.5 cm wide); apices usually acute or acuminate; bases cuneate; margins entire; adaxially dark green, glabrous to strigose; abaxially glaucous-whitened to appressed-pilose, occasionally with reddish hairs; **lateral vein pairs 4–7**, evenly spaced, curved toward the apices; petioles (0.5–2.5 cm long) flattened above, reddish

stems: **twigs purplish-red** from summer to winter, earlier greenish-red; glabrous to appressed or spreading pubescent; lenticels raised, elliptical, light in color; leaf scars crescent-shaped, dark; terminal buds ovoid, compressed; trunk bark red to occasionally greenish, smooth

HISTORICAL, FOOD, AND MEDICINAL USES: some Native Americans smoked the inner bark layers; a bark extract was used as an emetic in treating coughs and fevers; a red dye was made by boiling the small roots; a tea from the leaves has been used as a quinine substitute

LIVESTOCK LOSSES: none

FORAGE VALUES: not browsed by livestock, nearly worthless to big game, fruits are eaten by birds and small mammals

HABITATS: moist soil along streams, meadows, and lakeshores in wooded or open areas

Oneseeded juniper
Juniperus monosperma (Engelm.) Sarg.

SYN = *Juniperus gymnocarpa* (Lemmon) Cory, *Juniperus mexicana* var. *monosperma* (Engelm.) Cory, *Sabina monosperma* (Engelm.) Rydb.

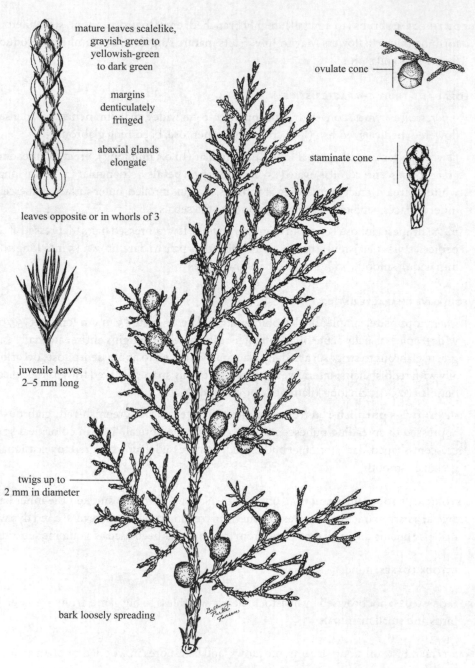

mature leaves scalelike, grayish-green to yellowish-green to dark green

margins denticulately fringed

abaxial glands elongate

leaves opposite or in whorls of 3

juvenile leaves 2–5 mm long

twigs up to 2 mm in diameter

bark loosely spreading

ovulate cone

staminate cone

FAMILY:	CUPRESSACEAE
SPECIES:	*Juniperus monosperma* (Engelm.) Sarg.
COMMON NAME:	Oneseed juniper (táscate, cherrystone juniper, enebro, sabina)
LIFE SPAN:	Perennial
ORIGIN:	Native
SEASON:	Evergreen

GROWTH FORM: dioecious shrubs or small trees (to 10 m tall); crown rounded to flattened, branching near the ground, often with several trunks; pollinates March to April, cones mature September, reproduces from seeds

CONE CHARACTERISTICS

cones: **axillary** and terminal, borne on branches of the previous year; staminate cones solitary, oblong-ovoid (3–4 mm long, 2 mm in diameter), brownish or yellowish; ovulate cones clustered

mature ovulate cones: berrylike; subglobose (3–7 mm diameter), somewhat fleshy, reddish-blue to brownish or copper-colored, infrequently blue, glaucous, maturing in 1 year, seeds 1 or occasionally 2; seeds 4–5 mm long

VEGETATIVE CHARACTERISTICS

leaves: opposite or in pairs or whorls of 3, scalelike; imbricate and appressed against the branches, ovate (1–3 mm long, 0.5–1.2 mm wide), fleshy, thickened, rounded; apices acute or acuminate; margins denticulately fringed; grayish-green to yellowish-green to dark green, old leaves reddish-brown; **abaxial glands elongate**

stems: twigs slender (up to 2 mm in diameter), reddish-brown; **bark loosely spreading**, exfoliating in flat plates; plates gray to brown

other: leaves on juvenile branchlets acerose, often longer (2–5 mm long) than on mature branchlets

HISTORICAL, FOOD, AND MEDICINAL USES: berrylike cones were gathered and eaten by some Native Americans; wood used for prayer sticks and bows; green dye was obtained from bark; cones were ground to make flour; fibrous bark was used for mats, saddles, and breechcloths; cones of *Juniperus* species are used to flavor gin

LIVESTOCK LOSSES: may cause abortion in livestock

FORAGE VALUES: worthless to cattle and sheep, seldom consumed; occasionally browsed by goats; mature cones are eaten by deer, quail, foxes, coyotes, small mammals, and birds

HABITATS: hillsides, canyons, arroyos, flats, and exposed ledges; adapted to a broad range of soil types

Rocky Mountain juniper
Juniperus scopulorum Sarg.

SYN = *Sabina scopulorum* (Sarg.) Rydb.

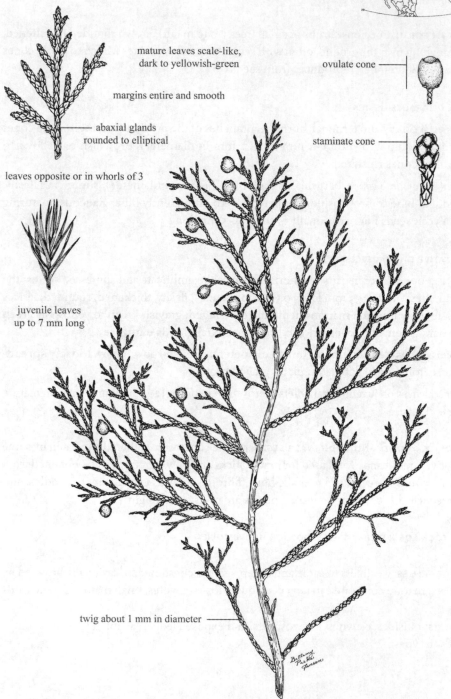

mature leaves scale-like,
dark to yellowish-green

margins entire and smooth

abaxial glands
rounded to elliptical

leaves opposite or in whorls of 3

juvenile leaves
up to 7 mm long

ovulate cone

staminate cone

twig about 1 mm in diameter

FAMILY:	CUPRESSACEAE
SPECIES:	*Juniperus scopulorum* Sarg.
COMMON NAME:	Rocky Mountain juniper (Rocky Mountain cedar)
LIFE SPAN:	Perennial
ORIGIN:	Native
SEASON:	Evergreen

GROWTH FORM: dioecious shrubs or small trees (to 20 m tall); crown rounded to pyramidal, irregular, branching above and near the ground; pollinates April to May, cones mature October to December of the second year following pollination, reproduces from seeds

CONE CHARACTERISTICS

cones: staminate cones oblong (2–4 mm long, 1–2 mm in diameter), inconspicuous, solitary at tips of branchlets, sessile, yellowish-brown; ovulate cones globose

mature ovulate cones: berrylike; subglobose (5–8 mm in diameter), fleshy, resinous-juicy, bright blue to purplish, glaucous, maturing in 2 years; seeds mostly 2

VEGETATIVE CHARACTERISTICS

leaves: opposite or in whorls of 3, scalelike; closely appressed, not or slightly imbricate, rhombic-ovate to ovate-elliptical (1–4 mm long, 0.7–1.5 mm wide), fleshy, thickened, rounded; apices obtuse to broadly acute; margins entire and smooth; dark to yellowish-green, often pale and glaucous, **abaxial glands rounded to elliptical**, conspicuous

stems: twigs slender (about 1 mm diameter), scaly; flattened at first, then becoming round; trunk bark grayish- to reddish-brown, **exfoliating in thin strips**, furrowed

other: leaves on juvenile branchlets acerose (to 7 mm long), often longer than on mature branchlets

HISTORICAL, FOOD, AND MEDICINAL USES: some Native Americans ate the cones raw or cooked, used as flavoring for meat, cones and young shoots were boiled for tea, cones were ground for mush and cakes, wax from berrylike cones used in candles; currently used as an ornamental, in shelterbelts, for fence posts, and for fuel; reported to have insect-repellent properties; cones of *Juniperus* species are used to flavor gin; cedar oil immersion is used to increase the resolving power of microscopes

LIVESTOCK LOSSES: may cause abortions in livestock

FORAGE VALUES: worthless to livestock, generally not consumed; important browse plant for pronghorn, mule deer, and bighorn sheep; birds and small mammals eat the cones

HABITATS: ridges, bluffs, canyons, hillsides, and wash areas; often on undeveloped, erodible soils; most abundant on calcareous and somewhat alkaline soils

Bracken fern
Pteridium aquilinum (L.) Kuhn

SYN = *Pteris aquilina* L., *Pteris capensis* Thunb., *Pteris lanuginosa* Bory *ex* Willd.

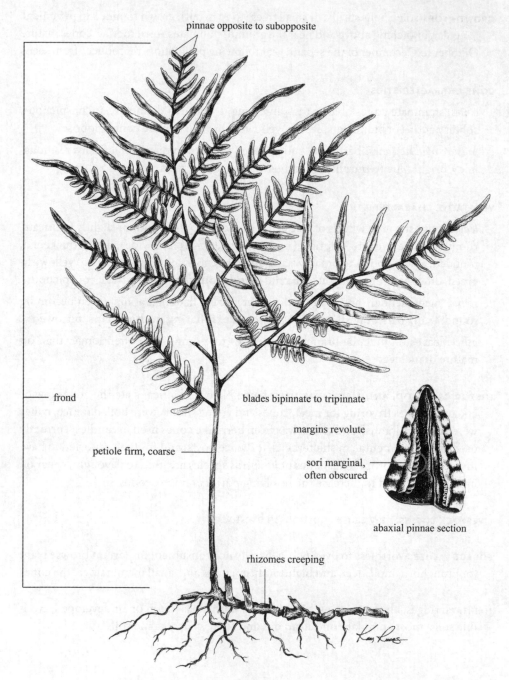

pinnae opposite to subopposite

frond

blades bipinnate to tripinnate

margins revolute

petiole firm, coarse

sori marginal, often obscured

abaxial pinnae section

rhizomes creeping

FAMILY:	DENNSTAEDTIACEAE
SPECIES:	*Pteridium aquilinum* (L.) Kuhn
COMMON NAME:	Bracken fern (helecho, western bracken, western brackenfern, brake)
LIFE SPAN:	Perennial
ORIGIN:	Native
SEASON:	Warm

GROWTH FORM: ferns (to 2 m tall); forming dense colonies from creeping rhizomes; spores produced July to September; reproduces from spores and rhizomes

SORI AND SPORE CHARACTERISTICS

sori: marginal, more or less continuous, borne on vascular strands connecting the vein ends of the leaf blades, **often obscured by recurved false outer indusium**; indusium double, outer false and reflexed; inner true and developed or obsolete

spores: minute, tetrahedral-globose, brown, without perispores

VEGETATIVE CHARACTERISTICS

fronds: widely spaced along rhizomes; blades bipinnate to tripinnate (20–80 cm long, 20–90 cm wide), broadly deltate, sparsely to densely hairy abaxially, rarely glabrous, **coriaceous**; pinnae opposite to subopposite, lower pair deltate to ovate, nearly as long as upper portion of blades, upper pinnae smaller, deltate to lanceolate; margins revolute, veins free or joined at margins; petioles firm, coarse, ascending to erect, dark at the bases; U-shaped in cross section

stems: **rhizomes** (to 1 cm thick), deeply subterranean, creeping and branched, **blackish, pilose to glabrate**; fronds erect to ascending from the rhizomes

other: variable plant with several varieties in North America

HISTORICAL, FOOD, AND MEDICINAL USES: some Native Americans ate young fronds raw or boiled and ground rhizomes into a meal; sometimes eaten to control internal parasites

LIVESTOCK LOSSES: toxic to horses and to some extent to cattle; toxicity due to a breakdown of thiamine in the blood; cumulative poison

FORAGE VALUES: cattle, elk, and deer may graze it in early spring, unpalatable throughout the rest of the year

HABITATS: pastures, open forests, cutover areas, and abandoned fields; poor soils of prairies and woodlands; less frequent in rich, moist woodlands

Russet buffaloberry
Shepherdia canadensis (L.) Nutt.

SYN = *Elaeagnus canadensis* (L.) A. Nelson, *Lepargyrea canadensis* (L.) Greene

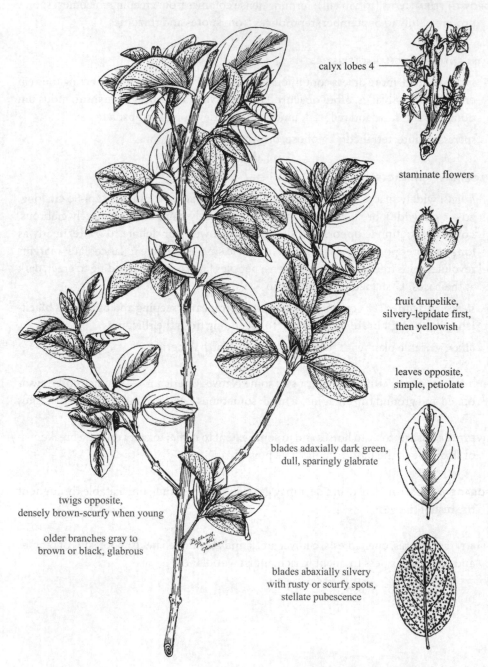

calyx lobes 4

staminate flowers

fruit drupelike,
silvery-lepidate first,
then yellowish

leaves opposite,
simple, petiolate

blades adaxially dark green,
dull, sparingly glabrate

twigs opposite,
densely brown-scurfy when young

older branches gray to
brown or black, glabrous

blades abaxially silvery
with rusty or scurfy spots,
stellate pubescence

FAMILY:	ELAEAGNACEAE
SPECIES:	*Shepherdia canadensis* (L.) Nutt.
COMMON NAME:	Russet buffaloberry (Canadian buffaloberry, rabbitberry)
LIFE SPAN:	Perennial
ORIGIN:	Native
SEASON:	Cool

GROWTH FORM: dioecious shrubs (to 3 m tall); much-branched, spreading, crowns rounded; flowers April to June, fruits mature July to September, reproduces from seeds

FLORAL AND FRUIT CHARACTERISTICS

inflorescences: flowers solitary or in clusters of 2–4; appearing before the leaves at nodes on 1-year-old twigs

flowers: unisexual, apetalous; calyx lobes of staminate flowers 4 (1–2 mm long), adaxial surfaces yellow, abaxial surfaces brown; stamens usually 8; pistillate flowers with united urn-shaped calyces enclosing the ovaries; lobes 4, spreading (0.8–1.5 mm long); veins 3, prominent

fruits: drupelike; oval to ellipsoidal (3–8 mm long), enveloped by mealy base of hypanthium, silvery-lepidote first then yellowish as the scales fall away, insipid, bitter; achenes 1–2

VEGETATIVE CHARACTERISTICS

leaves: opposite, simple; blades oval to ovate or elliptical (2–6 cm long), sizes variable on the same plant; apices obtuse; bases rounded to subcordate; margins entire; adaxially dark green, dull, sparingly glabrate; **abaxially silvery and with rusty or scurfy spots, stellate pubescence**; petioles (3–5 mm long)

stems: **twigs opposite; densely brown-scurfy when young**, older branches gray to brown or black, glabrous, unarmed

HISTORICAL, FOOD, AND MEDICINAL USES: fruits are edible, cooked or raw, although insipid; fruits are sometimes whipped with sugar for a dessert; may cause diarrhea

LIVESTOCK LOSSES: none

FORAGE VALUE: fair for sheep before frost, otherwise considered worthless; seldom browsed by cattle, horses, or big game animals; fruits are consumed by many kinds of birds and small mammals

HABITATS: limestone slopes, ledges, riverbanks, and open wooded slopes; adapted to a broad range of soils, most abundant in moist soils

Longleaf ephedra
Ephedra trifurca Torr. *ex* S. Watson

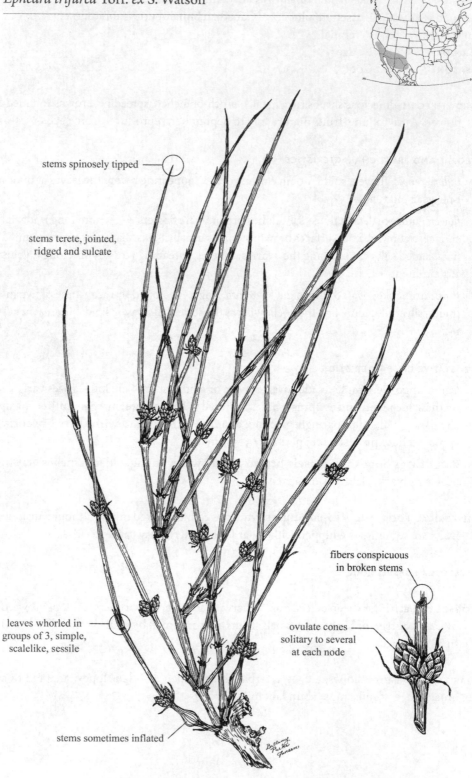

stems spinosely tipped

stems terete, jointed,
ridged and sulcate

fibers conspicuous
in broken stems

leaves whorled in
groups of 3, simple,
scalelike, sessile

ovulate cones
solitary to several
at each node

stems sometimes inflated

FAMILY:	EPHEDRACEAE
SPECIES:	*Ephedra trifurca* Torr. *ex* S. Watson
COMMON NAME:	Longleaf ephedra (cañatillo, Mormon tea, pitamo real, poptillo, jointfir, longleaf jointfir)
LIFE SPAN:	Perennial
ORIGIN:	Native
SEASON:	Evergreen

GROWTH FORM: dioecious shrubs (to about 2 m tall); erect, branches solitary or whorled at the nodes; pollinates March to April, seeds mature April to May, reproduces from seeds

CONE CHARACTERISTICS

cones: staminate cones solitary or several at stem nodes, obovate to obovoid (5–6 mm long), subtended by 2 bracts; **stamens 2–8, yellow, protruding beyond their united bracts**; ovulate cones solitary to several at each node, consisting of 2 erect ovules enclosed in urceolate involucres of scales (10–11 mm long); scales in many whorls of 3

mature ovulate cones: elliptical to obovate (7–12 mm long, 2–4 mm in diameter), reddish-brown; margins of the cone bracts conspicuously membranous

VEGETATIVE CHARACTERISTICS

leaves: **whorled in groups of 3, simple, scalelike** (5–13 mm long); united for one-half to three-fourths of their total length, persistent and spinose on older branches; apices subspinosely tipped; sheaths membranous, gray, becoming fibrous with age; sessile

stems: **rather rigid, terete, ridged and sulcate**, jointed (internodes 3–9 cm long), spinosely tipped, pale yellowish-green, glaucous; portions of stems may be inflated; fibers conspicuous in broken stems; bark gray, irregularly fissured

HISTORICAL, FOOD, AND MEDICINAL USES: some Native Americans and Mexicans used stem decoctions as a cooling beverage and ate the cones; common name Mormon tea is derived from use as a beverage by pioneer Latter-day Saints in the American West; contains the drug ephedrine

LIVESTOCK LOSSES: none

FORAGE VALUES: poor for livestock, heavily browsed in emergency situations, may be important winter forage; browsed by bighorn sheep and jackrabbits; cones are eaten by quail, other birds, and small mammals

HABITATS: mesas, foothills, open sites in valleys, and dry creek beds; most abundant in coarse soils

Pointleaf manzanita
Arctostaphylos pungens Kunth

SYN = *Arbutus ferruginea* Sessé & Moc., *Arbutus mucronata* Sessé & Moc., *Arbutus myrtifolia* Willd. *ex* Steud., *Arbutus rigida* Willd. *ex* Steud., *Daphnidostaphylis pungens* (Kunth) Klotzsch, *Uva-ursi pungens* (Kunth) Abrams

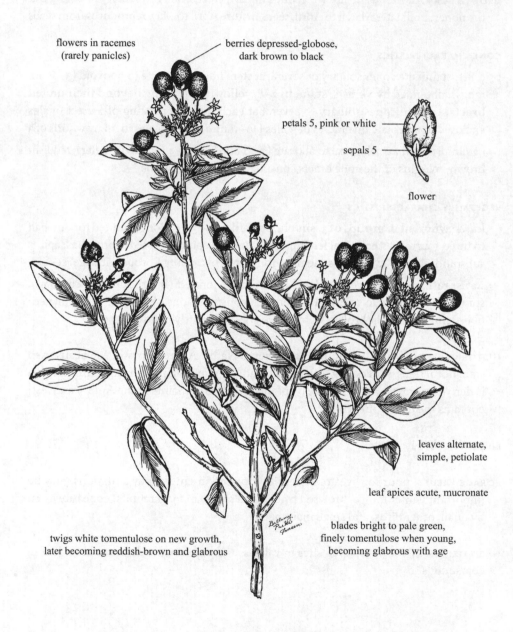

flowers in racemes (rarely panicles)

berries depressed-globose, dark brown to black

petals 5, pink or white

sepals 5

flower

leaves alternate, simple, petiolate

leaf apices acute, mucronate

twigs white tomentulose on new growth, later becoming reddish-brown and glabrous

blades bright to pale green, finely tomentulose when young, becoming glabrous with age

FAMILY:	ERICACEAE
SPECIES:	*Arctostaphylos pungens* Kunth
COMMON NAME:	Pointleaf manzanita (manzanilla, wíchari, Mexican manzanita, pingüica, madroño)
LIFE SPAN:	Perennial
ORIGIN:	Native
SEASON:	Evergreen

GROWTH FORM: shrubs (to 3 m tall); **erect or spreading**, branched from the base; forming extensive, dense thickets; flowers January to March, fruits mature April to July, reproduces from seeds and rootsprouts

FLORAL AND FRUIT CHARACTERISTICS

inflorescences: racemes (rarely panicles), spicate; cylindrical, congested, nodding; rachises often distinctly thickened and club-shaped at the apices

flowers: perfect, regular; calyces persistent; sepals 5, short-deltoid (2–3 mm long, 1–2 mm wide), firm, thick, pubescent or tomentose; **corollas urceolate** (5–7 mm long); **petals 5, pink or white**, recurved at summit; stamens 10, included; ovaries superior

fruits: berries; depressed-globose (5–8 mm in diameter), fleshy, dark brown to black, reddish when immature, glabrous; nutlets 4–10, ridged dorsally, each seeds 1

VEGETATIVE CHARACTERISTICS

leaves: alternate, simple, persistent (evergreen); blades oblong to oblanceolate (1.5–3 cm long, 9–15 mm wide); **apices acute and mucronate or mucronulate**; bases rounded or subcuneate; margins entire; **surfaces bright to pale green**, finely tomentulose when young, becoming glabrous with age; petiolate (to 1 cm long)

stems: twigs rigid, white tomentulose on new growth, **later becoming reddish-brown and glabrous; bark persistent on older stems**; sometimes rooting at the nodes where branches touch the ground

HISTORICAL, FOOD, AND MEDICINAL USES: fruits are sold in Mexican markets and used for jellies; leaves and fruits are used as a diuretic and household remedies for dropsy, bronchitis, and venereal diseases

LIVESTOCK LOSSES: none

FORAGE VALUES: worthless for cattle and sheep, poor to fair for goats; goats consume leaves and browse young twigs; goats will peel back bark in the spring, presumably for sap; fruits eaten by deer, small mammals, bears, coyotes, quail, grouse, and other birds

HABITATS: rocky mesas and mountain slopes; most abundant on dry, well-drained soils

Bearberry
Arctostaphylos uva-ursi (L.) Spreng.

SYN = *Arbutus buxifolia* Stokes, *Arbutus uva-ursi* L., *Arctostaphylos adenotricha*
(Fernald & J.F. Macbr.) Á. Löve, D. Löve & Kapoor, *Arctostaphylos*
coloradensis Rollins, *Arctostaphylos officinalis* Wimm. & Grab.,
Arctostaphylos procumbens E. Mey., *Daphnidostaphylis fendleri* Klotzsch,
Mairania uva-ursi (L.) Desv., *Uva-ursi procumbens* Moench

leaves alternate, simple, petiolate

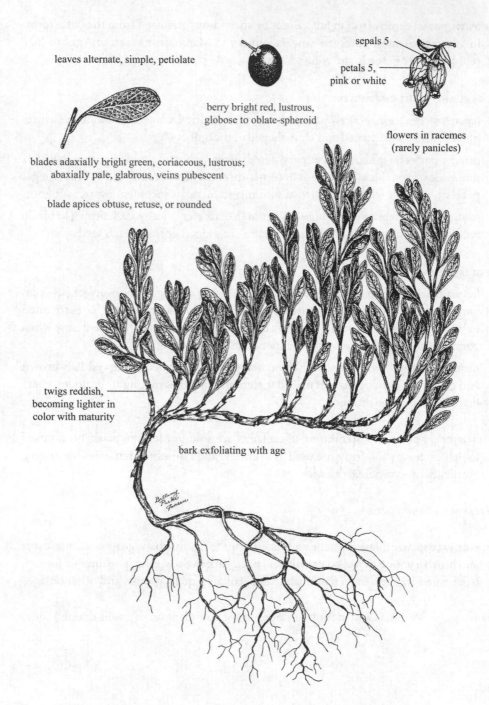

berry bright red, lustrous,
globose to oblate-spheroid

sepals 5

petals 5,
pink or white

flowers in racemes
(rarely panicles)

blades adaxially bright green, coriaceous, lustrous;
abaxially pale, glabrous, veins pubescent

blade apices obtuse, retuse, or rounded

twigs reddish,
becoming lighter in
color with maturity

bark exfoliating with age

FAMILY:	ERICACEAE
SPECIES:	*Arctostaphylos uva-ursi* (L.) Spreng.
COMMON NAME:	Bearberry (manzanita, kinnikinnick)
LIFE SPAN:	Perennial
ORIGIN:	Native
SEASON:	Evergreen

GROWTH FORM: shrubs with depressed or trailing stems forming prostrate mats (to 2 m diameter); much-branched, **terminal portions erect** (to 20 cm tall); flowers April to July, fruits mature July to October, reproduces from seeds

FLORAL AND FRUIT CHARACTERISTICS

inflorescences: racemes (rarely panicles); dense, terminal, scarcely exceeding the leaves, flowers few to several; flowers pendulous

flowers: perfect, regular; calyces persistent; sepals 5 (1–1.5 mm long), pink or white, distinct; **corollas ovoid to urceolate** (4–8 mm long); **petals 5, pink or white**, reflexed; stamens 10, shorter than the corollas; ovaries superior; pedicels short, recurved

fruits: berries; globose to oblate-spheroid (4–10 mm in diameter), fleshy, bright red, lustrous, somewhat persistent; nutlets 5, ridged dorsally

VEGETATIVE CHARACTERISTICS

leaves: alternate, simple, evergreen, crowded; blades spatulate to obovate (1–3 cm long, 5–10 mm wide); **apices obtuse, retuse, or rounded**; bases cuneate or narrowed; margins entire, revolute; **adaxially bright green, coriaceous**, lustrous; **abaxially pale**, glabrous, veins pubescent; petioles to 1 cm long

stems: young twigs reddish, becoming lighter in color with maturity; rooting at the nodes; **bark exfoliating with age**

HISTORICAL, FOOD, AND MEDICINAL USES: fruits insipid but edible if cooked; fruits used by early settlers for treating urinary disorders, leaves were used as a tobacco substitute

LIVESTOCK LOSSES: none

FORAGE VALUES: worthless to poor for livestock, browsed by bighorn sheep and deer; fruits provide food for grouse, wild turkeys, and other birds

HABITATS: open woodlands, hillsides, and mountain slopes to slightly above timberline; grows in well-drained soils in either open or shaded areas; most abundant in sandy and rocky soils

Guajillo
Acacia berlandieri Benth.

SYN = *Acacia emoryana* Benth., *Senegalia berlandieri* (Benth.) Britton & Rose

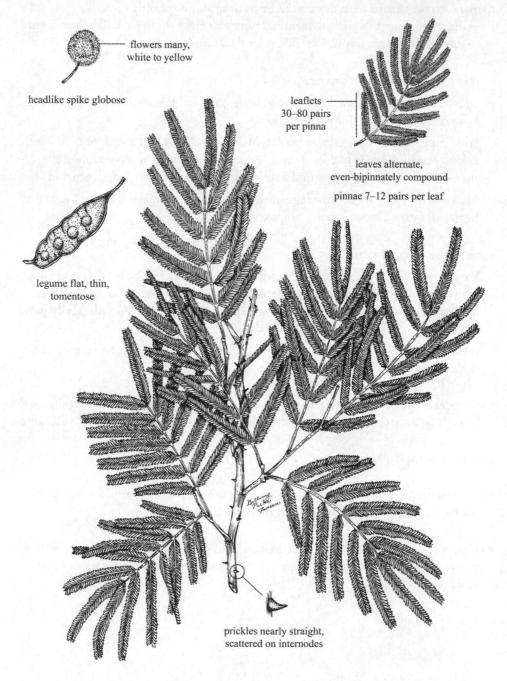

flowers many,
white to yellow

headlike spike globose

leaflets
30–80 pairs
per pinna

leaves alternate,
even-bipinnately compound

pinnae 7–12 pairs per leaf

legume flat, thin,
tomentose

prickles nearly straight,
scattered on internodes

FAMILY:	FABACEAE
SPECIES:	*Acacia berlandieri* Benth.
COMMON NAME:	Guajillo (Berlandier acacia)
LIFE SPAN:	Perennial
ORIGIN:	Native
SEASON:	Warm

GROWTH FORM: shrubs (to 4 m tall); several main stems branch from the base, ascending, sparingly branched; flowers November to March, fruits mature June to July, reproduces from seeds

FLORAL AND FRUIT CHARACTERISTICS

inflorescences: **headlike spikes; globose** (8–15 mm diameter), rarely elongated (2 cm long, 1 cm wide), axillary, solitary or in clusters, flowers many; peduncles pubescent (2–5 cm long)

flowers: perfect, regular; calyx lobes 5, pubescent; petals 5, white to yellow, pubescent; stamens numerous, exserted, distinct

fruits: legumes; oblong (8–15 cm long, 1.5–2.5 cm wide), **flat, thin**, straight or somewhat curved, apices obtuse or apiculate, margins thickened; one margin is generally straighter than the other; velvety-tomentose when mature, dark brown; seeds several

VEGETATIVE CHARACTERISTICS

leaves: alternate, even-bipinnately compound (9–18 cm long), delicate, almost fernlike in appearance; pinnae 7–12 pairs; **leaflets 30–80 pairs per pinna**, crowded; leaflets linear to oblong (2–6 mm long), oblique, apices acute, margins entire, tomentose when immature and glabrate when mature, veins prominent; petiole glands sessile; stipules small, caducous

stems: twigs gray to white; internodes striate; **generally armed with nearly straight prickles** (1–3 mm long) scattered on the internodes

HISTORICAL, FOOD, AND MEDICINAL USES: important honey plant, gums and dyes have been extracted from this shrub; used as an ornamental either as a hedge or specimen planting

LIVESTOCK LOSSES: may cause hydrocyanic acid poisoning in livestock when extremely large amounts are consumed

FORAGE VALUES: fair for livestock and wildlife, seeds furnish important food for birds and small mammals

HABITATS: limestone ridges and caliche hills; most abundant in sandy soils, uncommon in deep soils

Huisache
Acacia farnesiana (L.) Willd.

SYN = *Acacia acicularis* Humb. & Bonpl. *ex* Willd., *Acacia densiflora* (Alexander *ex* Small) Cory, *Acacia ferox* M. Martens & Galeotti, *Acacia smallii* Isely, *Mimosa farnesiana* L., *Pithecellobium minutum* M.E. Jones, *Poponax farnesiana* (L.) Raf., *Vachellia farnesiana* (L.) Wight & Arn.

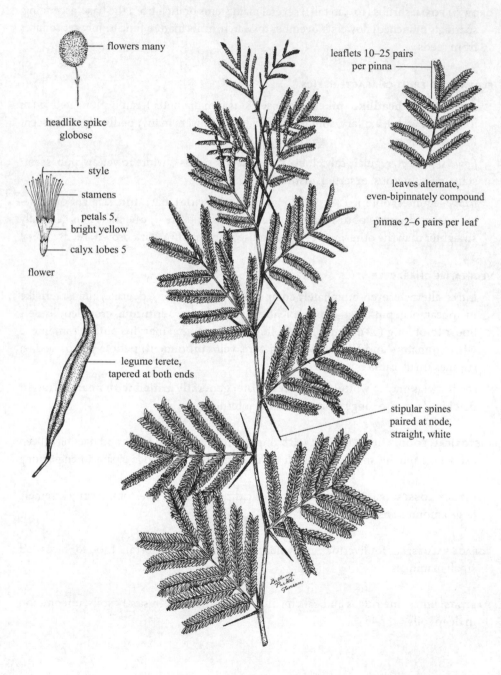

flowers many

headlike spike
globose

style

stamens

petals 5,
bright yellow

calyx lobes 5

flower

legume terete,
tapered at both ends

leaflets 10–25 pairs
per pinna

leaves alternate,
even-bipinnately compound

pinnae 2–6 pairs per leaf

stipular spines
paired at node,
straight, white

FAMILY:	FABACEAE
SPECIES:	*Acacia farnesiana* (L.) Willd.
COMMON NAME:	Huisache (huizache, sweet acacia)
LIFE SPAN:	Perennial
ORIGIN:	Native
SEASON:	Warm

GROWTH FORM: shrubs or small trees (to 9 m tall); often with several trunks, ascending; densely branched crown; flowers February to March, may flower whenever rain occurs during dry years; reproduces from seeds

FLORAL AND FRUIT CHARACTERISTICS

inflorescences: **headlike spikes; globose** (about 1 cm diameter), axillary, solitary to paired, flowers many; peduncles slender (1–4 cm long), pubescent

flowers: perfect, regular; calyx lobes 5, minute (1–2 mm long), pubescent; corollas funnelform; petals 5, bright yellow; stamens numerous, exserted

fruits: legumes; **terete** (2–8 cm long), tapered at both ends, straight or curved, woody and stout, coriaceous, reddish-brown to purple or black, pulpy within, very tardily dehiscent, seeds in 2 rows; seeds several

other: flowers intensely fragrant

VEGETATIVE CHARACTERISTICS

leaves: alternate, even-bipinnately compound (3–9 cm long); pinnae 2–6 pairs; **leaflets 10–25 pairs**, linear to oblong (3–5 mm long), apices acute or obtuse with minute mucros, bases unequal, margins entire, grayish-green; petiolate glands borne in the middle of the petioles or absent

stems: twigs rigid, slender, numerous, striate, glabrous or puberulent; **stipular spines** (6–14 mm long), **paired at the nodes**, straight, rigid, needlelike, white

HISTORICAL, FOOD, AND MEDICINAL USES: formerly a source of oils for perfumes; wood has been used for fence posts, farm tools, and other small wooden items; gummy sap can be used for manufacturing mucilage; ornamental; honey plant

LIVESTOCK LOSSES: spines may cause injury to mouths, eyes, and legs of browsing animals

FORAGE VALUES: poor for livestock and wildlife, young leaves are occasionally eaten; legumes and seeds furnish valuable food for wildlife

HABITATS: brushy areas, open woodlands, hummocks, and disturbed areas; found in a broad range of soils; most abundant in dry, sandy soils

Catclaw acacia
Acacia greggii A. Gray

SYN = *Acacia durandiana* Buckley, *Acacia wrightii* Benth., *Senegalia greggii* (A. Gray) Britton & Rose

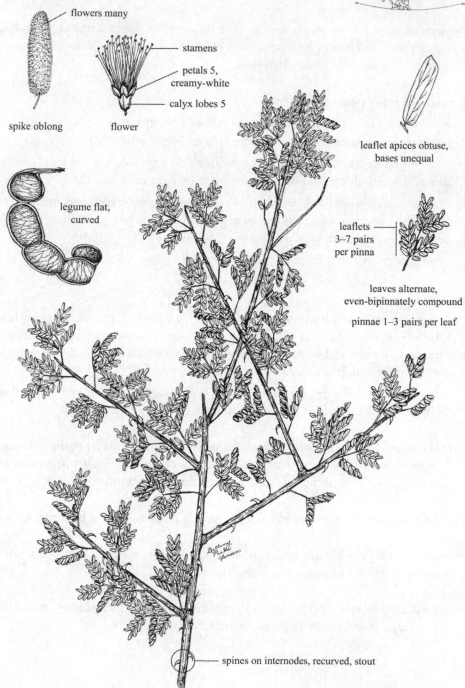

flowers many

spike oblong

stamens

petals 5, creamy-white

calyx lobes 5

flower

leaflet apices obtuse, bases unequal

legume flat, curved

leaflets 3–7 pairs per pinna

leaves alternate, even-bipinnately compound

pinnae 1–3 pairs per leaf

spines on internodes, recurved, stout

FAMILY:	FABACEAE
SPECIES:	*Acacia greggii* A. Gray
COMMON NAME:	Catclaw acacia (uña de gato, Gregg acacia, gatuño)
LIFE SPAN:	Perennial
ORIGIN:	Native
SEASON:	Warm

GROWTH FORM: shrubs or small trees (usually 1–2 m tall, rarely to 8 m); crowns rounded, forming thickets; flowers April to October, may flower later following rain in dry years, reproduces from seeds

FLORAL AND FRUIT CHARACTERISTICS

inflorescences: **spikes; oblong** (2–6 cm long, 1 cm wide), axillary, solitary or paired, flowers many

flowers: perfect, regular; calyx lobes 5, green, obscure (2–3 mm long), puberulent; petals 5, creamy-white; stamens numerous, long-exserted

fruits: legumes; **flat** (5–8 cm long, 1.5–2 cm wide), thin, curved or often curled and contorted, usually flexible, tardily becoming rigid, margins thickened, light brown to reddish-brown, seeds few to several, constricted between the seeds; seeds nearly orbicular, flat

VEGETATIVE CHARACTERISTICS

leaves: alternate, even-bipinnately compound; pinnae 1–3 pairs; **leaflets 3–7 pairs per pinna**, obovate to narrowly oblong (2–7 mm long), apices obtuse, bases unequal, petioles short, margins entire, pubescent, lightly reticulate

stems: twigs much-branched, pale or reddish-brown to gray; **spines (5–9 mm long) on internodes, recurved**; spines similar in appearance to cat's claws, stout, dark brown or gray

HISTORICAL, FOOD, AND MEDICINAL USES: wood used for fuel; some Pima and Papago made pinole flour from the legumes and prepared it as a mush; important honey plant; host plant for several insects that produce lac, a material used in varnish and shellac

LIVESTOCK LOSSES: spines may cause injury to soft tissues of browsing animals

FORAGE VALUES: poor for livestock and wildlife, foliage is seldom browsed; plants provide cover for wildlife, seeds are an important food for many species of wildlife, one of the main foods for some species of quail

HABITATS: valleys, ravines, mesas, washes, and canyon slopes; most abundant in dry, coarse-textured soils

Blackbrush acacia
Acacia rigidula Benth.

SYN = *Acacia amentacea* DC., *Acaciopsis rigidula* (Benth.) Britton & Rose, *Vachellia rigidula* (Benth.) Seigler & Ebinger

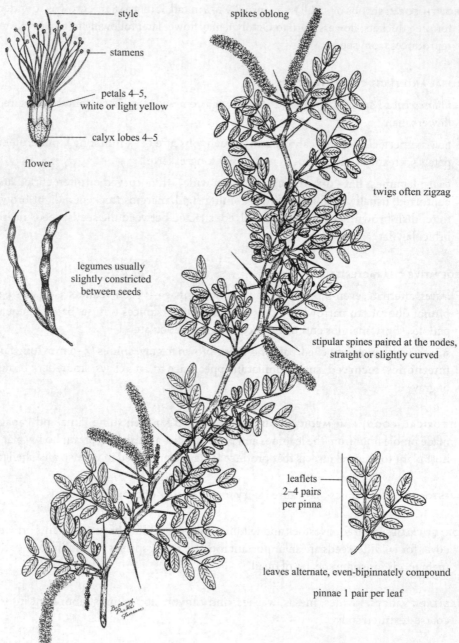

style

stamens

petals 4–5, white or light yellow

calyx lobes 4–5

flower

spikes oblong

twigs often zigzag

legumes usually slightly constricted between seeds

stipular spines paired at the nodes, straight or slightly curved

leaflets 2–4 pairs per pinna

leaves alternate, even-bipinnately compound

pinnae 1 pair per leaf

FAMILY:	FABACEAE
SPECIES:	*Acacia rigidula* Benth.
COMMON NAME:	Blackbrush acacia (chaparro prieto, blackbrush)
LIFE SPAN:	Perennial
ORIGIN:	Native
SEASON:	Warm

GROWTH FORM: shrubs (to 5 m tall); several branches from the base, branches divaricate; much-branched above, often forming impenetrable thickets; flowers April to May, reproduces from seeds

FLORAL AND FRUIT CHARACTERISTICS

inflorescences: **spikes; oblong** (2–6 cm long, 1 cm wide), numerous, axillary, flowers many

flowers: perfect, regular; calyx lobes 4–5; petals 4–5, white or light yellow; stamens several, exserted

fruits: **legumes; linear** (6–8 cm long, 4–7 mm wide), about as thick as broad, often falcate, apices acuminate, puberulent, reddish-brown to black, bivalved, dehiscent, **usually slightly constricted between seeds**; seeds several

other: flowers fragrant

VEGETATIVE CHARACTERISTICS

leaves: alternate, even-pinnately compound; usually only 1 pair of pinnae; **leaflets 2–4 pairs per pinna**, oblong (6–15 mm long, less than 7 mm wide), oblique, apices rounded and mucronate (sometimes notched), bases asymmetrical, margins entire, **dark green to brownish**, glabrous, glossy, veins conspicuous

stems: twigs rigid, often zigzag; bark glabrous, tight, reddish-gray to light or dark gray, some whitish bark on most branches; **stipular spines paired at the nodes**, straight or slightly curved (8–25 mm long), located immediately below the point of leaf attachment

HISTORICAL, FOOD, AND MEDICINAL USES: sometimes used as an ornamental, flowers a source for honey

LIVESTOCK LOSSES: spines may cause injury to the mouths, eyes, and legs of browsing animals

FORAGE VALUES: poor for livestock, fair for wildlife; plants provide cover for wildlife; seeds are an important food for some species of birds and small mammals

HABITATS: plains, mesas, and ridgetops; most abundant in sandy and limestone soils

Leadplant
Amorpha canescens Pursh

SYN = *Amorpha brachycarpa* E.J. Palmer

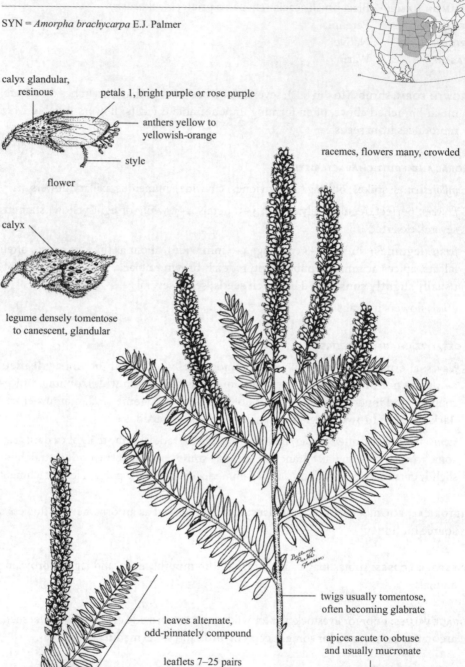

calyx glandular, resinous

petals 1, bright purple or rose purple

anthers yellow to yellowish-orange

style

flower

racemes, flowers many, crowded

calyx

legume densely tomentose to canescent, glandular

twigs usually tomentose, often becoming glabrate

leaves alternate, odd-pinnately compound

apices acute to obtuse and usually mucronate

leaflets 7–25 pairs

infructescence

leaflets adaxially woolly and dark green, abaxially gray-canescent

FAMILY:	FABACEAE
SPECIES:	*Amorpha canescens* Pursh
COMMON NAME:	Leadplant (prairie shoestring)
LIFE SPAN:	Perennial
ORIGIN:	Native
SEASON:	Warm

GROWTH FORM: shrubs (to 1.2 m tall, many reach 2 m in ungrazed areas), erect or ascending; stems 1 to several, often branched; flowers June through early August, reproduces from seeds and rhizomes

FLORAL AND FRUIT CHARACTERISTICS

inflorescences: racemes (5–15 cm long), spicate; **1 to several**, terminal and axillary, often clustered in upper leaf axils; **flowers many, crowded**; rachis densely villous

flowers: perfect, irregular; calyx tubes turbinate (3–5 mm long), pubescent to canescent, glandular-dotted, resinous; petals 1, broadly obovate (4–5 mm long, 3 mm wide), bright purple or rose-purple (occasionally light blue to violet blue); stamens 10, exserted; anthers yellow to yellowish-orange, conspicuous

fruits: legumes; curved (3–5 mm long, 1.6–2 mm wide), densely tomentose to canescent, glandular, seeds 1; seeds elliptical (2–3 mm long), with a slight beak, orangish-brown, smooth

other: flowers fragrant

VEGETATIVE CHARACTERISTICS

leaves: alternate, odd-pinnately compound (3.5–10 cm long); leaflets 7–25 pairs, crowded to imbricate, elliptical to oblong (7–18 mm long, 3–6 mm wide), apices acute to obtuse and usually mucronate, rounded at bases, margins entire, adaxially woolly and dark green, abaxially gray-canescent, sparingly glandular; petioles (0.5–3.5 mm long) pubescent; stipules subulate (1–3 mm long), caducous

stems: twigs usually tomentose, often becoming glabrate with age

HISTORICAL, FOOD, AND MEDICINAL USES: some Native Americans smoked dried leaves and made tea from leaves; they treated neuralgia and rheumatism by cutting stems into small pieces, attaching them to the skin by wetting them, and then lighting them and allowing them to burn down into the skin as a counterirritant; used as an ornamental

LIVESTOCK LOSSES: none

FORAGE VALUES: excellent, highly nutritive, and palatable for livestock and wildlife; commonly selected by livestock over most other species

HABITATS: prairies, open woodlands, dry plains, and hills; adapted to a broad range of soil types; rarely abundant on improperly managed rangelands

Woolly loco
Astragalus mollissimus Torr.

SYN = *Astragalus mogollonicus* Greene, *Astragalus simulans* Cockerell, *Astragalus thompsoniae* S. Watson, *Phaca villosa* S. Watson, *Tragacantha mollissima* (Torr.) Kuntze

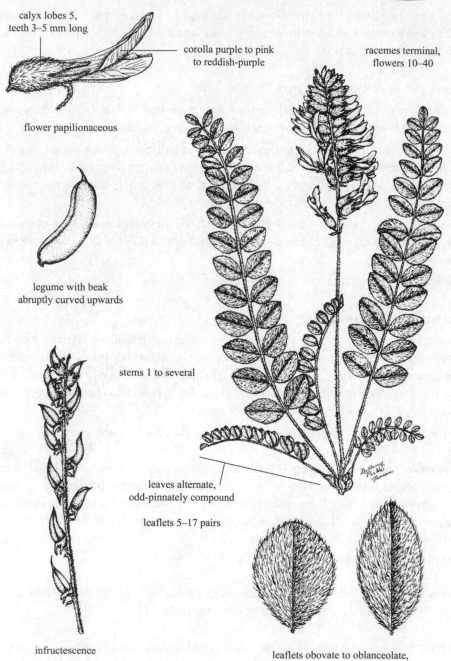

calyx lobes 5,
teeth 3–5 mm long

corolla purple to pink
to reddish-purple

racemes terminal,
flowers 10–40

flower papilionaceous

legume with beak
abruptly curved upwards

stems 1 to several

leaves alternate,
odd-pinnately compound

leaflets 5–17 pairs

infructescence

leaflets obovate to oblanceolate,
silky-tomentose, hairs basifixed

FAMILY:	FABACEAE
SPECIES:	*Astragalus mollissimus* Torr.
COMMON NAME:	Woolly loco (hierba loca, purple locoweed, poisonvetch, hierba plata, chinchin)
LIFE SPAN:	Perennial
ORIGIN:	Native
SEASON:	Cool

GROWTH FORM: forbs (10–30 cm tall, stems to 1 m long), stems 1 to several, prostrate to ascending from woody taproots; densely to loosely tufted, often robust and leafy; starts growth in March, flowers April to June, fruits mature July to August, reproduces from seeds

FLORAL AND FRUIT CHARACTERISTICS

inflorescences: racemes; oblong (4–10 cm long), terminal, dense, flowers 10–40; peduncles naked (5–20 cm long)

flowers: perfect, papilionaceous; calyx tubes cylindrical (5–10 mm long), somewhat oblique; lobes 5 (**teeth 3–5 mm long**), long-acuminate, silky; **corollas purple to pink to reddish-purple** (rarely yellow or white), often drying blue (1.7–2.2 cm long); banners moderately reflexed; keels rounded

fruits: legumes; plump-ellipsoid to oblong-ellipsoid (1–2.5 cm long, 4–9 mm wide), spreading or ascending, sessile, beaked; beaks abruptly curved upward; bilocular for most of the length, usually glabrous except at the apices, terete to dorsiventrally compressed, seeds many

VEGETATIVE CHARACTERISTICS

leaves: alternate, odd-pinnately compound (5–22 cm long); leaflets 5–17 pairs, **obovate to oblanceolate** (5–25 mm long, 2–15 mm wide), apices obtuse to acute and rarely mucronate, margins entire, silky-tomentose, hairs straight and curly, basifixed, often yellow in drying; **stipules distinct** (up to 1.5 cm long), triangular, silky; compare with *Oxytropis lambertii*

stems: young plants subcaulescent, **outer leaves and stems often prostrate**; mature plants with inner stems ascending, caulescent

HISTORICAL, FOOD, AND MEDICINAL USES: famous in western history as one of the causes of "locoed" animals

LIVESTOCK LOSSES: poisonous; can cause loco disease in livestock; toxic principles are the alkaloid locoine and selenium, which are accumulated by the plants; large amounts must be consumed for a lethal dose; both poisons are cumulative; both green and dry plants are poisonous

FORAGE VALUES: poor to worthless for livestock and wildlife, unpalatable and consumed only when other forage is not available; some animals (especially horses) may become addicted and refuse to eat better forage

HABITATS: dry prairies, hillsides, roadsides, stream valleys, and uplands; most abundant in sandy or rocky soils of improperly grazed rangelands

Purple prairieclover
Dalea purpurea Vent.

SYN = *Dalea violacea* (Michx.) Willd., *Kuhnistera violacea* (Michx.) Aiton *ex*
Steud., *Petalostemon pubescens* A. Nelson, *Petalostemon purpureus* (Vent.)
Rydb., *Petalostemon standleyanus* Rydb., *Petalostemon violaceus* Michx.,
Psoralea purpurea (Vent.) MacMillan

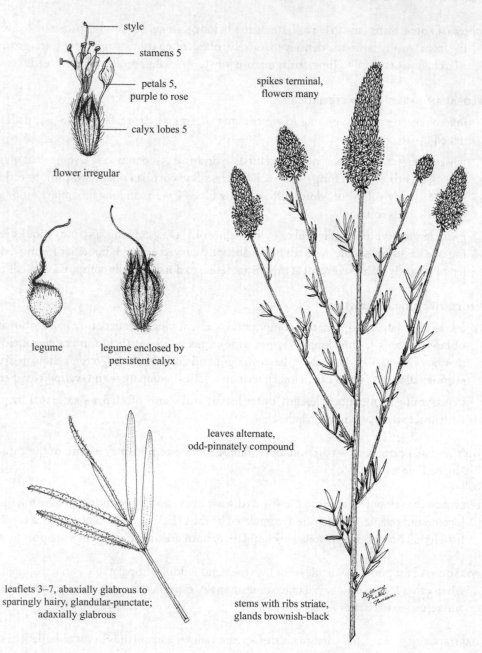

style

stamens 5

petals 5,
purple to rose

calyx lobes 5

flower irregular

spikes terminal,
flowers many

legume

legume enclosed by
persistent calyx

leaves alternate,
odd-pinnately compound

leaflets 3–7, abaxially glabrous to
sparingly hairy, glandular-punctate;
adaxially glabrous

stems with ribs striate,
glands brownish-black

FAMILY:	FABACEAE
SPECIES:	*Dalea purpurea* Vent.
COMMON NAME:	Purple prairieclover (violet prairieclover)
LIFE SPAN:	Perennial
ORIGIN:	Native
SEASON:	Warm

GROWTH FORM: forbs (20–90 cm tall); erect or ascending; stems few to many, simple or branched from thick caudices; flowers June to August, seeds mature July to September, reproduces from seeds and rootstocks

FLORAL AND FRUIT CHARACTERISTICS

inflorescences: spikes; oblong to ovate (1–7 cm long, 7–14 mm wide), terminal, dense, numerous, axis not visible, flowers many

flowers: perfect, irregular, **calyx tubes silky-villous** (2.5–4 mm long), lobes 5; lobes persistent, lanceolate; petals 5, purple to rose (5–7 mm long); 4 of the 5 petals and the 5 stamens joined to near the tip of the calyces, banners separate

fruits: legumes; obliquely ovate (2–2.5 mm long), enclosed by the persistent calyces, seeds 1

VEGETATIVE CHARACTERISTICS

leaves: alternate, odd-pinnately compound (1–4 cm long); **leaflets 3–7** (mostly 5), linear to linear-elliptical (5–25 mm long, 0.5–1.5 mm wide), margins entire, involute, adaxially glabrous, abaxially glabrous to sparingly hairy and **glandular-punctate**, midveins not visible on adaxial surfaces of leaflets; petioles similar in appearance to leaflets

stems: sparsely pilosulous to glabrous, ribs striate, glands brownish-black

HISTORICAL, FOOD, AND MEDICINAL USES: some Native Americans ate fresh leaves and boiled leaves, bruised leaves were steeped in water and applied to wounds, roots were chewed for their pleasant flavor, stems were used to make brooms

LIVESTOCK LOSSES: may cause bloat, but it is seldom abundant enough to be a problem

FORAGE VALUES: excellent for livestock and wildlife, important component of prairie hay; high in protein, highly palatable, and nutritious

HABITATS: prairies, plains, and hills; adapted to a broad range of soils, most abundant in dry soils

Tailcup lupine
Lupinus caudatus Kellogg

SYN = *Lupinus argentinus* Rydb., *Lupinus argophyllus* (A. Gray) Cockerell, *Lupinus cutleri* Eastw., *Lupinus montigenus* A. Heller

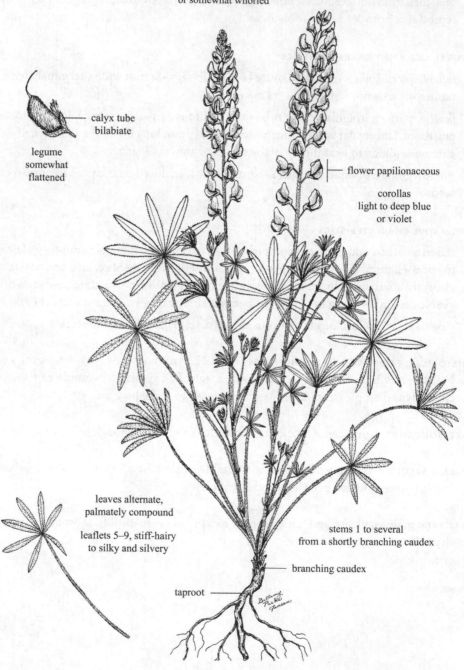

racemes with flowers scattered or somewhat whorled

calyx tube bilabiate

legume somewhat flattened

flower papilionaceous

corollas light to deep blue or violet

leaves alternate, palmately compound

leaflets 5–9, stiff-hairy to silky and silvery

stems 1 to several from a shortly branching caudex

branching caudex

taproot

FAMILY:	FABACEAE
SPECIES:	*Lupinus caudatus* Kellogg
COMMON NAME:	Tailcup lupine (lupino, altramuz, spurred lupine)
LIFE SPAN:	Perennial
ORIGIN:	Native
SEASON:	Cool

GROWTH FORM: forbs (20–60 cm tall); erect to ascending, stems 1 to several from shortly branching caudices surmounting taproots; branched or simple; flowers June to August, fruits mature July to September, reproduces from seeds

FLORAL AND FRUIT CHARACTERISTICS

inflorescences: racemes; oblong (6–20 cm long), terminal, flowers many; flowers scattered or somewhat whorled

flowers: perfect, papilionaceous; calyx tubes asymmetrical (3–4 mm long), bilabiate, spurred at the bases; corollas light to deep blue or violet (1–1.4 cm long); banner petals reflexed, often lighter blue or whitish in the middle, silky on back

fruits: legumes; somewhat flattened (2–3 cm long), **apices acuminate or long-acuminate**, silky, may be slightly constricted between the seeds; seeds several

VEGETATIVE CHARACTERISTICS

leaves: alternate, **palmately compound**; leaflets 5–9, oblanceolate (2–15 cm long, 5–15 mm wide), apices acute to rarely rounded, margins entire, **stiff-hairy to silky and silvery**; long-petiolate

stems: somewhat cespitose, strigose

HISTORICAL, FOOD, AND MEDICINAL USES: a drug has been extracted for management of cardiac arrhythmias

LIVESTOCK LOSSES: poisonous, especially to sheep, cattle, and horses, causing weakness and muscular trembling; deformed lambs and calves (crooked legs) may result from consumption of these plants 2 months into the pregnancy; alkaloids are concentrated in the seeds and occasionally in the young plants, plants are poisonous either green or dry, poisoning seldom occurs when other forage is adequate

FORAGE VALUES: poor for cattle and fair for sheep before the legumes develop; cattle may be attracted to the legumes and graze them selectively; fair to good for elk and deer

HABITATS: dry prairies, foothills, mesas, and stream valleys; adapted to a broad range of soil textures, but most abundant in coarse-textured and well-drained soils

Burclover
Medicago polymorpha L.

SYN = *Medicago apiculata* Willd., *Medicago denticulata* Willd., *Medicago hispida* Gaertn., *Medicago lappacea* Desr., *Medicago nigra* Krock.

corolla yellow

calyx

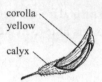

flower papilionaceous

seed

spines in a double row, with more or less hooked tips

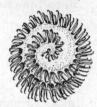

legume spirally coiled 2–5 times

leaflets obcordate to cuneate-obovate, sometimes with whitish or purplish splotches, margins serrulate

leaves alternate, pinnately trifoliate

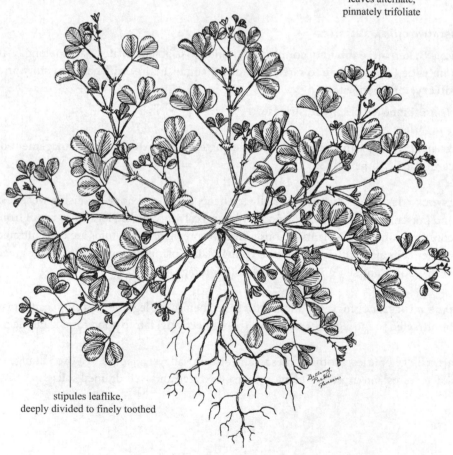

stipules leaflike, deeply divided to finely toothed

FAMILY:	FABACEAE
SPECIES:	*Medicago polymorpha* L.
COMMON NAME:	Burclover (carretilla, medic, toothed burclover)
LIFE SPAN:	Annual
ORIGIN:	Introduced (from Mediterranean region)
SEASON:	Cool

GROWTH FORM: forbs; procumbent to ascending (stems to 50 cm long), several stems from the bases surmounting taproots; flowers March to June, seeds mature in about 8 weeks, reproduces from seeds

FLORAL AND FRUIT CHARACTERISTICS

inflorescences: racemes; oblong to oval (5–15 mm long), axillary, flowers 2–8, borne on peduncles (1–3 cm long)

flowers: perfect, papilionaceous; pedicellate (pedicels 0.5–1 mm long); calyx tubes campanulate (about 1 mm long), pubescent; corollas yellow (3–5 mm long)

fruits: legumes; **spirally coiled 2–5 times** (4–6 mm in diameter, excluding the spines); spines (2–3 mm long) in a double row, with more or less hooked tips; seeds several

VEGETATIVE CHARACTERISTICS

leaves: alternate, pinnately trifoliate; **leaflets obcordate to cuneate-obovate** (6–15 mm long, mostly longer than broad), sometimes with whitish or purplish splotches, margins serrulate; petiolate; **stipules leaflike** (6–10 mm long) deeply divided to finely toothed

stems: nearly glabrous to puberulent

HISTORICAL, FOOD, AND MEDICINAL USES: young leaves may be used to garnish salads

LIVESTOCK LOSSES: excessive grazing of fresh herbage may cause bloat; fleece may become contaminated with fruits

FORAGE VALUES: excellent to good for livestock and wildlife, most valuable in spring; legumes are highly nutritious and are consumed during the dry season, somewhat resistant to heavy grazing; legumes are an important food for deer, birds, and small mammals

HABITATS: valleys, plains, lower slopes of foothills, waste ground, and a common weed in lawns and roadsides; thrives in rich soils and may occur in poor soils where few other species grow

Purple locoweed
Oxytropis lambertii Pursh

SYN = *Aragallus angustatus* Rydb., *Aragallus involutus* A. Nelson, *Astragalus lambertii* (Pursh) Spreng., *Oxytropis angustata* (Rydb.) A. Nelson, *Oxytropis involuta* (A. Nelson) K. Schum., *Spiesia lambertii* (Pursh) Kuntze

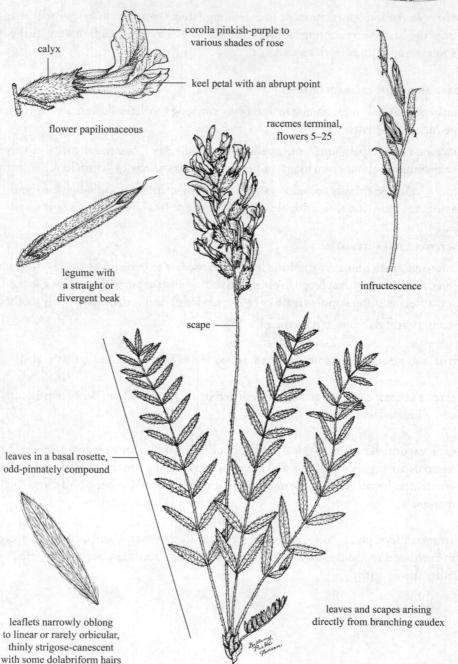

calyx

corolla pinkish-purple to
various shades of rose

keel petal with an abrupt point

flower papilionaceous

racemes terminal,
flowers 5–25

legume with
a straight or
divergent beak

infructescence

scape

leaves in a basal rosette,
odd-pinnately compound

leaflets narrowly oblong
to linear or rarely orbicular,
thinly strigose-canescent
with some dolabriform hairs

leaves and scapes arising
directly from branching caudex

360

FAMILY:	FABACEAE
SPECIES:	*Oxytropis lambertii* Pursh
COMMON NAME:	Purple locoweed (hierba loca, Lambert crazyweed, white loco, whitepoint loco)
LIFE SPAN:	Perennial
ORIGIN:	Native
SEASON:	Cool

GROWTH FORM: forbs acaulescent (scapes 10–35 cm tall); flowers April to August, fruits mature June to October, reproduces from seeds

FLORAL AND FRUIT CHARACTERISTICS

inflorescences: racemes; oblong (4–12 cm long, 2–7 cm wide), terminal, flowers 5–25, elevated above the leaves on scapes

flowers: perfect, papilionaceous; calyx tubes campanulate (5–9 mm long), lobes 5 (**teeth 1.5–3 mm long**), silky pilose; corollas (1.5–2.5 cm long) pinkish-purple to various shades of rose (white not uncommon), keel petals with an abrupt point (0.5–2.5 mm long)

fruits: legumes; oblong to oblong-ovoid (7–30 mm long, 5–6 mm wide), sessile or nearly so, with straight or divergent beaks (3–7 mm long), silky-strigose, becoming glabrous, seeds many, early dehiscent

VEGETATIVE CHARACTERISTICS

leaves: alternate in basal rosettes, odd-pinnately compound, dimorphic; principal leaves (4–20 cm long) with 7–19 leaflet pairs; **leaflets narrowly oblong to linear** or rarely orbicular (5–40 mm long, 2–7 mm wide), apices acute, margins entire, thinly strigose-canescent with some **dolabriform hairs**; petioles pubescent; **stipules adnate to the petioles;** compare with *Astragalus mollissimus*

stems: **leaves and scapes arising directly from branching caudices**

HISTORICAL, FOOD, AND MEDICINAL USES: famous in western history as one of the causes of "locoed" animals

LIVESTOCK LOSSES: poisonous; can cause loco disease in horses, cattle, sheep, and goats (see *Astragalus mollissimus* for a discussion of loco poisoning)

FORAGE VALUES: poor to worthless for livestock and wildlife; generally unpalatable and consumed only when other forages are not available, some animals may develop an addiction for the plants

HABITATS: dry upland prairies and plains, adapted to a broad range of soil types

Honey mesquite
Prosopis glandulosa Torr.

SYN = *Algarobia glandulosa* (Torr.) Torr. & A. Gray, *Neltuma glandulosa* (Torr.) Britton & Rose, *Neltuma neomexicana* Britton, *Prosopis chilensis* (Molina) Stuntz, *Prosopis juliflora* (Sw.) DC., *Prosopis odorata* Torr. & Frém.

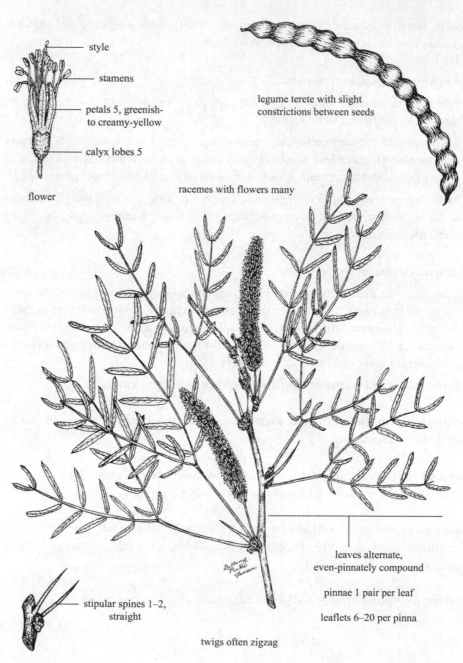

style

stamens

petals 5, greenish-
to creamy-yellow

calyx lobes 5

flower

legume terete with slight
constrictions between seeds

racemes with flowers many

leaves alternate,
even-pinnately compound

pinnae 1 pair per leaf

leaflets 6–20 per pinna

stipular spines 1–2,
straight

twigs often zigzag

FAMILY:	FABACEAE
SPECIES:	*Prosopis glandulosa* Torr.
COMMON NAME:	Honey mesquite (mezquite, mesquite, glandular mesquite)
LIFE SPAN:	Perennial
ORIGIN:	Native
SEASON:	Warm

GROWTH FORM: shrubs or small trees (to 6 m tall); single or multiple trunks, much-branched, crowns rounded; growth begins in late spring, flowers in May, fruits mature June to August, reproduces from seeds and basal shoots

FLORAL AND FRUIT CHARACTERISTICS

inflorescences: racemes (7–9 cm long); axillary, flowers many

flowers: perfect, regular; calyx tubes campanulate (1 mm long), pubescent, lobes 5; lobes triangular (0.3–0.4 mm long); petals 5 (3 mm long), greenish- to creamy-yellow, distinct, pubescent within; pedicels glandular (0.5 mm long)

fruits: legumes; straight or curved (10–20 cm long, 1 cm wide), terete, linear; often in clusters of 2–3, glabrous, **slight constrictions between seeds**, surfaces tan to golden, seeds several, dehiscent

other: flowers fragrant

VEGETATIVE CHARACTERISTICS

leaves: alternate, even-bipinnately compound (6–15 cm long); **pinnae usually 1 pair**, leaflets 6–20 pairs per pinna, linear to linear-oblong (2–6 cm long, 2–3 mm wide), terminal pair usually curved, apices acute and mucronate, margins entire, glabrous or nearly so; sessile or nearly so

stems: twigs rigid, **often zigzag**, reddish-brown or grayish-brown; **armed with 1–2 stipular spines** (to 5 cm long), above the point of leaf attachment, stout, straight, rarely spineless

HISTORICAL, FOOD, AND MEDICINAL USES: wood is used for fuel, lumber, and flooring; beans were ground into flour which was important in diets of some Native Americans, flour was fermented into alcohol; important to bees for production of honey

LIVESTOCK LOSSES: ingestion of large amounts by cattle may result in rumen stasis and impaction

FORAGE VALUES: poor to good for livestock, deer, and javelina; seeds are an important food for numerous wildlife species and as emergency feed for livestock

HABITATS: plains and prairies, on dry sandy or gravelly soils; especially abundant on abused rangelands and where fire has been eliminated

Slimflower scurfpea
Psoralidium tenuiflorum (Pursh) Rydb.

SYN = *Lotodes floribundum* (Nutt. *ex* Torr. & A. Gray) Kuntze, *Lotodes tenuiflora*
(Pursh) Kuntze, *Psoralea floribunda* Nutt. *ex* Torr. & A. Gray, *Psoralea*
obtusiloba Torr. & A. Gray, *Psoralea tenuiflora* Pursh, *Psoralidium batesii*
Rydb., *Psoralidium bigelovii* Rydb., *Psoralidium obtusilobum* (Torr. &
A. Gray) Rydb., *Psoralidium youngiae* (Tharp & F.A. Barkley)

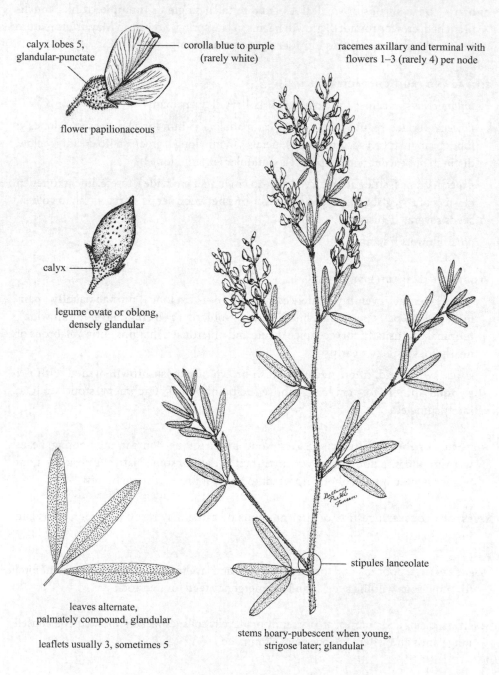

calyx lobes 5,
glandular-punctate

corolla blue to purple
(rarely white)

racemes axillary and terminal with
flowers 1–3 (rarely 4) per node

flower papilionaceous

calyx

legume ovate or oblong,
densely glandular

stipules lanceolate

leaves alternate,
palmately compound, glandular

leaflets usually 3, sometimes 5

stems hoary-pubescent when young,
strigose later; glandular

364

FAMILY:	FABACEAE
SPECIES:	*Psoralidium tenuiflorum* (Pursh) Rydb.
COMMON NAME:	Slimflower scurfpea (wild alfalfa, slender scurfpea)
LIFE SPAN:	Perennial
ORIGIN:	Native
SEASON:	Warm

GROWTH FORM: forbs (0.2–1.3 m tall); erect or ascending, stems 1 to several from caudices surmounting deep taproots, much-branched; flowers May to September, reproduces from seeds and rarely from rhizomes; above-ground portion disarticulates at the crowns following maturity and may tumble with the wind

FLORAL AND FRUIT CHARACTERISTICS

inflorescences: racemes; oblong (2–10 cm long), loose, axillary and terminal, flowers many, flowers per node 1–3 (rarely 4); peduncles (2–9 cm long) longer than the subtending leaves

flowers: perfect, papilionaceous; pedicellate (pedicels 1.5–3 mm long); calyx tubes campanulate (1.5–3 mm long), lobes 5; **lobes glandular-punctate**, strigose or villoushirsute; corollas (4–8 mm long) blue to purple (rarely white)

fruits: legumes; ovate or oblong (5–9 mm long), compressed, with a short and straight beak, glabrous, **densely glandular**, often asymmetrical, seeds 1

VEGETATIVE CHARACTERISTICS

leaves: alternate, palmately compound; **leaflets usually 3** (lower leaves sometimes with 5 leaflets); leaflets linear-oblanceolate to obovate (1–5 cm long, 4–12 mm wide), margins entire, **glandular**, adaxially glabrate, abaxially gray-strigose; petioles mostly shorter (4–20 mm long) than the leaflets; stipules lanceolate (2–3 mm long)

stems: hoary-pubescent when young, strigose later; glandular

HISTORICAL, FOOD, AND MEDICINAL USES: some Native Americans drank a tea made from stems and leaves for fever, a tea was made from the roots for headache, plants were burned to repel mosquitoes

LIVESTOCK LOSSES: may cause bloat; reported to be poisonous to cattle and horses, but no experimental evidence is available in support of this report

FORAGE VALUES: poor for livestock and wildlife, readily eaten after curing in hay; seeds are an important food for birds and small mammals

HABITATS: dry plains, prairies, and open woodlands; adapted to a broad range of soils

Gambel oak
Quercus gambelii Nutt.

SYN = *Quercus albifolia* C.H. Müll., *Quercus confusa* Wooton & Standl., *Quercus eastwoodiae* Rydb., *Quercus gunnisonii* (Torr. & A. Gray) Rydb., *Quercus leptophylla* Rydb., *Quercus marshii* C.H. Müll., *Quercus nitescens* Rydb., *Quercus submollis* Rydb., *Quercus utahensis* (A. DC.) Rydb., *Quercus vreelandii* Rydb.

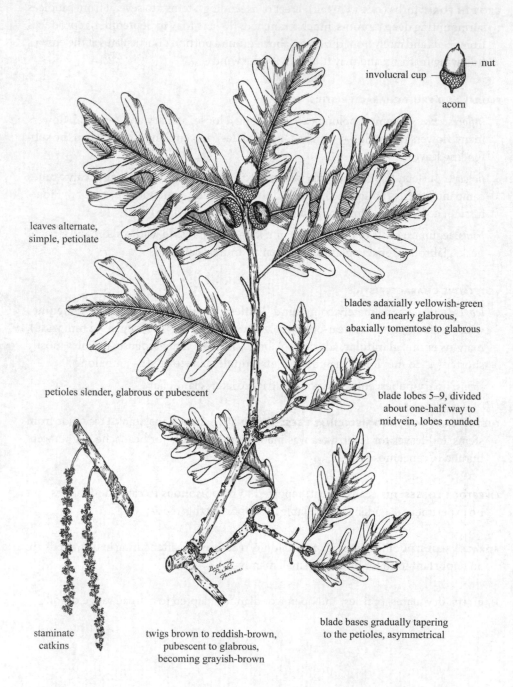

nut

involucral cup

acorn

leaves alternate, simple, petiolate

blades adaxially yellowish-green and nearly glabrous, abaxially tomentose to glabrous

petioles slender, glabrous or pubescent

blade lobes 5–9, divided about one-half way to midvein, lobes rounded

staminate catkins

twigs brown to reddish-brown, pubescent to glabrous, becoming grayish-brown

blade bases gradually tapering to the petioles, asymmetrical

FAMILY:	FAGACEAE
SPECIES:	*Quercus gambelii* Nutt.
COMMON NAME:	Gambel oak (encino blanco)
LIFE SPAN:	Perennial
ORIGIN:	Native
SEASON:	Cool

GROWTH FORM: shrubs or trees (to 20 m tall), monoecious; crowns rounded; grows in dense stands, often forming thickets; flowers March to April, fruits mature in autumn of the first year after flowering, reproduces from seeds

FLORAL AND FRUIT CHARACTERISTICS

inflorescences: staminate flowers in catkins (2.5–4 cm long), pendulous, flowers many; pistillate flowers 1–3 in reduced catkins at branch apices

flowers: unisexual, apetalous; staminate flower calyx lobes 5–6; stamens 5–10; pistillate flowers inconspicuous (3–8 mm long), calyx lobes 6

fruits: acorn; solitary or clustered; involucral cups (to 1.5 cm in diameter), enclose one-fourth to one-half of the nuts; scales appressed, ovate-acuminate, tomentulose, apices narrowed and rounded, bases rounded; nuts ovoid to ellipsoid (to 1.5 cm long), light brown, seeds 1; subsessile or on peduncles (to 10 mm long)

VEGETATIVE CHARACTERISTICS

leaves: alternate, simple; blades highly variable, ovate to obovate or oblong to elliptical (8–16 cm long, 4–7 cm wide), **lobes 5–9, usually divided to about one-half way to midveins**; lobes rounded, **middle lobes the largest**; bases gradually tapering to the petioles, asymmetrical; coriaceous; adaxially yellowish-green and nearly glabrous; abaxially tomentose to glabrous, sometimes glaucous; petioles slender (7–25 mm long), glabrous or pubescent

stems: twigs slender (1.5–2.5 mm diameter), glabrous to stellate pubescent; hairs brown to reddish-brown, becoming grayish-brown; bark gray or brown, deeply fissured, scaly; lenticels inconspicuous

HISTORICAL, FOOD, AND MEDICINAL USES: acorns are edible after tannic acid is removed; some Native Americans used ground or powdered acorns to thicken soup and make mush

LIVESTOCK LOSSES: shoots contain tannic and gallic acids, therefore, poisoning of cattle and occasionally sheep may occur from March to April

FORAGE VALUES: fair for all classes of livestock, deer, and porcupines; acorns eaten by livestock and wildlife

HABITATS: valleys, canyons, foothills, and lower mountain slopes in all soil textures

Post oak
Quercus stellata Wangenh.

SYN = *Quercus minor* (Marshall) Sarg., *Quercus obtusiloba* Michx.

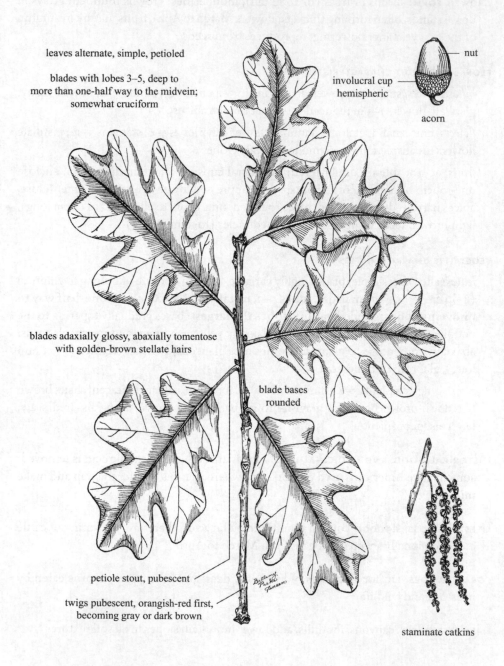

leaves alternate, simple, petioled

blades with lobes 3–5, deep to more than one-half way to the midvein; somewhat cruciform

nut

involucral cup hemispheric

acorn

blades adaxially glossy, abaxially tomentose with golden-brown stellate hairs

blade bases rounded

petiole stout, pubescent

twigs pubescent, orangish-red first, becoming gray or dark brown

staminate catkins

FAMILY:	FAGACEAE
SPECIES:	*Quercus stellata* Wangenh.
COMMON NAME:	Post oak (encino)
LIFE SPAN:	Perennial
ORIGIN:	Native
SEASON:	Warm

GROWTH FORM: trees (to 30 m tall), monoecious; limbs stout, crowns rounded or cylindrical; flowers March to May, fruits mature in the autumn of the first year after flowering, reproduces from seeds and basal shoots

FLORAL AND FRUIT CHARACTERISTICS

inflorescences: staminate flowers in **catkins (8–10 cm long)**, pendulous, in clusters from lateral buds, flowers many; pistillate flowers 2 to several, short-stalked or sessile

flowers: unisexual, apetalous; calyx of staminate flower lobes 5; lobes (1.2–1.4 mm long) yellow with brown tips, pubescent, acute; stamens 4–6; pistillate flowers inconspicuous

fruits: acorns; usually in clusters of 2–4, occasionally solitary; involucral cups hemispheric (to 2 cm in diameter), enclosing one-third (sometimes up to one-half) of the nuts; scales reddish-brown, tomentose, closely appressed, apices rounded or acute; nuts oval or obovoid-oblong (1–1.5 cm long), light brown; seeds 1; subsessile or on a peduncle (to 6 mm long)

VEGETATIVE CHARACTERISTICS

leaves: alternate, simple; blades broadly obovate to oblong-obovate (10–15 cm long, 7–10 cm wide), **lobes 3–5**, deep **to more than one-half way to the midveins, somewhat cruciform**; apices rounded; bases rounded to cordate or cuneate; adaxially glossy; golden-brown **tomentose abaxially; hairs stellate**; gray pubescence in axils of leaf veins; petioles stout (5–10 mm long), pubescent

stems: twigs rigid, densely stellate-pubescent, orangish-red first, becoming gray or dark brown; bark gray to reddish-brown, fissured with platelike scales

HISTORICAL, FOOD, AND MEDICINAL USES: wood used for furniture, fencing, flooring, fuel, railroad ties, and lumber

LIVESTOCK LOSSES: high content of tannic and gallic acids in the shoots, particularly following freezing temperatures, can poison cattle

FORAGE VALUES: poor to fair for cattle, fair for goats; buds, leaves, and twigs eaten by cattle and goats in early spring; acorns eaten by wildlife

HABITATS: open woodlands, hillsides, and ridges; adapted to a broad range of soil textures; most abundant on dry sites

Broadleaf filaree
Erodium botrys (Cav.) Bertol.

SYN = *Erodion botrydium* (Cav.) St.-Lag., *Geranium botrys* Cav.

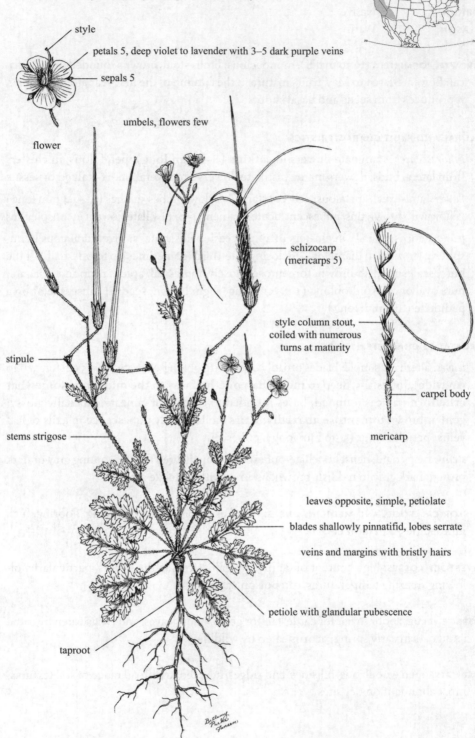

style

petals 5, deep violet to lavender with 3–5 dark purple veins

sepals 5

umbels, flowers few

flower

schizocarp
(mericarps 5)

style column stout,
coiled with numerous
turns at maturity

stipule

carpel body

stems strigose

mericarp

leaves opposite, simple, petiolate

blades shallowly pinnatifid, lobes serrate

veins and margins with bristly hairs

petiole with glandular pubescence

taproot

FAMILY:	GERANIACEAE
SPECIES:	*Erodium botrys* (Cav.) Bertol.
COMMON NAME:	Broadleaf filaree (longbeaked filaree, longbeaked storksbill)
LIFE SPAN:	Annual
ORIGIN:	Introduced (from Europe)
SEASON:	Cool

GROWTH FORM: forbs (stems 10–90 cm long); few to several stems from taproots, prostrate to somewhat ascending, first acaulescent then branching from the bases; flowers March to May, reproduces from seeds; seeds germinate in winter, forming rosettes, grows rapidly for 3–4 months before slowing significantly

FLORAL AND FRUIT CHARACTERISTICS

inflorescences: umbels; axillary or terminal on elongated peduncles, flowers few; peduncles glandular-hirsute

flowers: perfect, regular; sepals 5 (7–8 mm long, enlarging in fruit), glandular-pubescent; lobes 1–2, minute, with bristlelike hairs or awns (enlarging in fruit); petals 5, deep violet to lavender, cuneate (1–1.5 cm long), apices blunt; veins 3–5, dark purple; anther-bearing stamens 5, occurring alternately with 5 sterile stamens

fruits: schizocarps; **mericarps** 5, fusiform (**8–10 mm long**), pubescent, pits 2 at the base of each carpel body, each subtended by 2 folds forming smaller pits between; style columns stout (9–13 cm long), **coiled with numerous turns at maturity**

VEGETATIVE CHARACTERISTICS

leaves: opposite, simple; blades ovate to oblong-ovate (3–10 cm long), **shallowly pinnatifid**; lobes serrate, acute; basal leaves sometimes crenate; bristly hairs on veins and margins; petioles (8–24 mm long) with glandular pubescence; with stipules

stems: strigose, much-branched

HISTORICAL, FOOD, AND MEDICINAL USES: purple dyes have been extracted from the flowers

LIVESTOCK LOSSES: has been reported to cause bloat

FORAGE VALUES: good in winter and early spring for cattle, sheep, and wildlife; especially valuable on annual rangelands

HABITATS: pastures, plains, grassy lowlands, foothills, and waste places; adapted to a broad range of soils

Redstem filaree
Erodium cicutarium (L.) L'Hér. *ex* Aiton

SYN = *Erodium chaerophyllum* Steud., *Erodium millefolium* Kunth, *Erodium moranense* Kunth, *Erodium triviale* Jord., *Geranium arenicola* Steud., *Geranium chaerophyllum* Cav., *Geranium cicutariu*m L., *Geranium pimpinellifolium* With.

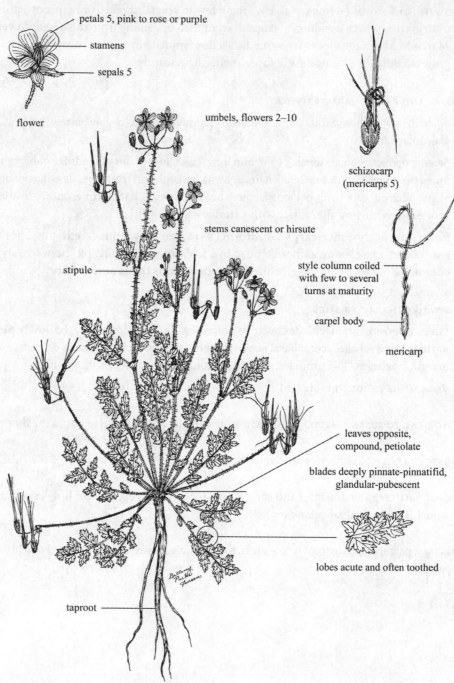

petals 5, pink to rose or purple

stamens

sepals 5

flower

umbels, flowers 2–10

schizocarp (mericarps 5)

stems canescent or hirsute

stipule

style column coiled with few to several turns at maturity

carpel body

mericarp

leaves opposite, compound, petiolate

blades deeply pinnate-pinnatifid, glandular-pubescent

lobes acute and often toothed

taproot

FAMILY:	GERANIACEAE
SPECIES:	*Erodium cicutarium* (L.) L'Hér. *ex* Aiton
COMMON NAME:	Redstem filaree (alfilaria, alfilerillo, redstem storksbill, heronbill, agujas de pastor)
LIFE SPAN:	Annual
ORIGIN:	Introduced (from Europe and the Mediterranean region)
SEASON:	Cool

GROWTH FORM: forbs (stems 10–50 cm long); few to several stems from taproots, decumbent to ascending; first appearing acaulescent as a winter rosette of basal leaves; flowers February to May (sometimes in September), reproduces from seeds; seeds germinate in autumn or early spring

FLORAL AND FRUIT CHARACTERISTICS

inflorescences: umbels; axillary on elongated peduncles, flowers 2–10; pedicels (5–20 mm long), often recurved from the bases, glandular-pubescent; hairs flattened

flowers: perfect, regular (1 cm in diameter); sepals 5, elliptical (2–6 mm long), with 1–2 short, white, bristlelike hairs or awns (0.1–0.5 mm long); petals 5, pink to rose or purple, elliptical to obovate (4–8 mm long), ciliate at the bases; anther-bearing stamens 5, occurring alternately with 5 sterile stamens

fruits: schizocarps; **mericarps** 5, fusiform (**4–5 mm long**), basal points sharp, tardily dehiscent, stiffly pubescent; style columns attached (2–5 cm long), **coiled with few to several turns at maturity**

VEGETATIVE CHARACTERISTICS

leaves: opposite, compound; blades elongate-oblanceolate (3–13 cm long, including the petiole), **deeply pinnate-pinnatifid**; lobes acute and often toothed; glandular-pubescent; hairs flattened, white; leaflets sessile; petioles 8–18 mm long; with stipules

stems: erect to decumbent to prostrate; canescent or hirsute, much-branched, reddish

other: leaves delicate, nearly fernlike; one of the first plants to bloom in the spring

HISTORICAL, FOOD, AND MEDICINAL USES: young leaves can be eaten raw or cooked; reputed in folklore to contain an antidote for strychnine

LIVESTOCK LOSSES: has been reported to cause bloat

FORAGE VALUES: excellent to good in spring for cattle, sheep, and wildlife; especially valuable on annual rangelands

HABITATS: annual rangelands, cultivated fields, waste places, roadsides, lawns, plains, mesas, and ravines; adapted to a broad range of soil types, most abundant on sandy soils

Wildwhite geranium
Geranium richardsonii Fisch. & Trautv.

SYN = *Geranium albiflorum* Hook., *Geranium gracilentum* Greene, *Geranium hookerianum* Walp., *Geranium loloense* H. St. John

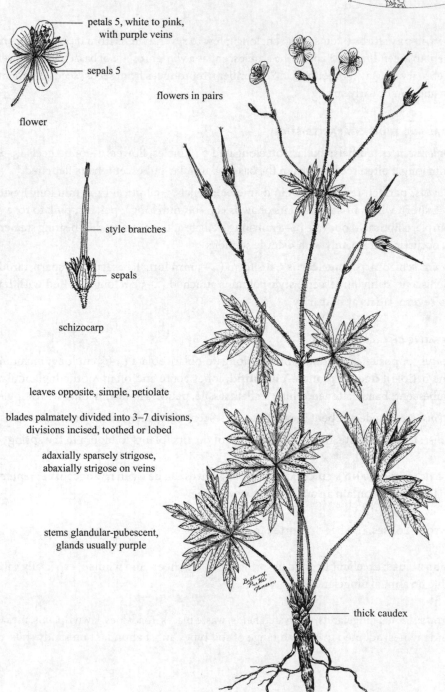

petals 5, white to pink,
with purple veins

sepals 5

flower

flowers in pairs

style branches

sepals

schizocarp

leaves opposite, simple, petiolate

blades palmately divided into 3–7 divisions,
divisions incised, toothed or lobed

adaxially sparsely strigose,
abaxially strigose on veins

stems glandular-pubescent,
glands usually purple

thick caudex

FAMILY:	GERANIACEAE
SPECIES:	*Geranium richardsonii* Fisch. & Trautv.
COMMON NAME:	Wildwhite geranium (Richardson cranesbill, white cranesbill)
LIFE SPAN:	Perennial
ORIGIN:	Native
SEASON:	Warm

GROWTH FORM: forbs (0.2–1 m tall), stems 1 to few, erect or ascending, from thick branched or unbranched caudices; sometimes rhizomatous; flowers May to August, fruits mature August to September; reproduces from seeds and occasionally from rhizomes

FLORAL AND FRUIT CHARACTERISTICS

inflorescences: **flowers in pairs**; pedicellate, peduncles (1–3 cm long) **on axillary branches**; branches purplish, glandular-pilose; gland tips often reddish-purple

flowers: perfect, regular; sepals 5, imbricate, ovate (6–11 mm long), tips setose; petals 5, imbricate, **white to pink, with purple veins**, obovate to obcordate (1–1.8 cm long), pilose at the bases for one-half their length on the inner surface; stamens 10

fruits: schizocarps; mericarps 5 (3–4.5 mm long); carpel stylar portions remaining attached at the apices, free portions coiling outward; style branches somewhat pubescent

VEGETATIVE CHARACTERISTICS

leaves: **opposite**, simple; **basal blades palmately divided into 3–7 divisions** (5–30 cm long, 4–15 cm wide), cleft three-fourths of their length or more; main divisions incised, toothed or lobed; **sparsely strigose adaxially** and on veins abaxially; petioles long (2–28 cm long); upper blades reduced, petioles shorter (2–12 cm long); with stipules

stems: erect to ascending, usually glandular-pubescent, but may be glabrous to rarely strigulose below; glands usually purple

HISTORICAL, FOOD, AND MEDICINAL USES: some Cheyenne pulverized leaves into a powder and snuffed it into their noses to control nosebleeds, roots were also powdered and made into a drink

LIVESTOCK LOSSES: none

FORAGE VALUES: poor to fair for sheep and deer, poor to worthless for cattle, rarely eaten by horses; worthless for livestock after reaching maturity

HABITATS: shaded and moist wooded hillsides, ravines, mountains, and foothills; most abundant in rich soils

Wax currant
Ribes cereum Douglas

SYN = *Ribes reniforme* Nutt.

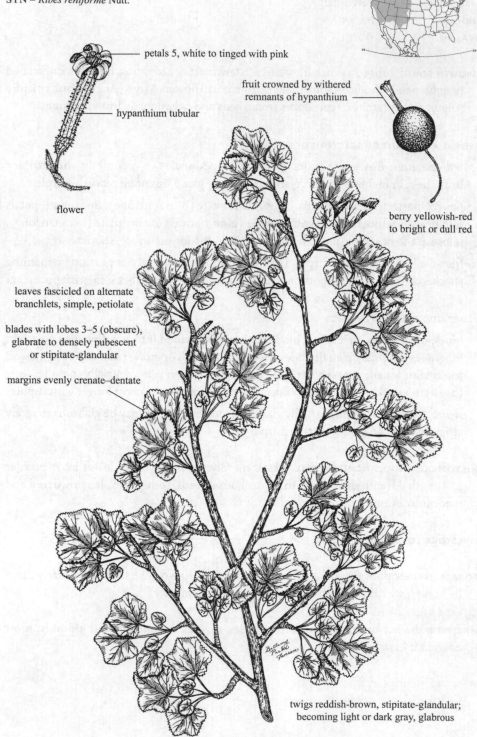

petals 5, white to tinged with pink

hypanthium tubular

flower

fruit crowned by withered remnants of hypanthium

berry yellowish-red to bright or dull red

leaves fascicled on alternate branchlets, simple, petiolate

blades with lobes 3–5 (obscure), glabrate to densely pubescent or stipitate-glandular

margins evenly crenate–dentate

twigs reddish-brown, stipitate-glandular; becoming light or dark gray, glabrous

FAMILY:	GROSSULARIACEAE
SPECIES:	*Ribes cereum* Douglas
COMMON NAME:	Wax currant (capulincillo, western redcurrant)
LIFE SPAN:	Perennial
ORIGIN:	Native
SEASON:	Cool

GROWTH FORM: shrubs (to 2 m tall); much-branched; erect to spreading; branches crooked, flowers May to July, fruits ripen July to August, reproduces from seeds

FLORAL AND FRUIT CHARACTERISTICS

inflorescences: flowers solitary or in corymbiform racemes; raceme flowers 2–8; inconspicuous, drooping

flowers: perfect, regular; **hypanthiums tubular** (6–9 mm long, 2–3.5 mm in diameter); calyces narrowly cylindrical, sepals 5; sepals ovate, spreading or reflexed (1.5–3 mm long, 1–1.9 mm wide), white or greenish-white and tinged with pink, pubescent and stipitate-glandular; petals 5 (1.5–2 mm long, 1.2–2 mm wide), white to tinged with pink; stamens 5, shorter than the petals

fruits: **berries**; spherical to ovoid (5–10 mm in diameter), **yellowish-red to bright or dull red**, smooth to slightly glandular-hairy, seeds few to several; **crowned by withered remnants of the hypanthiums**

VEGETATIVE CHARACTERISTICS

leaves: fascicled on alternate branchlets, simple; blades reniform to orbiculate (5–25 mm long, 1–4 cm wide); **lobes 3–5, obscure**; bases truncate to subcordate; **margins evenly crenate-dentate**; glabrate to densely pubescent or stipitate-glandular on both surfaces, waxy with age; petiolate (5–20 mm long)

stems: twigs stiff; young twigs reddish-brown, stipitate-glandular; eventually becoming light or dark gray (sometimes white), glabrous; lenticels transverse; aromatic

HISTORICAL, FOOD, AND MEDICINAL USES: some Native Americans ate the tasteless fresh berries or dried the berries to be later eaten with animal fat, plant extracts were used to alleviate stomachache; currently used in jellies

LIVESTOCK LOSSES: none

FORAGE VALUES: poor to fair for livestock, fair to good for wildlife; often abundant, therefore, a rather important browse plant; fruits valuable source of food for many species of birds and small mammals

HABITATS: prairies, open slopes, hills, ridges, and openings in woodlands; most abundant in dry, rocky or sandy soils

Range ratany
Krameria erecta Willd.

SYN = *Krameria imparata* (J.F. Macbr.) Britton, *Krameria interior* Rose &
    J.H. Painter, *Krameria navae* Rzed., *Krameria palmeri* Rose, *Krameria
    rosmarinifolia* Pavon *ex* Chodat

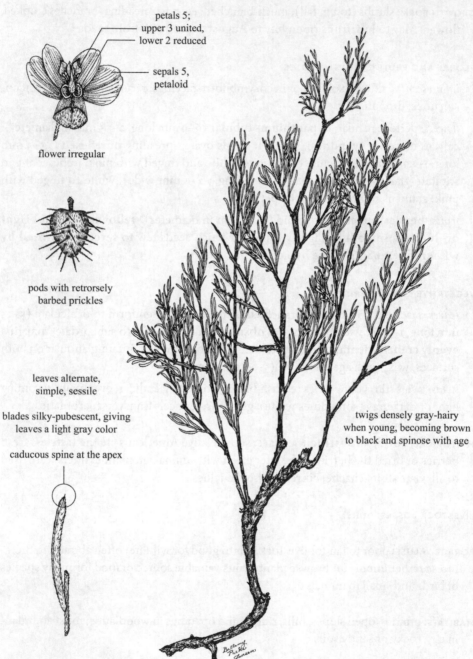

petals 5;
upper 3 united,
lower 2 reduced

sepals 5,
petaloid

flower irregular

pods with retrorsely
barbed prickles

leaves alternate,
simple, sessile

blades silky-pubescent, giving
leaves a light gray color

caducous spine at the apex

twigs densely gray-hairy
when young, becoming brown
to black and spinose with age

378

FAMILY:	KRAMERIACEAE
SPECIES:	*Krameria erecta* Willd.
COMMON NAME:	Range ratany (cósahui, tamichil, little ratany, littleleaf ratany)
LIFE SPAN:	Perennial
ORIGIN:	Native
SEASON:	Cool

GROWTH FORM: shrubs (to 70 cm tall); intricately branched, crowns rounded to irregular; flowers April to May and may flower a second time in late summer, reproduces from seeds

FLORAL AND FRUIT CHARACTERISTICS

inflorescences: flowers solitary; terminal and axillary, on glandular-hairy peduncles (4–9 mm long)

flowers: perfect, irregular; sepals 5 (sometimes 4), petaloid, unequal (5–10 mm long), ascending, oblong to oblanceolate to lanceolate, **strigose**; petals 5, oblong to obovate (2–5 mm long), upper 3 (4–5 mm long) united to form a short claw, remaining 2 reduced (2–3 mm long), elliptical, **purple**, showy, fragrant

fruits: pods; ovoid to globose (6–9 mm long), indehiscent, thick-walled, silky-hairy, **burlike with retrorsely barbed prickles; prickles slender (3–4 mm long), often curved, purplish or reddish**; seeds 1

VEGETATIVE CHARACTERISTICS

leaves: alternate, simple; **blades linear to oblanceolate** (3–15 mm long, 1–2 mm wide), caducous spine at the apices; margins entire; **silky-pubescent** giving leaves a light gray color; sessile; stipules absent

stems: twigs slender, stiff; much-branched above, spreading to divergent, **densely gray-hairy when young**, becoming brown to black and **spinose with age**

other: twigs give a bluish tint to the landscape where these plants are abundant

HISTORICAL, FOOD, AND MEDICINAL USES: some Native Americans made a decoction from the leaves to use as an eyewash, for diarrhea, and for sores; source for red and brown dyes; roots were used in manufacturing ink

LIVESTOCK LOSSES: none

FORAGE VALUES: fair to good for cattle, sheep, goats, and mule deer; furnishes valuable browse but cannot withstand heavy browsing because branches are brittle; burlike fruits are readily disseminated by livestock, important food for small mammals

HABITATS: foothills, mesas, hillsides, and plains; adapted to a broad range of soil textures; most abundant in dry, gravelly soils; least abundant in clay soils

Scarlet globemallow
Sphaeralcea coccinea (Nutt.) Rydb.

SYN = *Cristaria coccinea* (Nutt.) Pursh, *Malva coccinea* Nutt., *Malvastrum coccineum* (Nutt.) A. Gray, *Malvastrum cockerellii* A. Nelson, *Malvastrum dissectum* (Nutt. *ex* Torr. & A. Gray) Cockerell, *Sida coccinea* (Nutt.) DC., *Sida dissecta* Nutt. *ex* Torr. & A. Gray

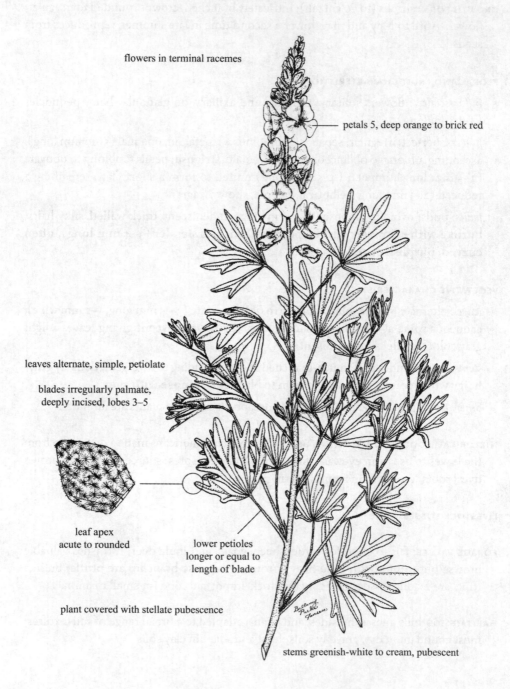

flowers in terminal racemes

petals 5, deep orange to brick red

leaves alternate, simple, petiolate

blades irregularly palmate, deeply incised, lobes 3–5

leaf apex acute to rounded

lower petioles longer or equal to length of blade

plant covered with stellate pubescence

stems greenish-white to cream, pubescent

FAMILY:	MALVACEAE
SPECIES:	*Sphaeralcea coccinea* (Nutt.) Rydb.
COMMON NAME:	Scarlet globemallow (hierba del negro, red falsemallow)
LIFE SPAN:	Perennial
ORIGIN:	Native
SEASON:	Warm

GROWTH FORM: forbs (stems 10–80 cm long); ascending to decumbent, stems simple or clustered from woody caudices that arise from deep horizontal roots; flowers April to August, reproduces from seeds; persists during dry periods by shedding leaves

FLORAL AND FRUIT CHARACTERISTICS

inflorescences: racemes; broadly or narrowly ovate (2–10 cm long), terminal; flowers several to many

flowers: perfect, regular; calyces conspicuously villous (3–10 mm long), persistent; sepals 5, connate, narrowly triangular to ovate, acuminate; petals 5 (1–2 cm long), **deep orange to brick red**, drying pinkish, emarginate; stamens numerous, united into a column

fruits: schizocarps; mericarps 10 or more (3–3.4 mm long and in diameter), differentiated into a smooth dehiscent portion at the apices and a roughened indehiscent basal portion; basal portion prominently reticulate, black tuberculate; seeds 1

VEGETATIVE CHARACTERISTICS

leaves: alternate, simple; blades **suborbicular to ovate** or deltate in outline (1–6 cm long, wider than long), **irregularly palmate, deeply incised, lobes 3–5 (rarely more)**; divisions again lobed and incised; segment apices acute to rounded, final segment oblong to obovate or oblanceolate to spatulate; bases cuneate; petioles of the lower leaves longer or equal (1–8 cm long) to the length of the blades

stems: decumbent or ascending to erect, greenish-white to cream, pubescent

other: entire plant is covered with stellate pubescence

HISTORICAL, FOOD, AND MEDICINAL USES: some Blackfoot chewed the leaves and stems and applied the paste to burns, scalds, and external sores as a cooling agent

LIVESTOCK LOSSES: none

FORAGE VALUES: excellent for deer and pronghorn, worthless to fair for cattle, fair to good for sheep; important forage in Southwest and West but infrequently grazed in the Great Plains; increases in abundance with heavy use and during dry periods

HABITATS: prairies, plains, hills, roadsides, and waste places; adapted to a broad range of soils, especially prevalent in sandy soils

Fireweed
Chamaenerion angustifolium (L.) Scop.

SYN = *Chamerion angustifolium* (L.) Holub, *Epilobium angustifolium* L.

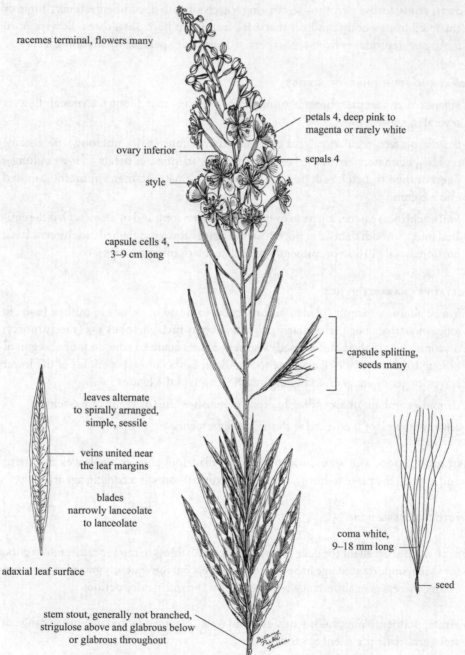

racemes terminal, flowers many

petals 4, deep pink to
magenta or rarely white

ovary inferior

sepals 4

style

capsule cells 4,
3–9 cm long

capsule splitting,
seeds many

leaves alternate
to spirally arranged,
simple, sessile

veins united near
the leaf margins

blades
narrowly lanceolate
to lanceolate

coma white,
9–18 mm long

adaxial leaf surface

seed

stem stout, generally not branched,
strigulose above and glabrous below
or glabrous throughout

FAMILY:	ONAGRACEAE
SPECIES:	*Chamaenerion angustifolium* (L.) Scop.
COMMON NAME:	Fireweed (willowherb, blooming Sally, great willowherb)
LIFE SPAN:	Perennial
ORIGIN:	Native
SEASON:	Warm

GROWTH FORM: forbs (0.3–2.5 m tall); erect; flowers June to September, reproduces from seeds and rhizomes (up to 8 m long)

FLORAL AND FRUIT CHARACTERISTICS

inflorescences: racemes; terminal, elongate, lax, drooping in bud, flowers many, spreading at anthesis; bracts resembling leaves

flowers: perfect, irregular; hypanthiums absent; sepals 4, narrowly lanceolate (7–16 mm long, 1.6–2.5 mm wide), acute, often purple-tinged, canescent; petals 4, slightly asymmetrical, obovate (8–20 mm long, 6–11 mm wide) tapering to a short claw, deep pink to magenta or rarely white; stamens 8, subequal, shorter than petals; stigma lobes 4, long and slender; ovary inferior

fruits: capsules; cells 4 (3–9 cm long), splitting into 4 valves; ascending to erect; canescent, often purplish; seeds many, fusiform (0.9–1.4 mm long); **coma white** (9–18 mm long); pedicellate

VEGETATIVE CHARACTERISTICS

leaves: alternate to spirally arranged, simple; **blades narrowly lanceolate to lanceolate** (2–20 cm long, 5–40 mm wide); apices acuminate; bases acute to acuminate; margins subentire or obscurely denticulate; midveins strong abaxially, glabrous or with strigulose hairs; **veins united near the leaf margins**; sessile to subsessile

stems: stout, becoming indurate at the base, **generally not branched**; glabrous below, throughout, or strigulose on upper stem

HISTORICAL, FOOD, AND MEDICINAL USES: used as a potherb, young shoots can be cooked like asparagus, young leaves used in salads and steeped for tea, pith of stems can be used to flavor and thicken stews and soups; important honey plant; grown as an ornamental

LIVESTOCK LOSSES: none

FORAGE VALUES: fair to good for sheep; poor to fair for cattle; grazed to a minor extent by horses, deer, moose, bighorn sheep, and elk; becomes unpalatable with maturity

HABITATS: open woodlands, along streams, roadsides, and disturbed areas; adapted to dry and moist soils, grows in a broad range of soil types; especially abundant following fire

Pinyon pine
Pinus edulis Engelm.

SYN = *Caryopitys edulis* (Engelm.) Small, *Pinus cembroides* Zucc., *Pinus monophylla* Torr. & Frém

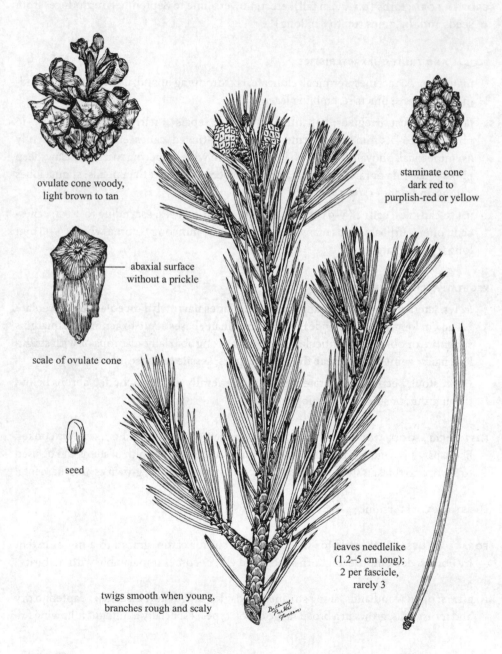

ovulate cone woody,
light brown to tan

staminate cone
dark red to
purplish-red or yellow

abaxial surface
without a prickle

scale of ovulate cone

seed

leaves needlelike
(1.2–5 cm long);
2 per fascicle,
rarely 3

twigs smooth when young,
branches rough and scaly

FAMILY:	PINACEAE
SPECIES:	*Pinus edulis* Engelm.
COMMON NAME:	Pinyon pine (pino piñonero, piñon pine, nut pine)
LIFE SPAN:	Perennial
ORIGIN:	Native
SEASON:	Evergeen

GROWTH FORM: trees (to 20 m tall), monoecious; crowns pyramidal, becoming round-topped with age; branches persistent to near trunk bases; flowers April to June, reproduces from seeds

CONE CHARACTERISTICS

cones: unisexual, in clusters at ends of branches; staminate cones ovoid (6–8 mm long), dark red to purplish-red or yellow, in clusters of 20–40; ovulate cones ovoid (2–6 cm long, 2–5 cm in diameter), light brown to tan, purplish to brown at maturity, sessile

mature cones: ovulate cones woody, scales thickened near apices, tips incurved, **without prickles** (compare to *Pinus ponderosa*), seeds borne in cavities at the base of middle scales, maturing the first season; seeds ovate (9–16 mm long), thick-shelled, wingless

VEGETATIVE CHARACTERISTICS

leaves: needlelike, in fascicles arranged in spirals; **needles 2 per fascicle** (rarely 3) (compare to *Pinus ponderosa*); **needles linear (1.2–5 cm long)**, often curved upward, persisting 4–6 years, aromatic; apices acuminate; margins entire; abaxially rounded, with whitish lines, new growth bluish-green turning yellowish-green; **fascicle sheaths mostly deciduous** (4–7 mm long), pale; sessile

stems: twigs smooth when young; branches rough and scaly; bark thin, gray to reddish-brown or nearly black with age; trunks frequently crooked and twisted; bark irregularly furrowed with small scales, yellowish- to reddish-brown, resinous

HISTORICAL, FOOD, AND MEDICINAL USES: seeds (nuts) are harvested, sold, and used in making candies, cakes, and cookies; seeds were a staple food in some Native American diets and were eaten raw, roasted, or ground into flour; needles were steeped for tea; inner bark served as starvation food for some Native Americans; wood is used for fuel and fence posts

LIVESTOCK LOSSES: none

FORAGE VALUES: worthless to cattle and sheep; seeds important wildlife food for birds, small mammals, black bears, mule deer, and goats

HABITATS: mountain slopes, foothills, mesas, and plateaus; most abundant on dry and rocky soils

Ponderosa pine
Pinus ponderosa P. Lawson & C. Lawson

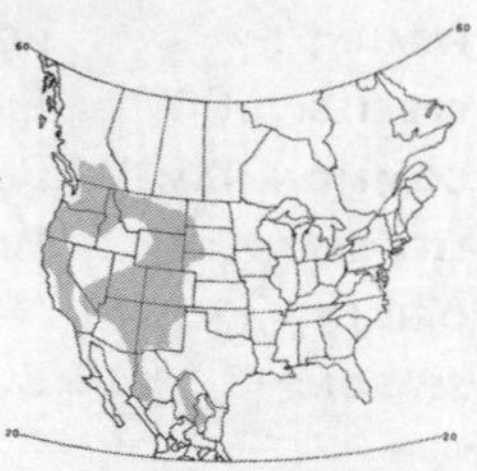

SYN = *Pinus arizonica* Engelm.

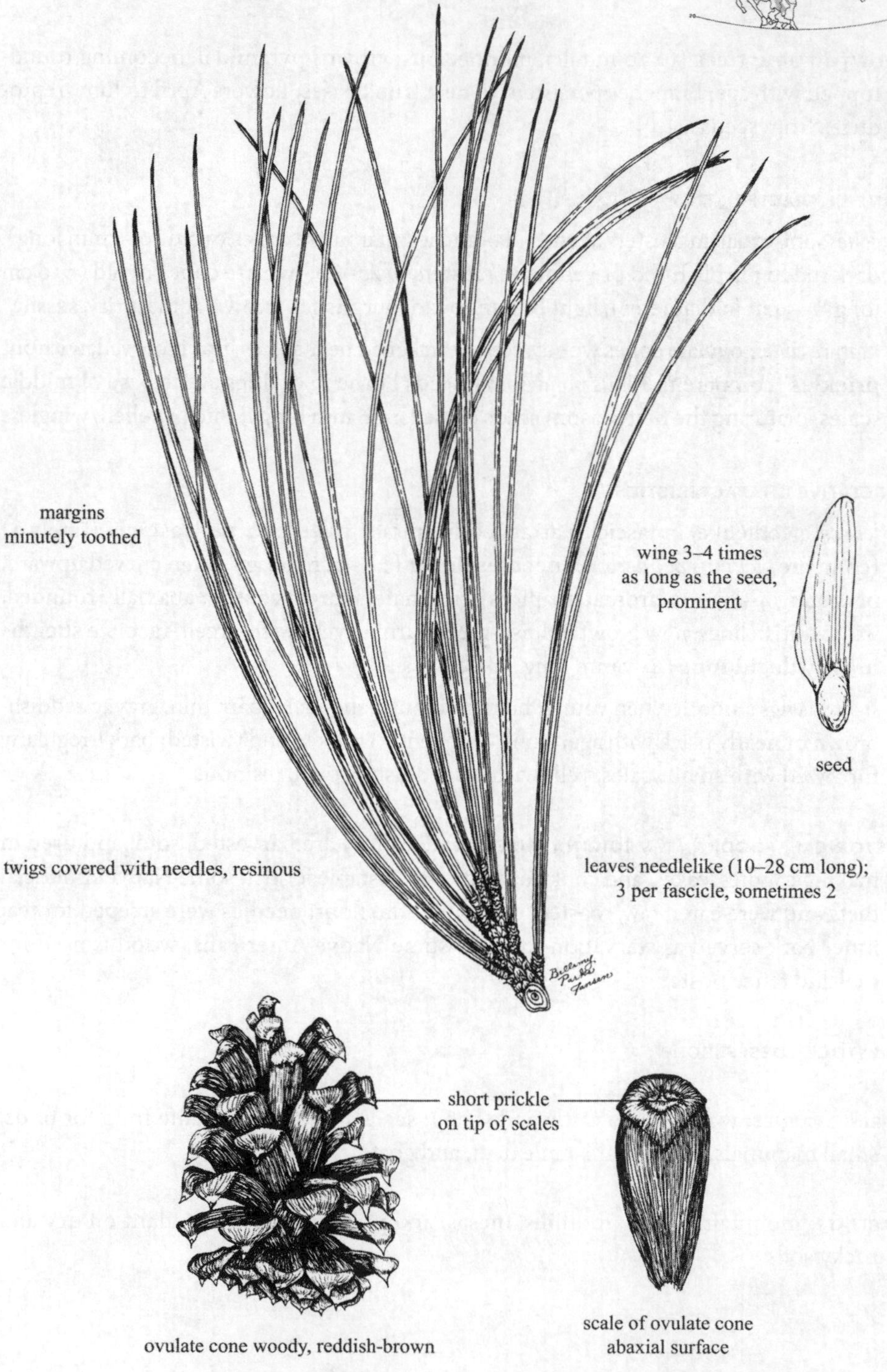

margins
minutely toothed

wing 3–4 times
as long as the seed,
prominent

seed

twigs covered with needles, resinous

leaves needlelike (10–28 cm long);
3 per fascicle, sometimes 2

short prickle
on tip of scales

ovulate cone woody, reddish-brown

scale of ovulate cone
abaxial surface

FAMILY:	PINACEAE
SPECIES:	*Pinus ponderosa* P. Lawson & C. Lawson
COMMON NAME:	Ponderosa pine (pino ponderosa, pino real, western yellowpine)
LIFE SPAN:	Perennial
ORIGIN:	Native
SEASON:	Evergreen

GROWTH FORM: trees (to 70 m tall); crowns open, pyramidal when immature and becoming round- to flat-topped with age; branches spreading to ascending or descending; mature trees generally without branches on the lower trunks; flowers April to June, reproduces from seeds

CONE CHARACTERISTICS

cones: unisexual; staminate cones cylindrical (1.5–3.5 cm long, 6–8 mm in diameter), yellowish-orange to deep purple, in clusters of 10–20; ovulate cones in clusters or pairs at base of new growth, broadly ovoid (6–15 cm long, 6–9 cm in diameter), reddish-brown, woody, sessile

mature cones: ovulate cones woody, tip of scales rounded and **often with a short prickle** (compare to *Pinus edulis*); seeds 2 per scale, maturing the summer of the second season; seeds ellipsoid to obovoid, dark brown to purplish-mottled (6–9 mm long), with a prominent papery wing 3–4 times as long as the seed

VEGETATIVE CHARACTERISTICS

leaves: needlelike, in fascicles arranged in spirals; **typically needles 3 per fascicle** (sometimes 2) (compare to *Pinus edulis*), fascicles clustered or whorled near the tips of the branches; **needles linear** (10–28 cm long, 1–1.5 mm wide), resinous, aromatic, one-third to one-half round in cross section; apices abruptly acuminate; margins minutely toothed; yellowish-green to dark green; fascicle sheaths dark brown to black, persisting 4–6 years; sessile

stems: twigs covered with needles and occasionally with old lanceolate leaf scales, resinous; buds brown; bark rough, thick, with deep fissures, **scales yellowish- to reddish-brown, large with intricate margins**; trunk straight

HISTORICAL, FOOD, AND MEDICINAL USES: some Native Americans used cones to help start fires; one of the principal lumber trees in North America; wood used commercially for boxes, crates, construction, and mill products; sometimes planted as an ornamental and in shelterbelts

LIVESTOCK LOSSES: browsing on needles may cause abortion in cattle

FORAGE VALUES: worthless to livestock; seeds are eaten by several species of birds and small mammals; browsed by deer, elk, and bighorn sheep

HABITATS: rocky hillsides, mountains, and plateaus; grows in moist or dry soils and in a wide range of soil types

Blue penstemon
Penstemon glaber Pursh

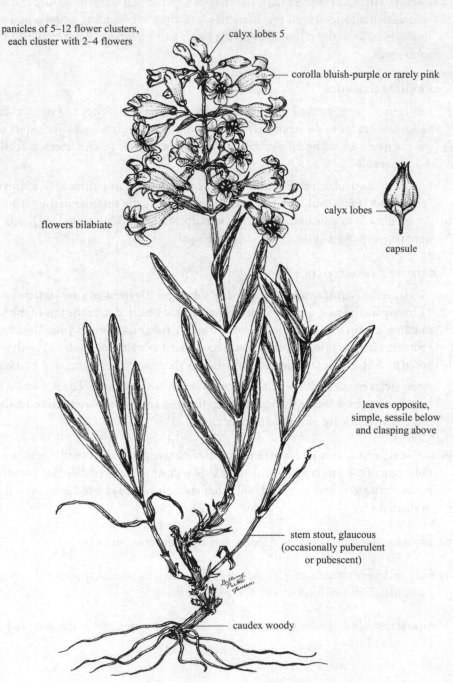

panicles of 5–12 flower clusters,
each cluster with 2–4 flowers

calyx lobes 5

corolla bluish-purple or rarely pink

calyx lobes

capsule

flowers bilabiate

leaves opposite,
simple, sessile below
and clasping above

stem stout, glaucous
(occasionally puberulent
or pubescent)

caudex woody

FAMILY:	PLANTAGINACEAE
SPECIES:	*Penstemon glaber* Pursh
COMMON NAME:	Blue penstemon (sawsepal penstemon, smooth beardtongue)
LIFE SPAN:	Perennial
ORIGIN:	Native
SEASON:	Cool

GROWTH FORM: forbs (15–80 cm tall); erect to ascending from woody caudices; flowers June to September, seeds mature August to October, reproduces from seeds

FLORAL AND FRUIT CHARACTERISTICS

inflorescences: panicles (8–30 cm long); flowers in 5–12 clusters, congested, each floral cluster with 2–4 flowers

flowers: perfect, **bilabiate**; calyx lobes 5, lanceolate to lance-ovate, rounded or short-acuminate (2–10 mm long, 1.5–4 mm wide), margins scarious and erose, glabrous to puberulent; corollas bluish-purple or rarely pink (2.5–4 cm long), pale internally and lined on the interior with reddish-purple nectar guides, lobes rounded; sterile filaments bearded, exserted

fruits: **capsules**; ovoid (1–1.5 cm long), **valves 2**, thin, pliable, brown at maturity; seeds several

VEGETATIVE CHARACTERISTICS

leaves: **opposite**, simple; basal leaves wanting or highly reduced, oblanceolate, apices acute to obtuse, occasionally mucronate; cauline leaves linear-lanceolate to lanceolate (3–15 cm long, 7–45 mm wide), thick, apices acute to obtuse, margins entire; glabrous (rarely pubescent), **glaucous; sessile below and clasping above**

stems: ascending, 1 to many, stout, glaucous (occasionally puberulent or pubescent)

HISTORICAL, FOOD, AND MEDICINAL USES: some Native Americans made a wet dressing from the leaves to treat snakebite, others made tea from the leaves to stop vomiting; grown as an ornamental

LIVESTOCK LOSSES: may accumulate selenium, but has not been substantiated to cause poisoning

FORAGE VALUES: fair for deer, pronghorn, elk, and sheep, becomes less palatable with maturity; worthless for cattle; seeds are eaten by birds and small mammals

HABITATS: plains, open woodlands, hills, and mountains; most abundant in sandy to gravelly soils

Woolly plantain
Plantago patagonica Jacq.

SYN = *Plantago purshii* Roem. & Schult.

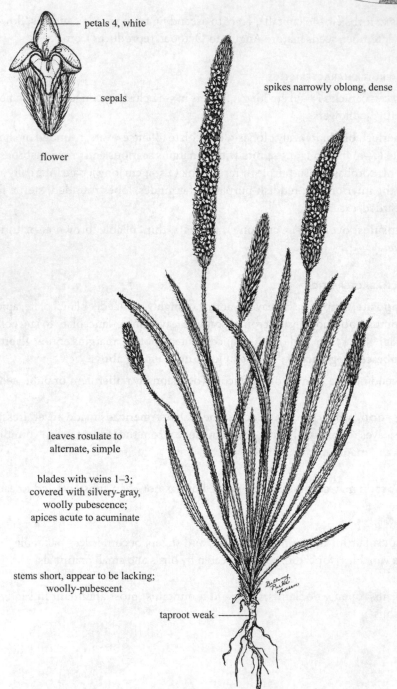

petals 4, white

sepals

flower

spikes narrowly oblong, dense

leaves rosulate to
alternate, simple

blades with veins 1–3;
covered with silvery-gray,
woolly pubescence;
apices acute to acuminate

stems short, appear to be lacking;
woolly-pubescent

taproot weak

FAMILY:	PLANTAGINACEAE
SPECIES:	*Plantago patagonica* Jacq.
COMMON NAME:	Woolly plantain (lanté, peludilla, woolly indianwheat, tallowweed)
LIFE SPAN:	Annual
ORIGIN:	Native
SEASON:	Cool

GROWTH FORM: forbs (5–35 cm tall), appearing acaulescent from weak taproots; forming winter rosettes; starts growth in autumn or winter in southern regions, flowers May to August, reproduces from seeds

FLORAL AND FRUIT CHARACTERISTICS

inflorescences: spikes; narrowly oblong (1–15 cm long, less than 1 cm wide), 1–20 per plant, extending above the foliage, **cylindrical, dense, borne on scapes** (to 30 cm long); scapes erect

flowers: perfect, regular; sepals narrowly obovate (1.4–2.5 mm long), margins scarious; petals 4 (1–2 mm long), suborbicular to ovate-lanceolate, spreading, white; stamens 4, included to slightly exserted

fruits: capsules (3–4 mm long); dehiscent at or just below the middle; seeds 2, brownish-black

other: **silvery-gray villous throughout**

VEGETATIVE CHARACTERISTICS

leaves: mostly basal, rosulate to alternate, simple; blades of rosettes oblanceolate (5–30 cm long); principal blades linear to oblanceolate (2–20 cm long, 0.5–15 mm wide), ascending, apices acute to acuminate, gradually tapering to the petiolar bases, veins 1–3; margins entire; **covered with silvery-gray woolly pubescence, hairs capillary**

stems: short, appear to be lacking, mostly unbranched, woolly pubescent

HISTORICAL, FOOD, AND MEDICINAL USES: some Native Americans chewed and swallowed leaves for internal hemorrhage, leaves were also chewed for toothache

LIVESTOCK LOSSES: none

FORAGE VALUES: good for sheep and poor to fair for cattle and wildlife; a major forage species on lambing ranges, most important in the Southwest where it is often regarded as a spring opportunist, abundance is generally considered an indicator of rangeland deterioration

HABITATS: prairies, pastures, waste places, and roadsides; most abundant in sandy soils

Low larkspur
Delphinium bicolor Nutt.

SYN = *Plectrornis bicolor* (Nutt.) Lunell

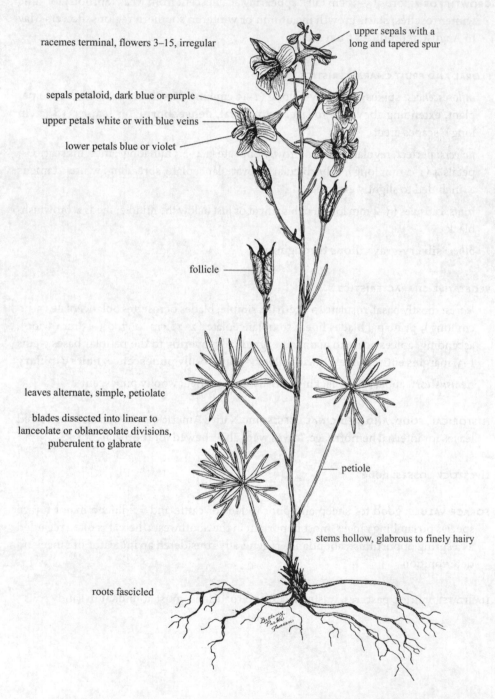

racemes terminal, flowers 3–15, irregular

upper sepals with a
long and tapered spur

sepals petaloid, dark blue or purple

upper petals white or with blue veins

lower petals blue or violet

follicle

leaves alternate, simple, petiolate

blades dissected into linear to
lanceolate or oblanceolate divisions,
puberulent to glabrate

petiole

stems hollow, glabrous to finely hairy

roots fascicled

FAMILY:	RANUNCULACEAE
SPECIES:	*Delphinium bicolor* Nutt.
COMMON NAME:	Low larkspur (little larkspur)
LIFE SPAN:	Perennial
ORIGIN:	Native
SEASON:	Cool

GROWTH FORM: forbs (**10–60 cm tall**), erect to weak-stemmed, usually unbranched; from a branching cluster of fascicled roots in early spring; flowers May to July, seeds mature June to July, reproduces from seeds

FLORAL AND FRUIT CHARACTERISTICS

inflorescences: racemes; oblong (usually less than 15 cm long), terminal, flowers usually 3–15; pedicels spreading

flowers: perfect, irregular; sepals 5, petaloid; lower sepals (1.5–2 cm long) dark blue or purple (rarely white or pink), lobes wavy; uppermost sepals with a long and tapered spur (1–2 cm long); corollas of 2 sets of 2 petals, white or with blue veins; petals smaller than the sepals; upper pair of petals prolonged at bases into the spurs; lower pair of petals clawed, blue or violet, concealing the stamens, sinuses absent; stamens numerous

fruits: follicles (1.5–2.5 cm long); usually divergent, brown, viscid-pubescent to glabrous, seeds many

VEGETATIVE CHARACTERISTICS

leaves: alternate, simple; leaves mostly basal to evenly spaced, usually 3–6; blades orbicular (2–4 cm wide), **dissected into linear to lanceolate or oblanceolate divisions**; puberulent to **glabrate**; lowest petioles long, abruptly reduced above

stems: hollow, glabrous to finely hairy

HISTORICAL, FOOD, AND MEDICINAL USES: some Native Americans crushed plants of this genus and applied it to their hair to control lice and other insects; used as an ornamental

LIVESTOCK LOSSES: poisonous to cattle throughout growth cycle, contains alkaloids which act on the nervous system, death may result following paralysis of breathing; bloat is common; seeds are the most poisonous part

FORAGE VALUES: fair to good for sheep and some wildlife; palatable to cattle even though other plants are available, unpalatable to horses

HABITATS: open woodlands, roadsides, hills, and meadows; most abundant on dry soils

Tall larkspur
Delphinium occidentale (S. Watson) S. Watson *ex* J.M. Coult.

SYN = *Delphinium cucullatum* A. Nelson

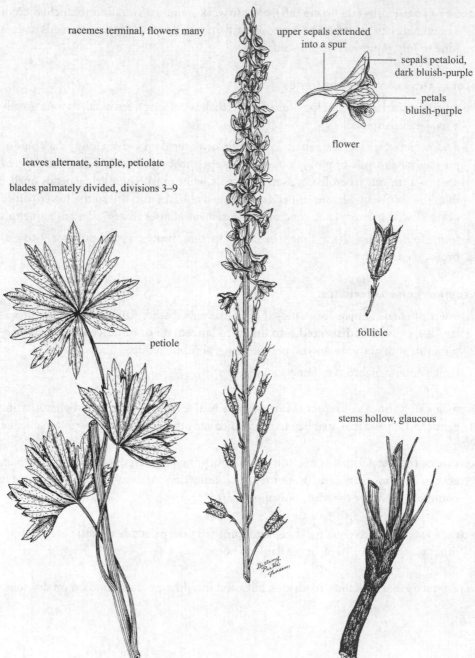

racemes terminal, flowers many

upper sepals extended
into a spur

sepals petaloid,
dark bluish-purple

petals
bluish-purple

flower

leaves alternate, simple, petiolate

blades palmately divided, divisions 3–9

petiole

follicle

stems hollow, glaucous

FAMILY:	RANUNCULACEAE
SPECIES:	*Delphinium occidentale* (S. Watson) S. Watson *ex* J.M. Coult.
COMMON NAME:	Tall larkspur (duncecap larkspur)
LIFE SPAN:	Perennial
ORIGIN:	Native
SEASON:	Warm

GROWTH FORM: forbs (**0.6–2 m tall**); erect from a taproot; growth begins in late spring; flowers July to August, seeds mature from August to September, reproduces from seeds

FLORAL AND FRUIT CHARACTERISTICS

inflorescences: racemes; narrowly oblong (usually more than 15 cm long), dense, spiciform to loosely paniculiform with age, terminal, flowers many, central axis glandular-hairy

flowers: perfect, irregular; sepals 5, petaloid, dark bluish-purple (rarely white); upper sepals with a spur (9–12 mm long), straight or curved near the tips; lower sepals ovate-oblong (6–12 mm long), apices rounded or acute, light or usually gray-canescent on the back; corollas of 2 sets of 2 petals; upper pair of petals bluish-purple, spurlike; lower pair of petals with broad, wavy-margined lobes, bluish-purple, sinuses 1–2 mm long; stamens numerous

fruits: follicles; short-oblong (9–12 mm long), glabrous to glandular-pubescent, seeds many

VEGETATIVE CHARACTERISTICS

leaves: alternate, simple; blades orbicular (10–18 cm long, 8–12 cm wide), **palmately divided into 3–9 divisions, divisions cleft**; pubescent to puberulent; petiolate

stems: hollow, somewhat stramineous at the base, glaucous

HISTORICAL, FOOD, AND MEDICINAL USES: some Native Americans crushed plants of this genus and applied to their hair to control lice and other insects; used as an ornamental

LIVESTOCK LOSSES: poisonous to cattle, chance of poisoning decreases after seed dispersal, contains alkaloids which act on the nervous system, death may result following paralysis of breathing, bloat is common; seeds are the most poisonous part

FORAGE VALUES: fair to good for sheep and some wildlife; palatable to cattle even though other plants are available, seldom eaten by horses

HABITATS: meadows, thickets, stream banks, and open woodlands; most abundant where the snowpack lasts the longest

Wedgeleaf buckbrush
Ceanothus cuneatus (Hook.) Nutt.

SYN = *Ceanothus ramulosus* (Greene) McMinn

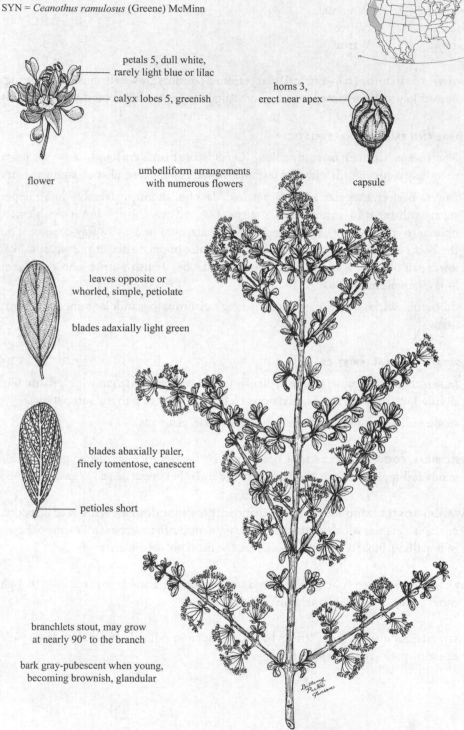

petals 5, dull white,
rarely light blue or lilac

calyx lobes 5, greenish

horns 3,
erect near apex

flower

umbelliform arrangements
with numerous flowers

capsule

leaves opposite or
whorled, simple, petiolate

blades adaxially light green

blades abaxially paler,
finely tomentose, canescent

petioles short

branchlets stout, may grow
at nearly 90° to the branch

bark gray-pubescent when young,
becoming brownish, glandular

FAMILY:	RHAMNACEAE
SPECIES:	*Ceanothus cuneatus* (Hook.) Nutt.
COMMON NAME:	Wedgeleaf buckbrush (raciliullo, narrowleaf buckbrush, wedgeleaf ceanothus)
LIFE SPAN:	Perennial
ORIGIN:	Native
SEASON:	Warm

GROWTH FORM: shrubs (to 4 m tall); branches rigid and divaricate, may form dense thickets; flowers March to May, reproduces from seeds; seeds germinate best following fire

FLORAL AND FRUIT CHARACTERISTICS

inflorescences: **umbelliform arrangements** (1–2.5 cm wide); borne on spurlike axillary branchlets; flowers many; pedicels slender

flowers: perfect, regular; calyx lobes 5, united below into a cuplike base, bent inward above, somewhat greenish; petals 5, dull white (rarely light blue or lilac), pipe-shaped, long-clawed; stamens 5, opposite the petals; styles clefts 3

fruits: capsules; subglobose (5–6 mm in diameter), lobes 3, cells 3, **with 3 erect horns near the apices**; separating into 3 carpels each with 1 seed

other: flowers fragrant

VEGETATIVE CHARACTERISTICS

leaves: **opposite** or whorled, simple; blades oblong to cuneate to broadly obovate (5–20 mm long); apices obtuse or rounded; **bases cuneate**; margins entire to finely serrate at apices; coriaceous; adaxially light green, glabrous; **abaxially paler, finely tomentose, canescent**; petioles short (1–5 mm long)

stems: branchlets stout, **may grow at nearly 90° to the branch**; bark gray-pubescent when young, becoming brownish; glandular

other: leaves and twigs glandular, producing a balsamlike fragrance

HISTORICAL, FOOD, AND MEDICINAL USES: leaves and flowers may be boiled for tea; some Native Americans made an infusion from bark for a tonic; fresh flowers when crushed and rubbed in water make a perfumed, cleansing lather; branches were used to build fish dams

LIVESTOCK LOSSES: unsubstantiated reports of kidney injury to animals browsing this species

FORAGE VALUES: fair browse for sheep, goats, and deer; browsed by cattle only when other forage is unavailable; seeds eaten by small mammals and birds

HABITATS: ridges, slopes, semiarid valleys, and open rocky sites; most abundant in gravelly or other well-drained soils

Fendler ceanothus
Ceanothus fendleri A. Gray

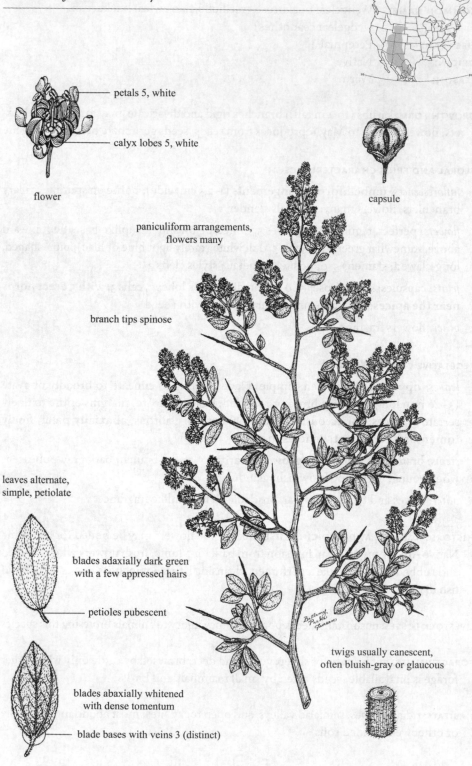

petals 5, white

calyx lobes 5, white

flower

capsule

paniculiform arrangements,
flowers many

branch tips spinose

leaves alternate,
simple, petiolate

blades adaxially dark green
with a few appressed hairs

petioles pubescent

blades abaxially whitened
with dense tomentum

blade bases with veins 3 (distinct)

twigs usually canescent,
often bluish-gray or glaucous

Bellamy
Parker
Jansson

FAMILY:	RHAMNACEAE
SPECIES:	*Ceanothus fendleri* A. Gray
COMMON NAME:	Fendler ceanothus (deerbriar, Fendler soapbloom)
LIFE SPAN:	Perennial
ORIGIN:	Native
SEASON:	Warm

GROWTH FORM: shrubs (to 1 m tall); stems clustered, loosely branched, forming low thickets; tardily deciduous; flowers April to October, fruits mature June to November, reproduces from seeds

FLORAL AND FRUIT CHARACTERISTICS

inflorescences: **paniculiform arrangements**, terminating main stem and branches, flowers many

flowers: perfect, regular; calyx lobes 5, white, ovate (1.5 mm long and wide), acute, glabrous, sharply incurved; petals 5, white, spreading, pipe-shaped (1.5–1.8 mm long), clawed; stamens 5, opposite the petals; style clefts 3 (1 mm long), white

fruits: capsules; subglobose (3–4 mm high, 5–6 mm diameter); surface rough, brown, lobes 3, cells 3, separating into 3 carpels each with 1 seed; capsule lobes slightly keeled, horns absent

VEGETATIVE CHARACTERISTICS

leaves: alternate, simple; blades ovate to elliptical (1–2.5 cm long, 2–10 mm wide); apices acute to rounded, mucronulate; bases cuneate; **veins 3, distinct**; margins usually entire, occasionally serrate; adaxially dark green with a few appressed hairs; abaxially whitened with dense tomentum; petioles pubescent (3–5 mm long)

stems: **twigs usually canescent**, often bluish-gray or glaucous; bark of old stems reddish-brown, often with blisterlike glands; **branch tips spinose**

HISTORICAL, FOOD, AND MEDICINAL USES: some Native Americans extracted a dye from the leaves and used the fruits for food; tea can be made from the dried leaves

LIVESTOCK LOSSES: none

FORAGE VALUES: good and important browse for goats, deer, and other wildlife; fair to good for cattle and sheep, frequently browsed by horses; highly variable in value between different areas in North America; heavily utilized by porcupines and jackrabbits in summer months

HABITATS: hillsides, open valleys, open woodlands, and rocky ledges in foothills and mountains; most abundant in well-drained soils

Deerbrush
Ceanothus integerrimus Hook. & Arn.

SYN = *Ceanothus andersonii* Parry, *Ceanothus californicus* Kellogg

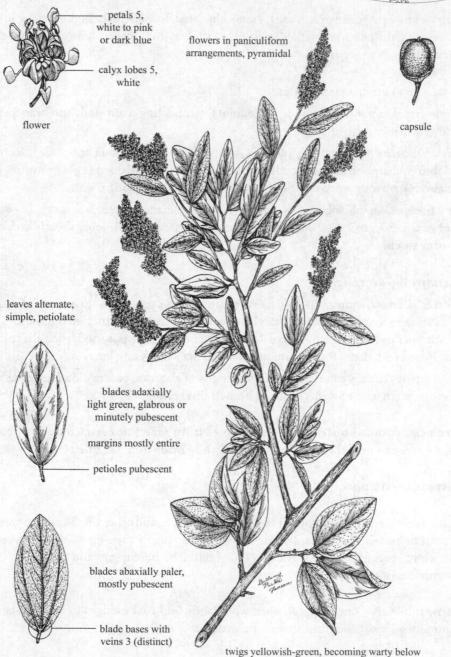

petals 5,
white to pink
or dark blue

calyx lobes 5,
white

flower

flowers in paniculiform
arrangements, pyramidal

capsule

leaves alternate,
simple, petiolate

blades adaxially
light green, glabrous or
minutely pubescent

margins mostly entire

petioles pubescent

blades abaxially paler,
mostly pubescent

blade bases with
veins 3 (distinct)

twigs yellowish-green, becoming warty below

FAMILY:	RHAMNACEAE
SPECIES:	*Ceanothus integerrimus* Hook. & Arn.
COMMON NAME:	Deerbrush (lila de California, bluebrush, mountain lilac, sweet birch, soapbush)
LIFE SPAN:	Perennial
ORIGIN:	Native
SEASON:	Warm

GROWTH FORM: shrubs (to 4 m tall, often broader than tall); widely and loosely branched, branches slender, often drooping; crowns rounded to irregular; flowers May to July, reproduces from seeds and rootstocks; seeds germinate best following fire

FLORAL AND FRUIT CHARACTERISTICS

inflorescences: **paniculiform arrangements** (4–15 cm long), **pyramidal, terminating branches**, showy; flowers many

flowers: perfect, regular; calyx lobes 5, united below, white, curved sharply between the petals; petals 5, white to pink or dark blue, pipe-shaped, long-clawed; stamens 5, opposite the petals; style clefts 3 near the apices

fruits: capsules; globose to widely ovate, viscid, **lobes 3**, cells 3, separating into 3 carpels each with 1 seed; oblong glands common on the backs of each capsule lobe, sometimes indistinct

other: flowers fragrant

VEGETATIVE CHARACTERISTICS

leaves: alternate, simple; blades ovate to oblong-ovate (2–8 cm long, 1–4 cm wide); apices acute to rounded; bases rounded; **veins 3, distinct**; margins mostly entire or occasionally denticulate near the apices; adaxially light green, glabrous or minutely pubescent; abaxially paler, mostly pubescent; petioles pubescent (3–15 mm long)

stems: twigs slender, often slightly drooping, yellowish-green, becoming warty below

HISTORICAL, FOOD, AND MEDICINAL USES: extract from the bark was used by some Native Americans as a tonic and for making a soapy lather, flexible stems were used in basket making; valuable for honey production

LIVESTOCK LOSSES: none

FORAGE VALUES: good to excellent for cattle, sheep, goats, and deer; fair to good for horses; seeds are an important source of food for small mammals; quail and other birds eat the seeds

HABITATS: mountain slopes and ridges; adapted to a broad range of soils, most abundant in well-drained and fertile soils; grows in the open or in partial shade, does not tolerate full shade

Buckbrush
Ceanothus velutinus Douglas

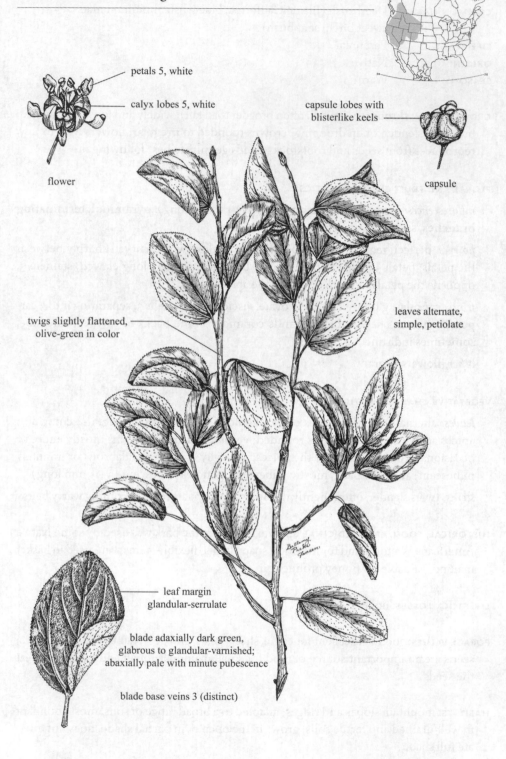

petals 5, white

calyx lobes 5, white

capsule lobes with
blisterlike keels

flower

capsule

twigs slightly flattened,
olive-green in color

leaves alternate,
simple, petiolate

leaf margin
glandular-serrulate

blade adaxially dark green,
glabrous to glandular-varnished;
abaxially pale with minute pubescence

blade base veins 3 (distinct)

FAMILY:	RHAMNACEAE
SPECIES:	*Ceanothus velutinus* Douglas
COMMON NAME:	Buckbrush (snowbrush, tobaccobrush, mountain balm)
LIFE SPAN:	Perennial
ORIGIN:	Native
SEASON:	Evergreen

GROWTH FORM: shrubs (to 2 m tall); several to many stems from the bases, diffusely spreading, often procumbent, forming dense colonies, crowns often rounded; flowers May to July, fruits mature in August and September, reproduces from seeds

FLORAL AND FRUIT CHARACTERISTICS

inflorescences: paniculiform arrangements (3–10 cm long, 3–4 cm wide); main branches 1–5, each with a lanceolate bract near the base, axes pubescent; on axillary and terminal peduncles (1–4 cm long); usually drying to chocolate-brown

flowers: perfect, regular; calyx lobes 5, united below, white, triangular (1.5–2 mm long, 1.5 mm wide), curved sharply between the petals; petals 5, white, pipe-shaped (2–3 mm long), long-clawed, usually curved downward; stamens 5, opposite the petals; style clefts 3 near the apices

fruits: capsules; subglobose (3–4 mm high, 5–6 mm diameter), sticky-glandular, lobes 3, cells 3, separating into 3 carpels each with 1 seed, each capsule lobe with a blisterlike keel

VEGETATIVE CHARACTERISTICS

leaves: alternate, simple; blades broadly ovate to elliptical (2.5–8 cm long, 1.5–5 cm wide); apices obtuse; bases rounded or subcordate, **veins 3, distinct; margins glandular-serrulate; coriaceous**; adaxially dark green, glabrous to glandular-varnished; abaxially pale, with minute pubescence; petiolate (5–21 mm long)

stems: twigs rigid, slightly flattened, olive-green, **unarmed** (without spines or thorns), puberulent; bark reddish-brown

other: plants produce a **strong cinnamonlike or balsamlike odor**

HISTORICAL, FOOD, AND MEDICINAL USES: some Native Americans used the leaves as a tobacco substitute

LIVESTOCK LOSSES: none

FORAGE VALUES: generally not considered a browse species, although it is occasionally browsed by deer and elk in winter; goats, sheep, and deer may consume flowers

HABITATS: open woodlands and mountain slopes; most abundant in rocky soils on logged or burned areas

Chamise
Adenostoma fasciculatum Hook. & Arn.

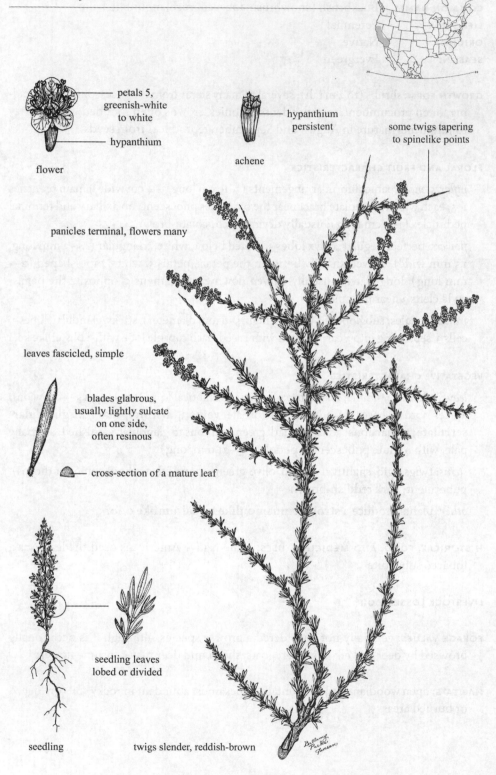

petals 5,
greenish-white
to white

hypanthium

flower

hypanthium
persistent

achene

some twigs tapering
to spinelike points

panicles terminal, flowers many

leaves fascicled, simple

blades glabrous,
usually lightly sulcate
on one side,
often resinous

cross-section of a mature leaf

seedling leaves
lobed or divided

seedling

twigs slender, reddish-brown

FAMILY:	ROSACEAE
SPECIES:	*Adenostoma fasciculatum* Hook. & Arn.
COMMON NAME:	Chamise (chamizo, greasewood, hierba del pasmo)
LIFE SPAN:	Perennial
ORIGIN:	Native
SEASON:	Evergreen

GROWTH FORM: shrubs (to 3.5 m tall); diffusely branched, forming dense thickets; growth starts in January and ends after flowering in June, reproduces from basal sprouts and seeds, seedlings rapidly establish following fire

FLORAL AND FRUIT CHARACTERISTICS

inflorescences: panicles; terminal (4–12 cm long), crowded, flowers many

flowers: perfect, regular; hypanthiums obconical (1–2 mm long), striate, veins 10, with glands inside the margins; sepals 5; petals 5, greenish-white to white, rounded, spreading; stamens 10–15, in groups of 2–3, alternating with the petals; sessile or nearly so

fruits: achenes; obovoid, small, hard, 1 per flower; contained within the persistent hypanthiums

VEGETATIVE CHARACTERISTICS

leaves: **fascicled, simple; blades subulate** to spatulate (4–15 mm long); margins entire; **glabrous, usually lightly sulcate on one side**, often resinous, rigid; stipules minute; **seedling leaves lobed** or divided

stems: twigs slender, stiff, some tapering to spinelike points, **reddish-brown**; old bark grayish- or reddish-brown, exfoliating as fibers with age

other: thick canopies and numerous roots provide erosion control

HISTORICAL, FOOD, AND MEDICINAL USES: used by some Native Americans for various ceremonial purposes

LIVESTOCK LOSSES: none

FORAGE VALUES: mature plants are worthless to cattle, sheep, and horses; largely unpalatable; fair browse for deer in winter; seedlings are palatable to cattle and sheep

HABITATS: mountain slopes, ridges, and foothills; chaparral; adapted to a broad range of soils, common on infertile soils; most abundant at elevations of 200–1800 m; a known fire hazard

Serviceberry
Amelanchier alnifolia (Nutt.) Nutt. *ex* M. Roem.

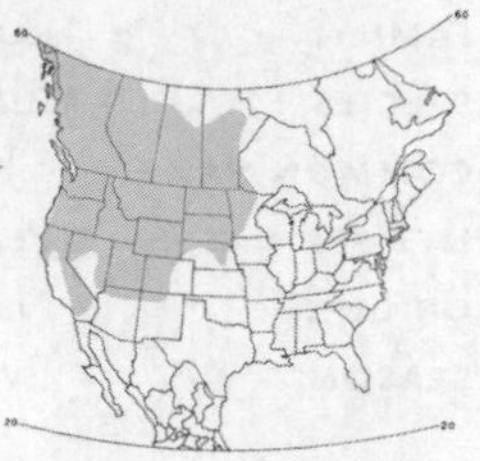

SYN = *Amelanchier canadensis* (L.) Medik., *Amelanchier sanguinea* (Pursh) DC., *Amelanchier utahensis* Koehne

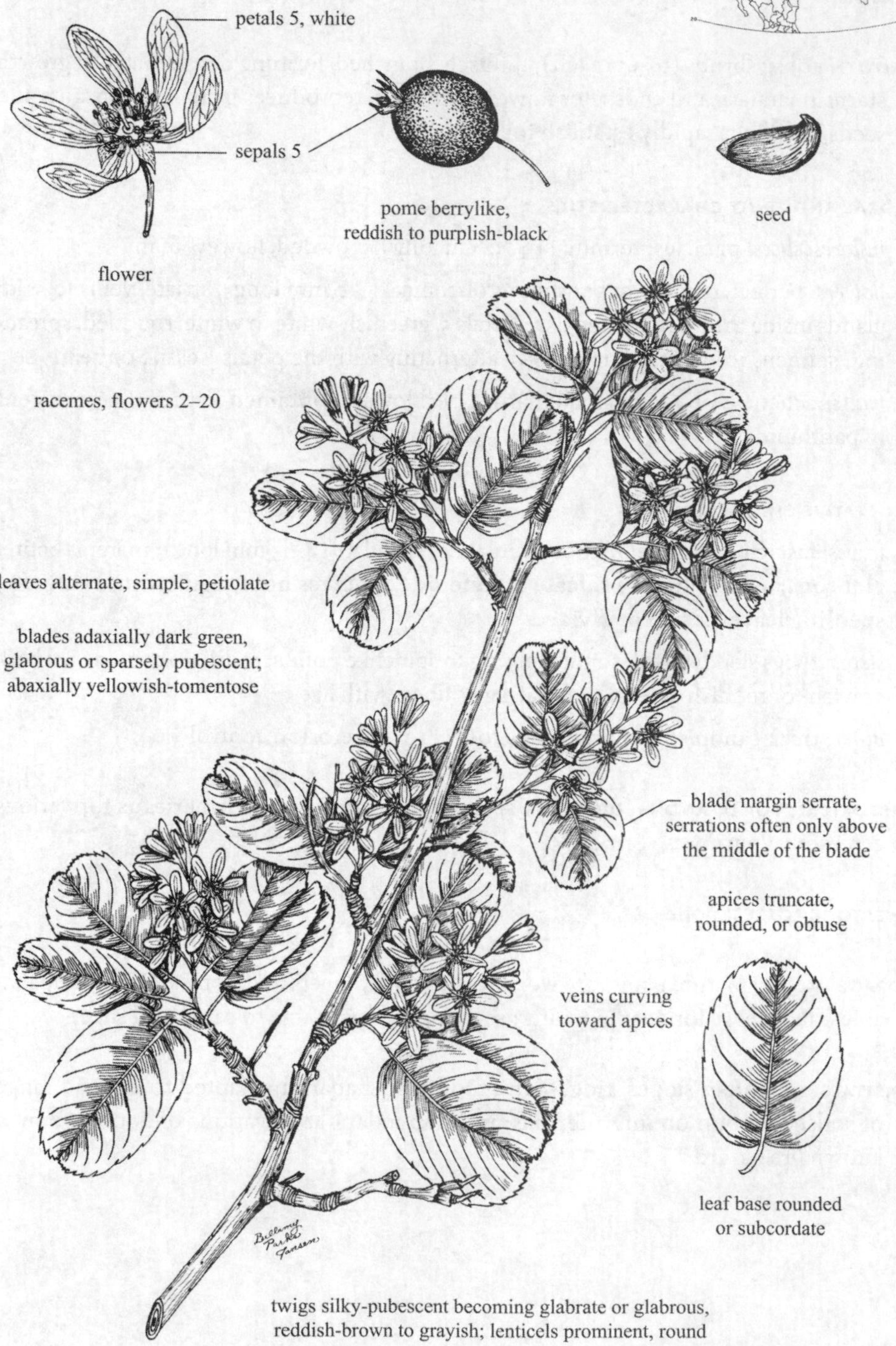

petals 5, white

sepals 5

pome berrylike, reddish to purplish-black

seed

flower

racemes, flowers 2–20

leaves alternate, simple, petiolate

blades adaxially dark green, glabrous or sparsely pubescent; abaxially yellowish-tomentose

blade margin serrate, serrations often only above the middle of the blade

apices truncate, rounded, or obtuse

veins curving toward apices

leaf base rounded or subcordate

twigs silky-pubescent becoming glabrate or glabrous, reddish-brown to grayish; lenticels prominent, round

FAMILY:	ROSACEAE
SPECIES:	*Amelanchier alnifolia* (Nutt.) Nutt. *ex* M. Roem.
COMMON NAME:	Serviceberry (Saskatoonberry, Saskatoon serviceberry, juneberry)
LIFE SPAN:	Perennial
ORIGIN:	Native
SEASON:	Cool

GROWTH FORM: shrubs or small trees (to 10 m tall); single or clustered trunks with branches near the bases; flowers April to June, fruits mature in July and August, reproduces from seeds and stolons, forming colonies that may become thickets

FLORAL AND FRUIT CHARACTERISTICS

inflorescences: racemes; widely ovate to ovate (2–4 cm long), erect, terminal on new growth, flowers 2–20; axes silky-pubescent, becoming glabrous or glabrate

flowers: perfect, regular; hypanthiums campanulate (3–4 mm in diameter), brownish-white; sepals 5, triangular (1.5–4 mm long), reflexed at anthesis; petals 5 (5–16 mm long, 2–3 mm wide), white, obovate to spatulate, spreading; apices rounded or acute; stamens 10–20

fruits: pomes; berrylike, globose to obovoid (8–15 mm long, 7–13 mm in diameter), fleshy, reddish at first to purplish-black with age, seeds 3–10

other: flowers are ill-scented

VEGETATIVE CHARACTERISTICS

leaves: alternate, simple; blades oval or obovate to oblong (2–6 cm long, 2.5–4 cm wide); **apices truncate or rounded or obtuse**; bases rounded or subcordate; margins serrate, **serrations often only above the middle, main lateral veins curving toward apices, entering the teeth** (compare to *Cercocarpus montanus*); adaxially dark green, glabrous or sparsely pubescent; abaxially yellowish-tomentose; petioles (7–20 mm long); stipules narrowly lanceolate (1–3 mm long), caducous

stems: young twigs silky-pubescent, becoming glabrate or glabrous and reddish-brown to grayish, smooth, rigid; lenticels round

HISTORICAL, FOOD, AND MEDICINAL USES: fresh fruit is dry, mealy, and not very palatable; fruit can be used to make jams, pies, and wines; some Native Americans used stems for arrow shafts and tepee stakes

LIVESTOCK LOSSES: none

FORAGE VALUES: young growth is fair to good for livestock, excellent browse for deer, elk, and moose; fruits are an important food for small mammals, black bears, and birds; bark is eaten by beavers and marmots

HABITATS: brushy hillsides, open woodlands, canyons, and creek banks; usually growing in well-drained soils but occasionally grows around bogs

Curlleaf mountainmahogany
Cercocarpus ledifolius Nutt.

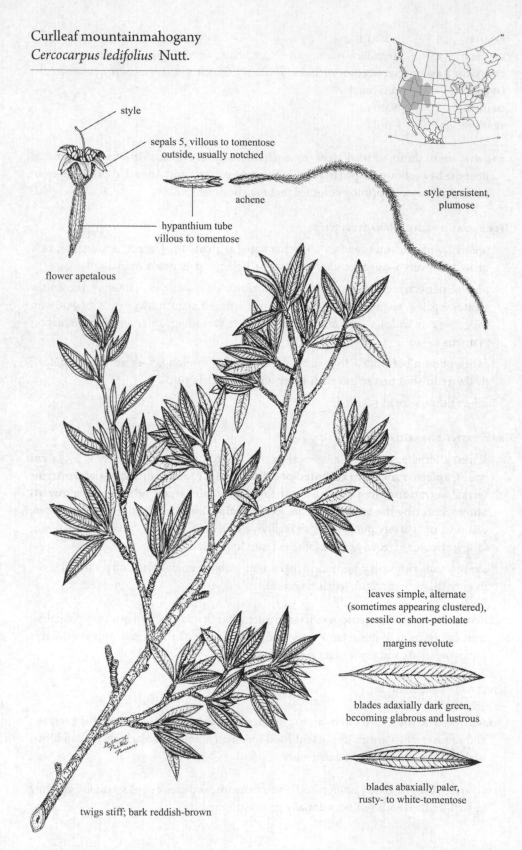

style

sepals 5, villous to tomentose outside, usually notched

achene

style persistent, plumose

hypanthium tube villous to tomentose

flower apetalous

leaves simple, alternate (sometimes appearing clustered), sessile or short-petiolate

margins revolute

blades adaxially dark green, becoming glabrous and lustrous

blades abaxially paler, rusty- to white-tomentose

twigs stiff; bark reddish-brown

FAMILY:	ROSACEAE
SPECIES:	*Cercocarpus ledifolius* Nutt.
COMMON NAME:	Curlleaf mountainmahogany (desert mahogany)
LIFE SPAN:	Perennial
ORIGIN:	Native
SEASON:	Evergreen

GROWTH FORM: shrubs or small trees (to 8 m tall); trunks 1 to several; characteristically grows in scattered patches; flowers May to July, reproduces from seeds

FLORAL AND FRUIT CHARACTERISTICS

inflorescences: flowers solitary or in clusters of 2–3; sessile or subsessile **in axils**

flowers: perfect, apetalous; hypanthium tubes (3–12 mm long) villous to tomentose; sepals 5, ovate to triangular (1–2.5 mm long), spreading, usually notched, **villous to tomentose outside (more hair than *Cercocarpus montanus*)**, glabrous within; stamens 10–30; pistils single, styles elongating in fruit

fruits: **achenes; tailed, solitary**, terete (5–9 mm long), hard, narrow, sharply pointed; styles terminal, elongate (4–8 cm long), exserted, persistent, plumose

VEGETATIVE CHARACTERISTICS

leaves: alternate (sometimes appearing clustered); blades evergreen, narrowly lanceolate to oblanceolate (1–4 cm long, 5–12 mm wide); apices acute; bases cuneate; **margins entire, revolute**; midveins prominent; coriaceous, resinous, aromatic; adaxially dark green, becoming glabrous and lustrous; **abaxially paler, rusty- to white-tomentose**; sessile or short-petiolate; petioles to 6 mm long

stems: twigs stiff; bark reddish-brown, finely villous or tomentose with white hairs, becoming gray and deeply sulcate with age

HISTORICAL, FOOD, AND MEDICINAL USES: wood is hard and dense (will not float), provides excellent fuel, producing intense heat and burning for long periods; the Gosiute of present-day Utah made bows from this wood

LIVESTOCK LOSSES: none

FORAGE VALUES: fair for sheep and cattle in autumn and winter, good for big game in winter, not readily eaten at other times; provides cover and some browse for deer, elk, bighorn sheep, and pronghorn

HABITATS: hills, canyons, rocky slopes, and rocky ridges at altitudes of 1200–3000 m; usually on south- and west-facing slopes; adapted to a wide range of soil textures, most abundant in dry coarse-textured soils

Birchleaf mountainmahogany
Cercocarpus montanus Raf.

SYN = *Cercocarpus breviflorus* A. Gray

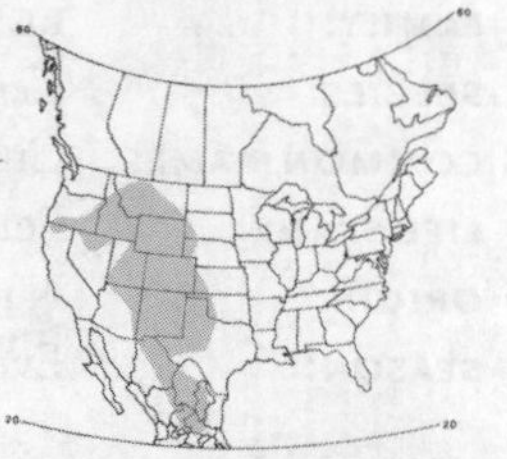

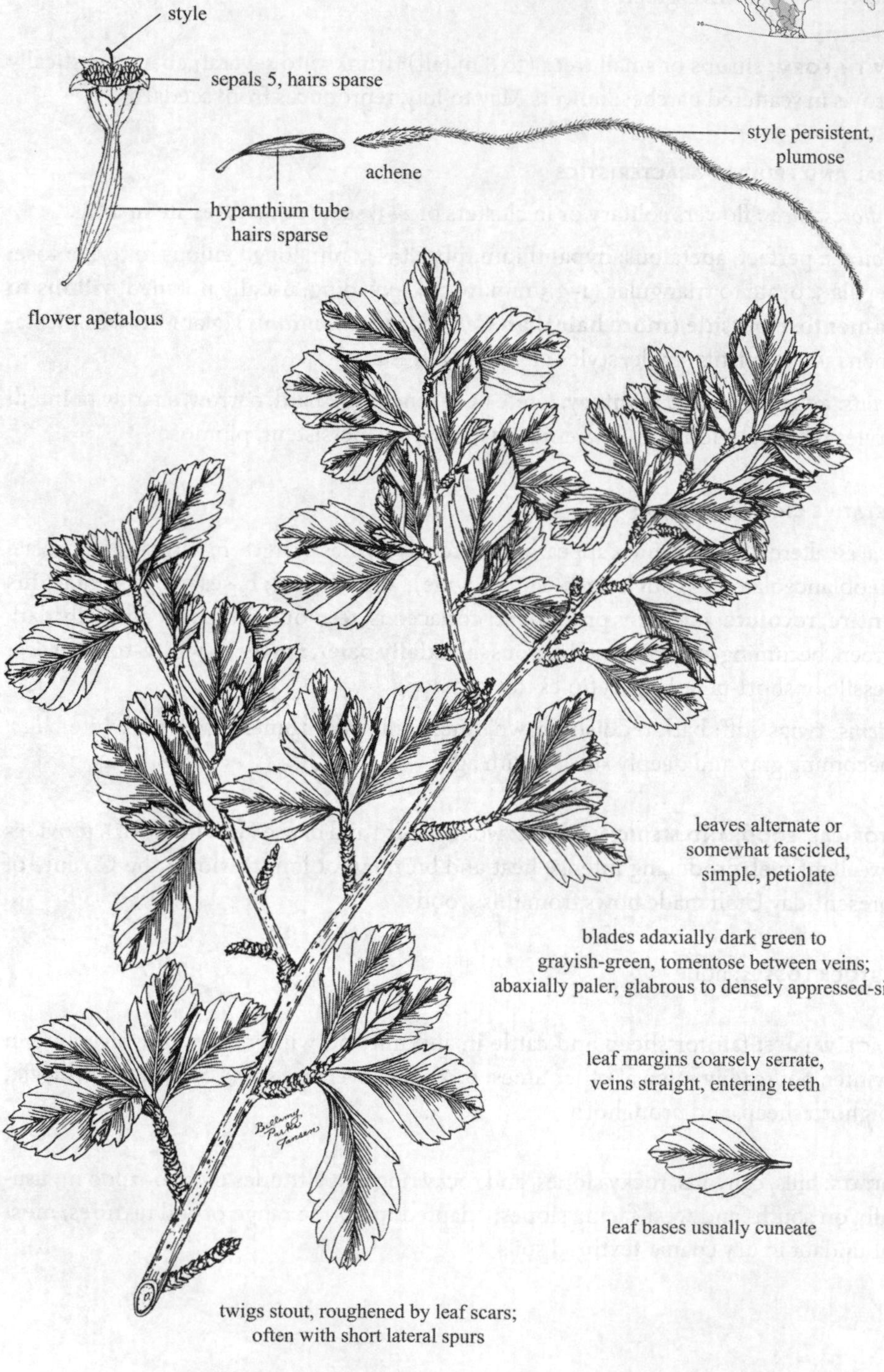

style

sepals 5, hairs sparse

achene

style persistent, plumose

hypanthium tube hairs sparse

flower apetalous

leaves alternate or somewhat fascicled, simple, petiolate

blades adaxially dark green to grayish-green, tomentose between veins; abaxially paler, glabrous to densely appressed-silky

leaf margins coarsely serrate, veins straight, entering teeth

leaf bases usually cuneate

twigs stout, roughened by leaf scars; often with short lateral spurs

FAMILY:	ROSACEAE
SPECIES:	*Cercocarpus montanus* Raf.
COMMON NAME:	Birchleaf mountainmahogany (palo duro, lintisco, true mountainmahogany)
LIFE SPAN:	Perennial
ORIGIN:	Native
SEASON:	Cool

GROWTH FORM: shrubs or small trees (to 6 m tall), much-branched; branches ascending or spreading; flowers June to July (some southern varieties flower as late as November), fruits usually mature in August, reproduces from seeds

FLORAL AND FRUIT CHARACTERISTICS

inflorescences: flowers solitary or clustered in groups of 2–3; axillary, sometimes crowded in groups of 5–15 on short spurlike branchlets

flowers: perfect, apetalous; inconspicuous; hypanthium tubes cylindrical (5–11 mm long), hairs sparse; sepals 5 (1–2 mm long, 3–5 mm wide), greenish-yellow, becoming reddish-brown, elongate; **hairs sparse, spreading-villous** or appressed-silky (less hairy than *Cercocarpus ledifolius*); stamens 22–44; pistils single, elongate in fruit

fruits: **achenes; tailed, solitary,** cylindrical-fusiform (8–12 mm long), appressed-silky; styles terminal, elongate (5–10 cm long), twisted, exserted, persistent, plumose

VEGETATIVE CHARACTERISTICS

leaves: alternate or somewhat fascicled at spur tips, simple; blades usually ovate to oval or obovate (2–5 cm long, 1.5–3.5 cm wide); apices acute to rounded; bases usually cuneate; margins coarsely serrate; adaxially dark green to grayish-green, tomentose between the veins; abaxially paler, varying from glabrous to densely appressed-silky; veins 3–10, prominent, straight, entering teeth of serration (compare to *Amelanchier alnifolia*); petioles short (3–6 mm long)

stems: twigs stout, rigid, roughened by leaf scars; often with numerous short lateral spurs; branches without elongated internodes; bark thin, young branches reddish-brown, older branches gray to brown

HISTORICAL, FOOD, AND MEDICINAL USES: some Native Americans used wood from these plants to make tools and war clubs, the Hopi used the bark to make a reddish-brown dye for leather

LIVESTOCK LOSSES: leaves may contain a cyanogenic glycoside, which can cause hydrocyanic acid poisoning

FORAGE VALUES: good to very good for cattle, sheep, and goats; extremely valuable winter browse for deer, elk, and bighorn sheep

HABITATS: rocky bluffs, mountain sides, canyons, rimrock, breaks, and open woodlands; most abundant in dry soils

Blackbrush
Coleogyne ramosissima Torr.

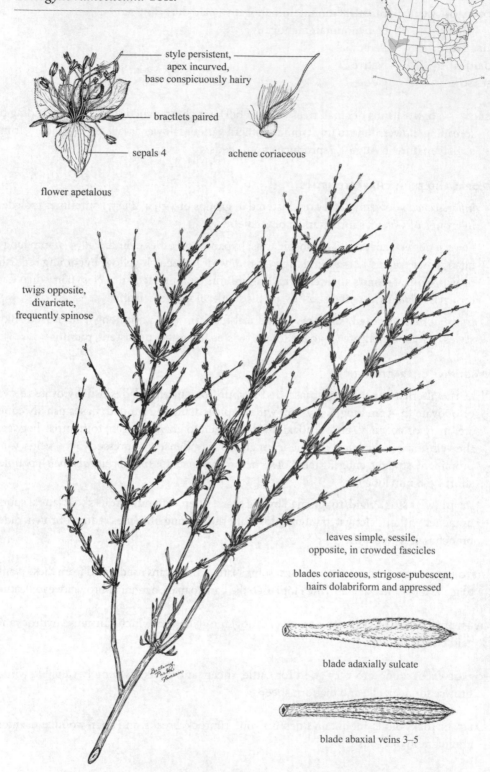

style persistent,
apex incurved,
base conspicuously hairy

bractlets paired

sepals 4

achene coriaceous

flower apetalous

twigs opposite,
divaricate,
frequently spinose

leaves simple, sessile,
opposite, in crowded fascicles

blades coriaceous, strigose-pubescent,
hairs dolabriform and appressed

blade adaxially sulcate

blade abaxial veins 3–5

412

FAMILY:	ROSACEAE
SPECIES:	*Coleogyne ramosissima* Torr.
COMMON NAME:	Blackbrush (burrobrush)
LIFE SPAN:	Perennial
ORIGIN:	Native
SEASON:	Evergreen

GROWTH FORM: shrubs (to 3 m tall); much-branched; branches opposite, short, rigid; forms nearly pure stands on large areas; flowers March to May, reproduces from seeds

FLORAL AND FRUIT CHARACTERISTICS

inflorescences: flowers solitary; subtended by 1–2 pairs of bractlets; bractlet lobes 3; terminating short branchlets

flowers: perfect, apetalous; hypanthium tubes very short, coriaceous; sepals 4 (8–12 mm long), arranged in 2 pairs, petaloid; adaxially yellow to orangish-yellow (sometimes greenish or brownish); **abaxially green to purplish**, strigose; 2 outer sepals lanceolate to oblong (4–7 mm long), inner ones ovate and acute (5–8 mm long), margins scarious; stamens 20–40; ovaries enclosed in tubular sheaths

fruits: achenes; somewhat compressed (3–5 mm long), glabrous, coriaceous, reddish-brown; with long, twisted, and threadlike styles from 1 side; styles persistent, apices incurved, **bases conspicuously hairy**; hair long, dense

VEGETATIVE CHARACTERISTICS

leaves: **opposite, in crowded fascicles**, simple; blades linear to oblanceolate (4–15 mm long, 1–5 mm wide); apices obtuse to mucronate; margins entire; coriaceous, **strigose-pubescent, hairs dolabriform and appressed**; adaxially sulcate, longitudinal, **abaxially veins** 3–5; sessile, bases often clasping

stems: twigs opposite and divaricate, frequently spinose at the tips; **bark gray to ashy-gray, becoming black with age, striate**

other: gray bark turns black when wet

HISTORICAL, FOOD, AND MEDICINAL USES: occasionally planted as an ornamental

LIVESTOCK LOSSES: none

FORAGE VALUES: fair for cattle, sheep, goats, elk, and deer during the winter; survives on heavily used rangelands because of the spiny character of the branches

HABITATS: desert mesas, open plains, and foothills in dry and well-drained soils; most abundant in sandy, gravelly, and rocky soils

Shrubby cinquefoil
Dasiphora fruticosa (L.) Rydb.

SYN = *Dasiphora floribunda* (Pursh) Raf., *Pentaphylloides fruticosa* (L.)
O. Schwarz, *Potentilla floribunda* Pursh, *Potentilla fruticosa* L.

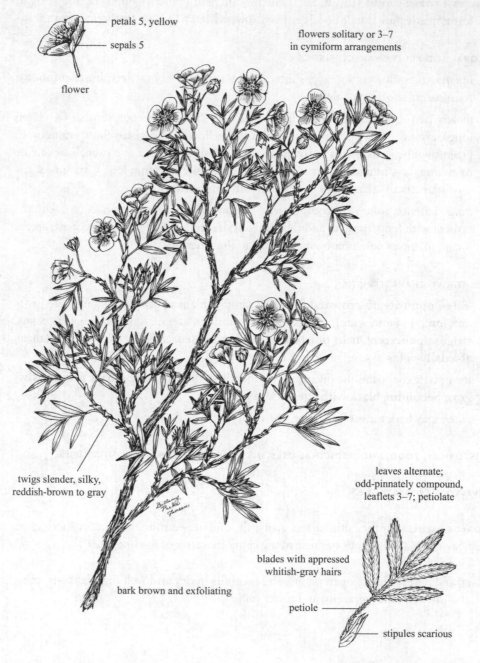

petals 5, yellow

sepals 5

flower

flowers solitary or 3–7
in cymiform arrangements

twigs slender, silky,
reddish-brown to gray

leaves alternate;
odd-pinnately compound,
leaflets 3–7; petiolate

bark brown and exfoliating

blades with appressed
whitish-gray hairs

petiole

stipules scarious

FAMILY:	ROSACEAE
SPECIES:	*Dasiphora fruticosa* (L.) Rydb.
COMMON NAME:	Shrubby cinquefoil (bush cinquefoil, yellow rose)
LIFE SPAN:	Perennial
ORIGIN:	Native
SEASON:	Cool (partially evergreen)

GROWTH FORM: shrubs (to 1.5 m tall); usually erect or ascending, much-branched, stems leafy; flowers April to June, seeds mature June to September, reproduces from seeds

FLORAL AND FRUIT CHARACTERISTICS

inflorescences: solitary or flowers 3–7 in cymiform arrangements, terminating branches or solitary flowers in leaf axils

flowers: perfect, regular; hypanthiums saucer-shaped, long-hairy; sepals 5, ovate (4–7 mm long), acuminate, alternating with 5 lanceolate and pubescent bractlets; petals 5 (5–15 mm long), yellow (rarely white), nearly orbicular; stamens 15–25; carpels numerous, angular, pubescent

fruits: achenes (1.5–2.5 mm long), light brown; pubescent; hairs ascending

VEGETATIVE CHARACTERISTICS

leaves: alternate, odd-pinnately compound; **leaflets 3–7, crowded,** upper 3 leaflets sometimes confluent at base, narrowly elliptical (5–20 mm long), tapering at each end; margins entire and often revolute; **appressed whitish-gray hairs, especially abaxially;** petioles variable (5–12 mm long); stipules broadly lanceolate to ovate, scarious

stems: twigs slender, silky and reddish-brown to gray; **bark later brown and exfoliating**

HISTORICAL, FOOD, AND MEDICINAL USES: steeped leaves were sometimes used as a tea by some Native Americans and pioneers; popular ornamental with many horticultural varieties; sometimes planted for erosion control

LIVESTOCK LOSSES: none

FORAGE VALUES: poor for cattle, good browse for sheep and goats in the Southwest and Intermountain areas; browsed extensively by elk and mule deer, less so by white-tailed deer

HABITATS: alpine meadows, bogs, rocky grounds at higher elevations (2000–3000 m), and subarctic regions; adapted to a broad range of soils, most abundant in moist soils

Apache plume

Fallugia paradoxa (D. Don) Endl. *ex* Torr.

SYN = *Fallugia micrantha* Cockerell, *Sieversia paradoxa* D. Don

petals 5, white

style persistent, elongated, plumose, pinkish to purplish or reddish with age

flowers solitary or 2–6 in corymbiform arrangements

achene

infructescence of achenes appearing as a plumose ball

twigs slender, white when young

leaves alternate and fascicled or clustered on short spurs, simple, sessile

blades pinnately divided; lobes 3–7 narrow and long; adaxially dark green; margins revolute; abaxially yellowish- to rust-tomentose;

bark dark, exfoliating in flakes

FAMILY:	ROSACEAE
SPECIES:	*Fallugia paradoxa* (D. Don) Endl. *ex* Torr.
COMMON NAME:	Apache plume (poñil, pluma de Apache, feather rose)
LIFE SPAN:	Perennial
ORIGIN:	Native
SEASON:	Cool

GROWTH FORM: shrubs (to 2.5 m tall); branchlets slender, crowns irregular, rhizomatous, forming thickets; flowers May to October, reproduces from seeds and rhizomes

FLORAL AND FRUIT CHARACTERISTICS

inflorescences: flowers solitary or 2–6 in corymbiform arrangements; peduncles elongated, nearly leafless

flowers: mostly unisexual, sometimes perfect, regular; hypanthiums cupulate; sepals 5, ovate, long-acuminate or trifid alternating with 5 linear-lanceolate or bifid bractlets; corollas showy (2.5–3.5 cm in diameter); **petals 5, white**, rounded (9–16 mm long), spreading; stamens numerous, in 3 series; style persistent

fruits: achenes; tailed, **numerous**, appearing as plumose balls, obovoid-fusiform (2.5–3.5 mm long); styles terminal, elongate (2.5–4 cm long), **pinkish to purplish or reddish with age**; achenes with plumose styles

VEGETATIVE CHARACTERISTICS

leaves: alternate and fascicled or clustered on short spurs, simple; blades obovate (1–2 cm long), **pinnately divided into 3–7 (sometimes 9) narrow and long lobes**; apices obtuse; margins entire and revolute, thick; adaxially dark green; **abaxially yellowish- to rust-tomentose**; midveins prominent; sessile; stipules triangular, small, persistent

stems: twigs slender; bark white when young, turning dark with age, **exfoliating in flakes**

other: mostly evergreen

HISTORICAL, FOOD, AND MEDICINAL USES: some Native Americans used bundles of twigs as brooms and older stems for arrow shafts; a decoction from leaves was used to stimulate hair growth; currently used as an ornamental and for erosion control

LIVESTOCK LOSSES: none

FORAGE VALUES: fair for cattle and goats, good winter forage for sheep; furnishes important browse for big game

HABITATS: arroyos, foothills, plains, and mesas in deserts and chaparral; most abundant in rocky or gravelly slopes and alluvial plains

Chokecherry
Prunus virginiana L.

SYN = *Cerasus virginiana* (L.) Michx., *Padus virginiana* (L.) Mill.

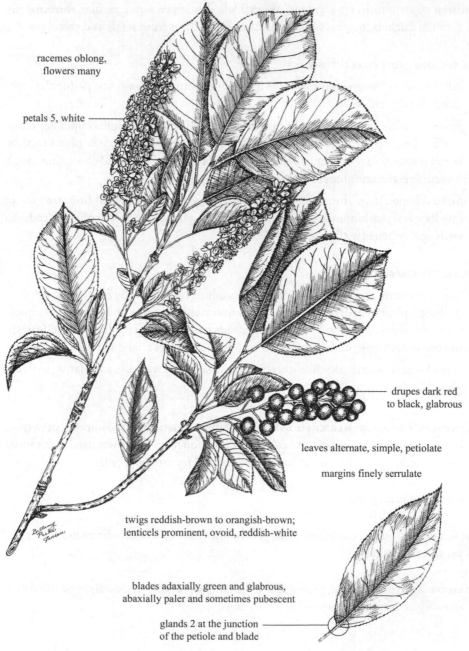

racemes oblong,
flowers many

petals 5, white

drupes dark red
to black, glabrous

leaves alternate, simple, petiolate

margins finely serrulate

twigs reddish-brown to orangish-brown;
lenticels prominent, ovoid, reddish-white

blades adaxially green and glabrous,
abaxially paler and sometimes pubescent

glands 2 at the junction
of the petiole and blade

418

FAMILY:	ROSACEAE
SPECIES:	*Prunus virginiana* L.
COMMON NAME:	Chokecherry (capulín, western chokecherry, black chokecherry)
LIFE SPAN:	Perennial
ORIGIN:	Native
SEASON:	Cool

GROWTH FORM: shrubs or small trees (to 10 m tall); erect with horizontal branches, rhizomatous, forming dense thickets; thicket crowns rounded; flowers April to July, fruits mature July to September; reproduces from seeds, rhizomes, and basal sprouts

FLORAL AND FRUIT CHARACTERISTICS

inflorescences: **racemes; oblong** (5–15 cm long), cylindrical, terminating lateral shoots, dense, flowers many, erect to drooping

flowers: perfect, regular; hypanthiums 2–3 mm deep; sepals 5 (0.7–1.4 mm long), obtuse at apices, glandular-lacinate on the margins; corollas 2–6 mm in diameter; petals 5, white, suborbicular (3–5 mm long); stamens 20–30

fruits: drupes; globose (6–12 mm in diameter), pendant to spreading, dark red to black, glabrous, lustrous, skin thick, flesh juicy, astringent and acidulous

other: flowers fragrant

VEGETATIVE CHARACTERISTICS

leaves: alternate, simple; blades elliptical to broadly elliptical (2–10 cm long, 1–6 cm wide); apices acute to acuminate; **margins finely serrulate,** teeth acuminate; adaxially green, glabrous; abaxially paler, sometimes pubescent; **glands 2 at the junction of the petioles and blades**; petioles 1–2 cm long

stems: twigs slender, greenish first and then becoming reddish- to orangish-brown; bark gray to black; **lenticels prominent**, ovoid (compare to *Amelanchier alnifolia*), reddish-white

HISTORICAL, FOOD, AND MEDICINAL USES: fruits are used for jelly; bark is sometimes used as a flavoring agent in cough syrup; some Native Americans used bark extract to cure diarrhea; fruits were used to treat canker sores and added to pemmican; wood was used for arrows, and pipestems; grown as an ornamental

LIVESTOCK LOSSES: may contain toxic quantities of hydrocyanic acid in leaves, stems, and seeds; poisonous to all classes of livestock, poisoning generally occurs when plant has been stressed from drought or freezing

FORAGE VALUES: poor to fair for cattle and sheep, twigs are good winter browse for wildlife; fruits are an important food sources for wildlife

HABITATS: prairies, fencerows, roadsides, hillsides, and canyons; most abundant in moist soils; adapted to a broad range of soil types

Mexican cliffrose
Purshia stansburyana (Torr.) Henrickson

SYN = *Cowania mexicana* D. Don, *Purshia mexicana* (D. Don) S.L. Welsh

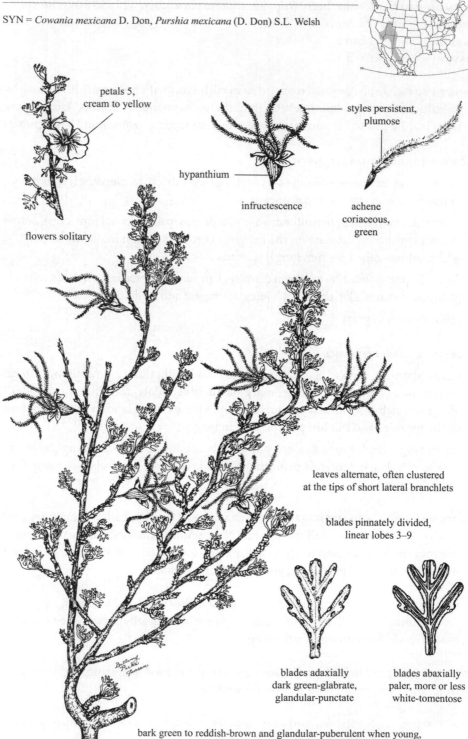

petals 5,
cream to yellow

styles persistent,
plumose

hypanthium

infructescence

achene
coriaceous,
green

flowers solitary

leaves alternate, often clustered
at the tips of short lateral branchlets

blades pinnately divided,
linear lobes 3–9

blades adaxially
dark green-glabrate,
glandular-punctate

blades abaxially
paler, more or less
white-tomentose

bark green to reddish-brown and glandular-puberulent when young,
later becoming dark gray to black, scaly, and exfoliating

FAMILY:	ROSACEAE
SPECIES:	*Purshia stansburyana* (Torr.) Henrickson
COMMON NAME:	Mexican cliffrose (rosa mexicana, quininebush, Stansbury cliffrose)
LIFE SPAN:	Perennial
ORIGIN:	Native
SEASON:	Evergreen

GROWTH FORM: shrubs or small trees (1–6 m tall); much-branched, spreading, crowns rounded; flowers April to June, fruits mature September to October, reproduces from seeds

FLORAL AND FRUIT CHARACTERISTICS

inflorescences: flowers solitary; **borne at ends of small lateral branches**

flowers: perfect, regular; hypanthium tubes obconic (4–6 mm long); sepals 5, broadly ovate, glandular-tomentose; corollas showy (1.2–1.8 cm in diameter); petals 5, **cream to yellow**, broadly ovate to obovate (7–15 mm long); stamens numerous, in 2 series; pistils 4–10, styles elongating in fruit

fruits: **achenes; 4–10 per fruit**, body narrowly oblong (4–8 mm long, 1.5-2.5 mm wide), coriaceous, glabrous at maturity; ribs 12, striate; green (compare to *Fallugia paradoxa*); styles terminal, elongate (1.2–5 cm long), persistent, plumose, silvery

other: flowers fragrant

VEGETATIVE CHARACTERISTICS

leaves: alternate, often clustered at the tips of short lateral branchlets, simple; blades obovate to narrowly spatulate (6–15 mm long, 3–8 mm wide); **pinnately divided, lobes 3–9; lobes linear**; margins revolute; adaxially dark green-glabrate, glandular-punctate; **abaxially paler, more or less white-tomentose, scattered glands**; sessile to subsessile

stems: twigs erect, stiff; bark green to reddish-brown and glandular-puberulent when young, later becoming dark gray to black, scaly, and exfoliating

other: resinous with a strong odor

HISTORICAL, FOOD, AND MEDICINAL USES: leaves used by some Native Americans as a medicinal wash for wounds; wood used for arrow shafts; fibers were used for baskets, sandals, ropes, and clothing; grown as an ornamental

LIVESTOCK LOSSES: none

FORAGE VALUES: good for cattle and sheep, especially important browse in winter; staple feed for mule deer

HABITATS: cliffs, hillsides, mesas, and washes; most abundant in dry and rocky soils at altitudes of 1000–2500 m

Antelope bitterbrush
Purshia tridentata (Pursh) DC.

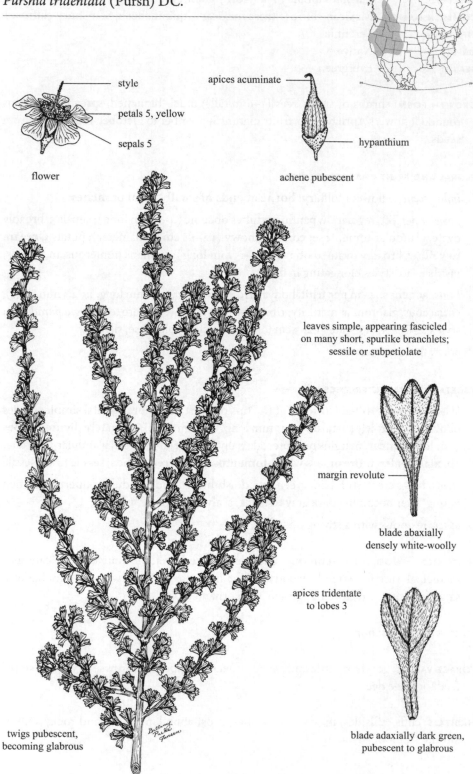

style

petals 5, yellow

sepals 5

flower

apices acuminate

hypanthium

achene pubescent

leaves simple, appearing fascicled
on many short, spurlike branchlets;
sessile or subpetiolate

margin revolute

blade abaxially
densely white-woolly

apices tridentate
to lobes 3

blade adaxially dark green,
pubescent to glabrous

twigs pubescent,
becoming glabrous

FAMILY:	ROSACEAE
SPECIES:	*Purshia tridentata* (Pursh) DC.
COMMON NAME:	Antelope bitterbrush (bitterbrush)
LIFE SPAN:	Perennial
ORIGIN:	Native
SEASON:	Evergreen

GROWTH FORM: shrubs (to 3 m tall); late-deciduous to evergreen, much-branched, crowns rounded; flowers April to August, fruits mature July to September, reproduces from seeds; trailing branches often root when in contact with the soil

FLORAL AND FRUIT CHARACTERISTICS

inflorescences: flowers solitary; subsessile; **terminating short branchlets**

flowers: perfect, regular; hypanthiums obconic (2–5 mm long), tomentose and sometimes glandular; sepals 5 (1.8–3 mm long), ovate to oblong; petals 5 (4–8 mm long), yellow, spatulate or obovate; stamens 20–25, exserted; styles beaklike (4–6 mm long), stout, persistent

fruits: achenes (7–12 mm long); exceeding the calices, **apices acuminate**, ribbed longitudinally, gray, pubescent

VEGETATIVE CHARACTERISTICS

leaves: alternate, appearing fascicled, simple; blades cuneate (5–26 mm long); **apices tridentate to lobes 3**; margins entire but revolute; adaxially dark green, pubescent to glabrous; **abaxially densely white-woolly**, glands absent (compare to *Purshia stansburyana*); sessile or subpetiolate; stipules small, triangular, persistent

stems: twigs gray to brown, with many short, spurlike branchlets, pubescent at first then becoming glabrous; buds small and scaly; bark exfoliating in long strips

HISTORICAL, FOOD, AND MEDICINAL USES: occasionally grown as an ornamental

LIVESTOCK LOSSES: none

FORAGE VALUES: good browse for cattle, sheep, and goats especially in late autumn and winter when the ground is snow-covered; usually not eaten by horses; excellent for mule deer; less valuable to pronghorn, white-tailed deer, and elk; seeds are important for birds and small mammals; withstands heavy browsing

HABITATS: plains, foothills, mountain slopes, mesas, and open woodlands; most abundant in well-drained sandy, gravelly, or rocky soils

Wild rose
Rosa woodsii Lindl.

SYN = *Rosa demareei* E.J. Palmer

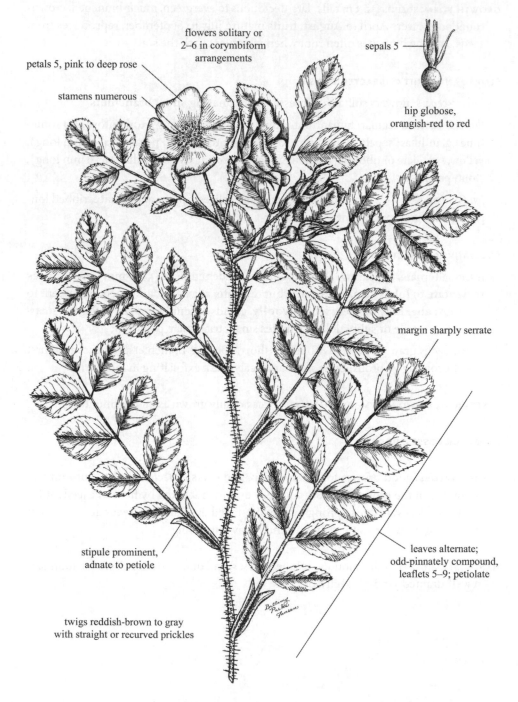

flowers solitary or
2–6 in corymbiform
arrangements

petals 5, pink to deep rose

stamens numerous

sepals 5

hip globose,
orangish-red to red

margin sharply serrate

stipule prominent,
adnate to petiole

leaves alternate;
odd-pinnately compound,
leaflets 5–9; petiolate

twigs reddish-brown to gray
with straight or recurved prickles

FAMILY:	ROSACEAE
SPECIES:	*Rosa woodsii* Lindl.
COMMON NAME:	Wild rose (rosa silvestre, Woods rose)
LIFE SPAN:	Perennial
ORIGIN:	Native
SEASON:	Cool

GROWTH FORM: shrubs (to 1.5 m tall), much-branched; usually forming thickets; growth starts in early spring, flowers May to July, reproduces from seeds

FLORAL AND FRUIT CHARACTERISTICS

inflorescences: flowers solitary or in corymbiform arrangements of 2–6; terminating branches of the current season

flowers: perfect, regular (to 3 cm diameter); hypanthiums glabrous (3–5 mm diameter); sepals 5 (1–2 cm long, 1.5–3.5 mm wide at the base), apices attenuate, often dilated, pubescent (rarely glandular), persistent, erect or spreading in fruit; petals 5, obovate (1.5–2.5 cm long), pink to deep rose; **stamens numerous**

fruits: achenes (3–4 mm long); hairy on one side, 15–35 contained inside an **orangish-red to red globose hips** (6–12 mm in diameter) formed by the hypanthiums

other: flowers fragrant

VEGETATIVE CHARACTERISTICS

leaves: alternate, **odd-pinnately compound; leaflets 5–9**, elliptical to oval or elliptical-ovate (2–5 cm long, 1–2.5 cm wide); **margins sharply serrate**; adaxially glabrous and shiny; abaxially puberulent or glandular to glabrous; petiolate; **stipules prominent**, adnate to the petioles at the bases (4–7 mm wide), glandular-pubescent on back

stems: twigs reddish-brown to gray, with **straight or recurved prickles**; infrastipular prickles usually more prominent than those on the internodes; prickles sometimes sparse, inconspicuous

HISTORICAL, FOOD, AND MEDICINAL USES: hips are consumed for a source of vitamin A and C, rose hip powder was used as a flavoring in soups and for making syrup; some Native Americans ate the young shoots as a potherb and steeped leaves for tea; raw petals are consumed today in salads, candied, or made into syrup; inner bark has been smoked like tobacco; dried petals were stored with items for their fragrance

LIVESTOCK LOSSES: prickles may injure soft tissue of browsing animals

FORAGE VALUES: fair to good for sheep and cattle; good browse for elk and deer; small mammals and birds feed on hips

HABITATS: prairies, open woodlands, plateaus, dry slopes, ravines, and thickets; in a broad range of soils

Quaking aspen
Populus tremuloides Michx.

SYN = *Populus aurea* Tidestr., *Populus tremula* L.

pistillate catkin

leaves alternate, simple, petiolate

blades adaxially dark green, glabrous, lustrous, main veins white and conspicuous; abaxially pale green, glabrous

margin finely crenate-serrate

petiole long and slender, flattened laterally near the blade

twigs reddish-brown to gray, lustrous and glabrous or slightly hairy

FAMILY:	SALICACEAE
SPECIES:	*Populus tremuloides* Michx.
COMMON NAME:	Quaking aspen (chopo temblón, trembling aspen, aspen, álamo blanco, álamo tremblón, alamillo)
LIFE SPAN:	Perennial
ORIGIN:	Native
SEASON:	Cool

GROWTH FORM: dioecious trees (to 35 m tall); trunks erect, long and slender; crowns rounded; flowers April to June, fruits ripen May to July, grows rapidly from basal sprouts and root sprouts, seldom reproduces from seeds

FLORAL AND FRUIT CHARACTERISTICS

inflorescences: **catkins** (2–8 cm long, may lengthen to 12 cm in fruit); drooping, flowers 30–100, dense, **rachis sparsely pubescent**

flowers: unisexual, apetalous; staminate floral disks oblique (1.2–1.9 mm in diameter), entire, stamens 6–12; scale lobes 3–5; lobes triangular-lanceolate, acute to acuminate, hairy-fringed; pistillate floral disks crenate, lobes 2–3; stigmas slender

fruits: **capsules; borne in catkins**; oblong-conic to lance-ovoid (3–6 mm long), light green to brown, thin walled, valves 2, seeds 3–6 per valve

VEGETATIVE CHARACTERISTICS

leaves: alternate, simple; blades highly variable, ovate to broadly ovate or reniform (2.5–7.5 cm long, 2.5–7 cm wide); **apices acute or acuminate**; bases truncate or rounded; **margins finely crenate-serrate**, teeth mostly rounded; adaxially dark green, glabrous, lustrous, main veins conspicuous and white; abaxially pale green, glabrous; **petioles slender (4–6.5 cm long), flattened laterally near the blade**, as long as the blades

stems: twigs slender, reddish-brown to gray, lustrous and glabrous or slightly hairy; **trunk bark whitish with dark markings, soft, becoming dark**

other: leaves change to bright yellow or yellowish-orange in autumn

HISTORICAL, FOOD, AND MEDICINAL USES: bark was used by pioneers and some Native Americans as a fever remedy and for scurvy, contains the anti-inflammatory glucoside salicin (similar to the active ingredient in aspirin); a substance similar to turpentine was extracted and used internally as an expectorant and externally as a counterirritant; wood is used for pulp and lumber

LIVESTOCK LOSSES: none

FORAGE VALUES: fair to good for sheep, fair for cattle; twigs, bark, and buds are browsed by horses, elk, and deer; seeds eaten by birds and small mammals

HABITATS: moist upland woodlands, mountain slopes, parklands, and stream banks; occurs on nearly all soil types

Beaked willow
Salix bebbiana Sarg.

SYN = *Salix cinerascens* (Wahlenb.) Flod., *Salix depressa* L., *Salix livida* Wahlenb., *Salix perrostrata* Rydb., *Salix rostrata* Richardson, *Salix starkeana* Willd., *Salix vagans* Hook. f. *ex* Andersson, *Salix xerophila* Flod.

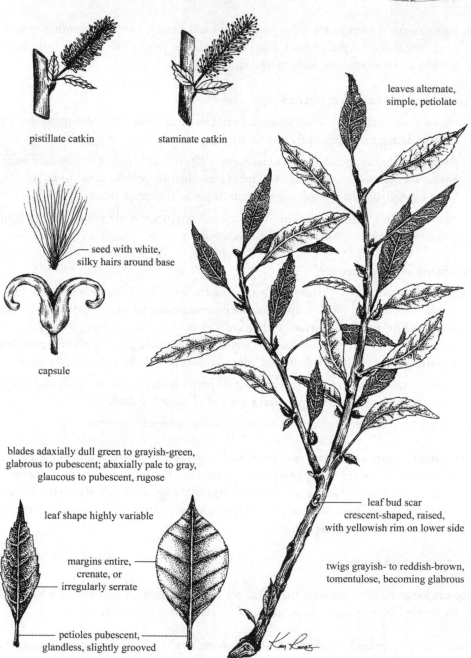

pistillate catkin

staminate catkin

leaves alternate, simple, petiolate

seed with white, silky hairs around base

capsule

blades adaxially dull green to grayish-green, glabrous to pubescent; abaxially pale to gray, glaucous to pubescent, rugose

leaf shape highly variable

margins entire, crenate, or irregularly serrate

petioles pubescent, glandless, slightly grooved

leaf bud scar crescent-shaped, raised, with yellowish rim on lower side

twigs grayish- to reddish-brown, tomentulose, becoming glabrous

FAMILY:	SALICACEAE
SPECIES:	*Salix bebbiana* Sarg.
COMMON NAME:	Beaked willow (Bebb willow, longbeaked willow)
LIFE SPAN:	Perennial
ORIGIN:	Native
SEASON:	Cool

GROWTH FORM: shrubs (to 4 m tall), dioecious, rhizomatous, forming thickets; flowers April to May, fruits ripen May to June; reproduces from seeds, rhizomes, and basal sprouts

FLORAL AND FRUIT CHARACTERISTICS

inflorescences: catkins; appearing before or with the leaves; staminate catkins (1.5–3 cm long) densely flowered on short new stems (usually with 2 leaves); pistillate catkins (1.5–6 cm long) loosely flowered on short new stems (usually with 2–4 leaves), **persistent after capsule dehiscence**

flowers: unisexual, apetalous; staminate scales ovate (1.5–2.5 mm long), blunt, pubescent, yellowish to brown, stamens 2; glands 1; pistillate scales ovate (2 mm long), pubescent, greenish-yellow, apices dark when immature, becoming yellowish-brown with age; ovary silky; stigmas 2

fruits: capsules; borne in catkins, ovoid-conic (5–8 mm long), silky pubescent, long-stipitate, long-beaked, valves 2; seeds obovoid (1 mm long) with white silky hairs (5–8 mm long) around base

VEGETATIVE CHARACTERISTICS

leaves: alternate, simple; **elliptical to narrowly ovate or obovate or oblanceolate (3–6 cm long, 1.5–2.5 cm wide)**, often widest above the middle; apices short acuminate to acute; bases cuneate; margins entire, crenate, or irregularly serrulate; adaxially dull green to grayish-green, glabrous to pubescent, somewhat coriaceous; abaxially pale to gray, glaucous to pubescent, rugose, veins raised; stipules small, caducous; petioles pubescent (5–15 mm long), glandless, slightly grooved

stems: twigs grayish- to reddish-brown (1–1.2 mm in diameter), flexuous, tomentulose, becoming glabrous; lenticels elliptical, light-colored, not obvious; leaf bud scars crescent-shaped, raised, with a yellowish rim on lower side; trunks gnarled, bark gray

other: reproduces from stem cuttings for revegetation

HISTORICAL, FOOD, AND MEDICINAL USES: some Native Americans made tea from the bark to treat fever and headache; contains salicin, an anti-inflammatory glucoside (similar to the active ingredient in aspirin)

LIVESTOCK LOSSES: none

FORAGE VALUES: fair for cattle and good for sheep; valuable browse for elk, deer, and moose

HABITATS: moist valleys, stream banks, wet meadows, seeps, moist to dry hillsides

Sandbar willow
Salix exigua Nutt.

SYN = *Salix fluviatilis* Nutt., *Salix interior* Rowlee, *Salix longifolia* Lam.

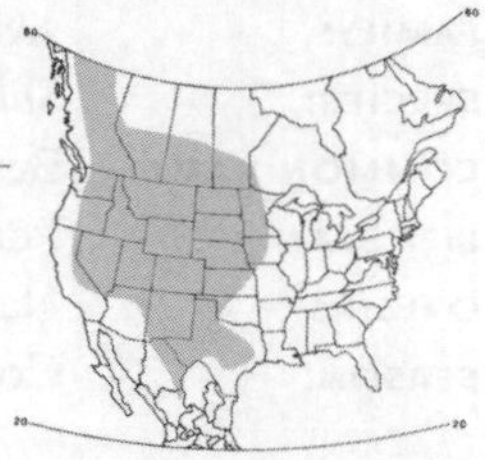

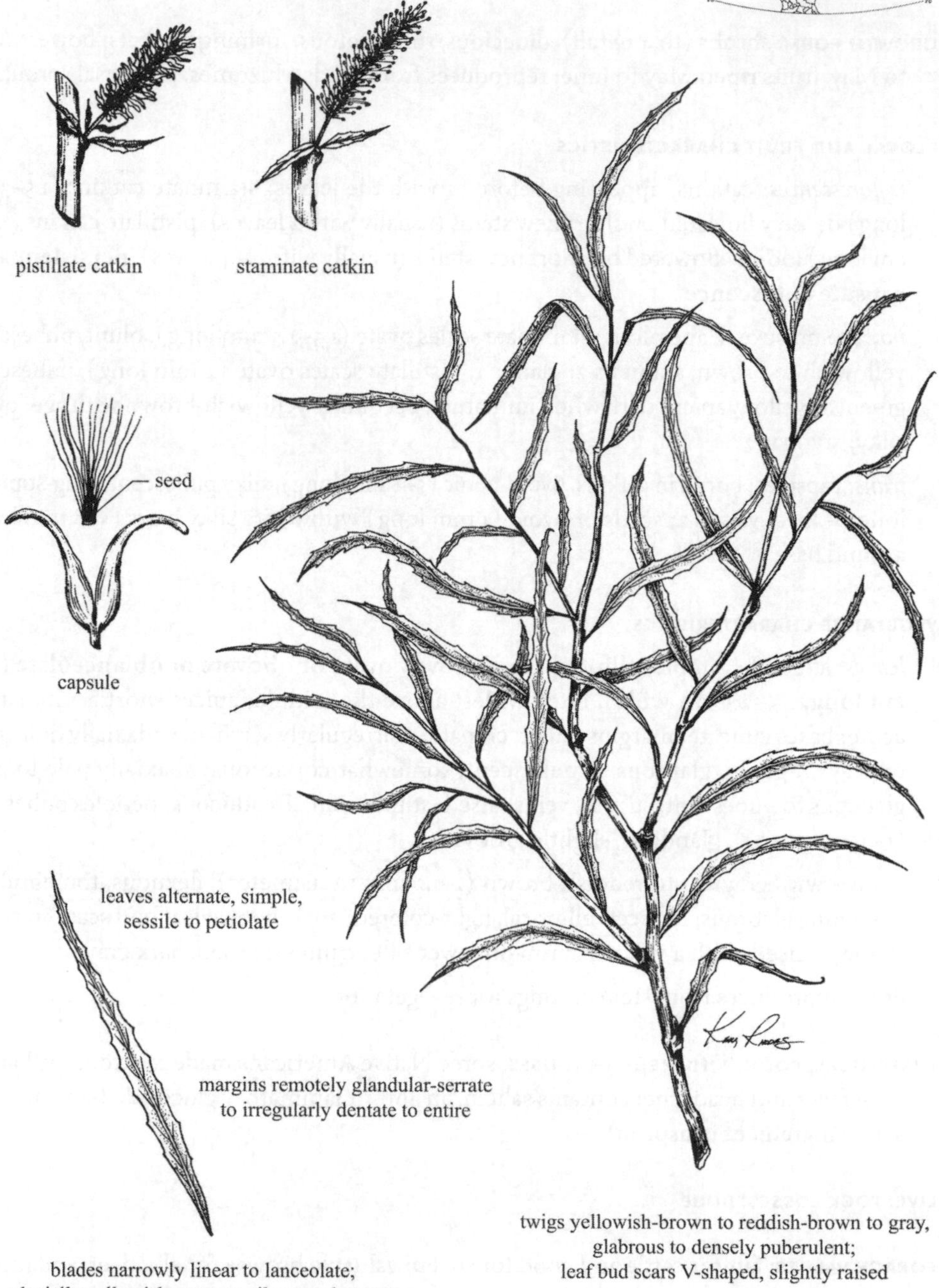

pistillate catkin

staminate catkin

— seed

capsule

leaves alternate, simple,
sessile to petiolate

margins remotely glandular-serrate
to irregularly dentate to entire

twigs yellowish-brown to reddish-brown to gray,
glabrous to densely puberulent;
leaf bud scars V-shaped, slightly raised

blades narrowly linear to lanceolate;
adaxially yellowish-green to silvery pubescent;
abaxially pale to gray with appressed pubescence

FAMILY:	SALICACEAE
SPECIES:	*Salix exigua* Nutt.
COMMON NAME:	Sandbar willow (sauz, narrowleaf willow, coyote willow, sauce, sauce coyote)
LIFE SPAN:	Perennial
ORIGIN:	Native
SEASON:	Cool

GROWTH FORM: dioecious shrubs to small trees (to 9 m tall); rhizomatous, forming thickets; stems and branches spreading to form rounded crowns; flowers May to June, fruits June to July, reproduces from seeds and rhizomes

FLORAL AND FRUIT CHARACTERISTICS

inflorescences: **catkins**; staminate catkins (1.5–4.5 cm long, 3–8 mm in diameter) emerging after the leaves on new leafy shoots, erect or ascending; pistillate catkins (1.5–8 cm long, 4–10 mm in diameter) on new shoots, loosely to densely flowered, ascending to slightly drooping

flowers: unisexual, apetalous; staminate scales ovate to obovate, acute, densely pubescent on both sides, greenish-yellowish, deciduous, stamens 2; pistillate scales rounded, greenish-yellow, rugose, pubescent on both sides, ovary sparsely pubescent

fruits: capsules; borne in catkins; capsules ovoid (4–7 mm long, 1.5 mm in diameter), blunt, glabrous or slightly pubescent, valves 2

VEGETATIVE CHARACTERISTICS

leaves: alternate, simple; **blades narrowly linear to lanceolate (4–12 cm long, 2–10 mm wide)**; apices acute; bases acuminate; margins remotely glandular-serrate to **irregularly dentate to entire**; adaxially yellowish-green to silvery pubescent, permanently pubescent to becoming glabrous, midveins usually raised; abaxially pale to gray with appressed pubescence; stipules minute; sessile to petiolate (1–5 mm long)

stems: twigs yellowish-brown to reddish-brown to gray, flexuous, glabrous to densely puberulent; new growth green, pubescent; lenticels elliptical, raised, scattered; leaf-bud scars V-shaped, slightly raised; trunk bark grayish-green to brown, smooth, shallowly furrowed with age

other: variable species with at least 2 subspecies

HISTORICAL, FOOD, AND MEDICINAL USES: some Native Americans used peeled stems to make baskets; a tea was made from the bark to treat fever and headache; contains salicin, an anti-inflammatory glucoside (similar to the active ingredient in aspirin)

LIVESTOCK LOSSES: none

FORAGE VALUES: fair for cattle and good for sheep; valuable browse for deer, elk, and moose

HABITATS: valleys, stream banks, marshy areas, ditch banks, occasionally on dry ground

Creosotebush
Larrea tridentata (DC.) Coville

SYN = *Covillea tridentata* (DC.) Vail, *Larrea glutinosa* Engelm., *Neoschroetera tridentata* (DC.) Briq., *Schroeterella tridentata* (DC.) Briq., *Zygophyllum tridentatum* DC.

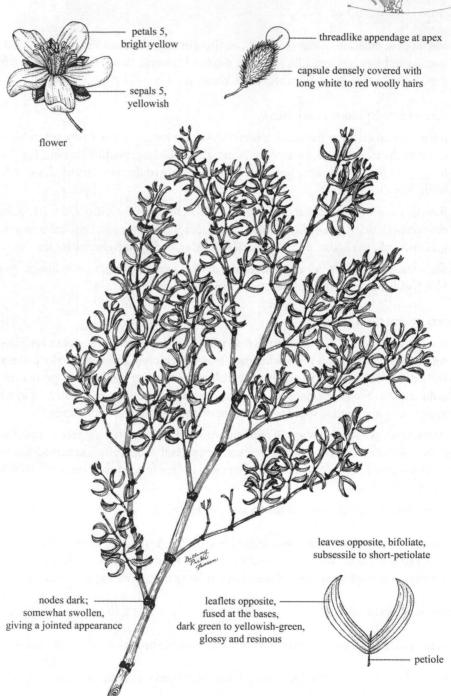

petals 5, bright yellow

sepals 5, yellowish

flower

threadlike appendage at apex

capsule densely covered with long white to red woolly hairs

leaves opposite, bifoliate, subsessile to short-petiolate

nodes dark; somewhat swollen, giving a jointed appearance

leaflets opposite, fused at the bases, dark green to yellowish-green, glossy and resinous

petiole

FAMILY:	ZYGOPHYLLACEAE
SPECIES:	*Larrea tridentata* (DC.) Coville
COMMON NAME:	Creosotebush (gobernadora, hediondilla)
LIFE SPAN:	Perennial
ORIGIN:	Native
SEASON:	Evergreen

GROWTH FORM: shrubs (to 3 m tall); no well-defined trunk, numerous limber stems from near ground level; often evenly spaced in nearly pure stands with little variation in size; flowers February to August, reproduces from rhizomes and seeds

FLORAL AND FRUIT CHARACTERISTICS

inflorescences: flowers solitary, axillary

flowers: perfect, regular; sepals 5 (5–8 mm long), unequal, yellowish, silky, early deciduous; petals 5 (7–10 mm long, 5 mm wide), bright yellow, obovate, concave, twisted following pollination; stamens 10; filaments winged

fruits: capsules; globose (7–8 mm long), **covered with dense white to red woolly long hairs and tipped by a threadlike appendage**, cells 5, each cell with seeds 1

VEGETATIVE CHARACTERISTICS

leaves: opposite, **bifoliate; leaflets opposite, fused at the bases**, ovate to oblong or obovate (5–10 mm long, 3–4 mm wide); margins entire; **dark green** to yellowish-green, glossy and **resinous**; subsessile to short-petiolate; stipules at base of petiole brown, **glandular-hairy**

stems: twigs slender, brown; **nodes dark, somewhat swollen**, conspicuous, **giving a jointed appearance**

other: foliage emits creosote odor, especially when wet or burned; some plants are estimated to be over 10,000 years old

HISTORICAL, FOOD, AND MEDICINAL USES: used by some Native Americans for fuel; decoctions were used as antiseptics and medicines; some Pima used lac from insect larvae found on the plants as glue and to waterproof baskets

LIVESTOCK LOSSES: may cause dermatitis in humans and animals; sheep, especially pregnant ewes, have been reported to die after eating the leaves; capsules may contaminate fleece

FORAGE VALUES: worthless to livestock; seldom consumed by livestock or wildlife, although jackrabbits will occasionally eat the leaves

HABITATS: alluvial plains, deserts, mesas, and hillsides; most abundant in sandy and gravelly soils; usually does not occupy saline soils

Glossary

A- prefix meaning without

ABAXIAL on the side away from the axis; lower surface

ABORTION failure of an organ or structure to develop; early termination of pregnancy resulting in the death of the embryo or fetus

ABRUPT changing sharply or quickly, rather than gradually

ABSENT not present; never developing

ACAULESCENT stemless; without an above-ground stem or apparently so

ACEROSE needle-shaped; acicular

ACHENE dry, 1-seeded, indehiscent fruit with a relatively thin wall in which the seed coat is not fused to the ovary wall except at one point at the base

ACICULAR needle-shaped; acerose

ACIDIC soil with a low pH (less than 5.5), may interfere with the growth of some plant species

ACIDULOUS acid in taste; sour

ACORN dry fruit seated in or surrounded by a hard, woody cupule of indurate bracts, as in the genus *Quercus*

ACTINOMORPHIC symmetrical, regular; divisible into equal parts in 2 or more planes (contrast with zygomorphic)

ACUMINATE gradually tapering to a sharp point and forming concave sides along the apex (compare with acute)

ACUTE tapering to a pointed apex with more or less straight sides; sharply pointed; angle less than 90°

ADAXIAL on the side nearest the axis; upper surface

ADDICTED having a compulsive need for a habit-forming substance

ADNATE attached or grown together; fusion of unlike parts, such as palea and caryopsis in the genus *Bromus*; attached or grown together

AGGREGATED densely clustered

ALKALI soil with a high pH (8.5 or higher) and high exchangeable sodium content (15% or more), may interfere with the growth of some species

ALKALOID any of numerous nitrogen-containing organic bases, many naturally occurring secondary metabolites in plants; may be toxic to animals

ALLUVIUM mineral or soil material deposited by running water

ALPINE mountainous regions or areas; zone above the tree line

ALTERNATE located singly at each node; not opposite or whorled

AMENT dense spike or raceme with many small, usually naked, flowers; a catkin, as in the genus *Populus*

ANDROGYNOUS with both pistillate and staminate flowers, the staminate flowers are borne above the pistillate flowers in the same inflorescence

-ANGLED suffix meaning a corner

ANGULAR forming an angle; with one or more angles

ANNUAL within one year; applied to plants that do not live more than one year or growing season

ANTERIOR in front of; in a flower, the side away from the axis and adjacent to the bract

ANTHER part of a stamen in which pollen develops

ANTHESIS period of flowering; period during which pollination occurs

ANTIDOTE remedy to counter the effects of a toxic substance

ANTRORSE directed upward or forward; opposed to retrorse

APETALOUS without petals

APEX, APICES tip or distal end

APICAL relating to the apex

APICULATE ending in an abrupt, sharp point

APOMIXIS process of setting seed without fusion of gametes

APPRESSED lying against an organ; flatly pressed against

AQUATIC growing in, on, or near water

ARCHING curved like an arch

ARCUATE curved like a bow or arch

ARISTATE awned or tapering; a very long, narrow apex

ARMED having sharp thorns, spines, or prickles

AROMATIC fragrant or having an odor; bearing essential oils

ARROYO gully or channel; a watercourse in an arid region

ARTICULATE jointed, provided with nodes; separating clearly at maturity

ARTICULATION joint or point of attachment

ASCENDING growing or angled upward; obliquely upward

ASTRINGENT able to draw together soft tissues; styptic

ASYMMETRICAL not symmetrical; not divisible into equal parts

ATTENUATE gradually narrowing to a slender apex or base

AURICLE ear-shaped lobes, such as those that occur at the base of leaf blades of some grasses

AWL-SHAPED narrow and sharp pointed; gradually tapering from a narrow base to a pointed apex; subulate

AWN in grasses, the extension of a vein beyond the leaflike tissue

AWN COLUMN undivided portion of the awn below their branches, such as in the genus *Aristida*

AWNLESS without awns

AXIL angle between an organ and its axis

AXILLARY growing in an axil

AXIS, AXES central or main longitudinal support upon which parts are attached

BALSAM aromatic and resinous substances from plants containing benzoic or cinnamic acid

BANNER upper petal (standard) of the papilionaceous flower in the FABACEAE

BARB short, rigid projection

BARBED furnished with retrorse projections

BARK exterior covering of a woody stem or root; tissues lying outside the cambium

BARREN unproductive sites, usually with shallow soils

BASAL located at or near the base of a structure, such as leaves arising from the base of the stem

BASIFIXED attached by the base (compare with dolabriform)

BEADLIKE sphere-shaped

BEAK narrow or prolonged tip; a hard point or projection (frequently the remnants of the style base)

BEARDED furnished with long, stiff hairs

BERRY fleshy fruit developing from a single pistil with several or many seeds; a fruit that is fleshy or pulpy throughout

BI- prefix meaning 2

BICOLORED having 2 colors

BIDENTATE having 2 teeth

BIENNIAL plant that lives for 2 years

BIFID 1-cleft or 2-toothed; applied to the summit of glumes, lemmas, paleas, petals, or leaflets

BIFOLIATE having 2 leaves or leaflets

BILABIATE having 2 lips, as in irregular flowers

BILATERAL having 2 sides

BILOCULAR having 2 compartments or locules in the ovary, anther, or fruit

BIPINNATE twice-pinnate

BITTER disagreeable taste sensation; 1 of the 4 dominant basic taste sensations

BIVALVED having 2 valves

BLADE part of the leaf above the sheath, petiole, or petiolule

BLEACHED having lost most of the original color

BLISTER enclosed raised spot on the surface

BLOAT digestive disturbance of livestock (especially cattle) marked by abdominal swelling due to a buildup of gas, potentially fatal

BLOTCHED spot or mark in an irregular shape

BLOWOUT depression in the surface of sand or sandy soil caused by wind erosion

BLUNT having a point or edge that is not sharp

BODIES parts of the whole

BOG poorly drained area; wet, spongy ground

BORNE attached to or carried by

BRACKISH salty water, with a saline content less than that of sea water

BRACT reduced leaves (frequently associated with the flowers)

BRACTEATE having bracts

BRACTEOLE bract borne on a secondary axis, diminutive; bractlet; scale

BRANCH lateral stem

BRANCHLET final, or ultimate, division of the branch

BREAKS rough land form usually caused by water erosion; canyons

BRISTLE stiff, slender appendage; a reduced inflorescence branch in the genus *Setaria*

BRISTLY covered with bristles

BRITTLE easily broken

BROOMLIKE shaped like a broom used for sweeping

BROWSE twigs, leaves, and other parts of woody plants consumed by herbivores; the act of consuming portions of woody plants

BUD underdeveloped shoot or flower

BULB underground bud with fleshy, thick scales

BULBOUS swollen at the base, like a bulb or corm

BULGE swelling; an abrupt expansion

BUR rough and prickly covering of a fruit

BURLAP coarse fabric of jute or hemp, commonly used to make bags

BUSHY resembling a bush; thick and spreading

CADUCOUS early deciduous; falling off early

CALCAREOUS soil containing sufficient calcium carbonate (often with magnesium carbonate) to effervesce when treated with hydrochloric acid

CALLOUS having the texture of a callus

CALLUS, CALLUSES the indurate downward extension of tissue from the mature lemma in the genera *Nassella, Hesperostipa, Aristida,* and others; hardened tissue

CALYCULUS, CALYCULI group of small bracts resembling a calyx

CALYX, CALYCES sepals of a flower considered collectively, usually green bracts

CAMPANULATE shaped like a bell

CANESCENT pale or gray-colored because of a dense, fine pubescence

CAPILLARY fine and slender; hairlike, such as a branch or awn

CAPITATE aggregated into a dense cluster; headlike; caplike

CAPSULE dry, dehiscent fruit of more than 1 carpel with more than 2 seeds

CARPEL modified leaf forming the ovary; may be compound or simple

CARYOPSIS fruit or grain of grasses; a small, dry, indehiscent fruit having a single seed with a thin, closely adherent pericarp

CATALYST substance that increases the speed of a reaction

CATHARTIC substance causing the evacuation of the bowels

CATKIN dense spike or raceme with many small, usually naked, flowers; an ament, as in the genus *Populus*

CAUDATE bearing a slender taillike projection or appendage

CAUDEX, CAUDICES short, usually woody, vertical stem located just below the soil surface

CAULESCENT having a stem

CAULINE pertaining to the stem or belonging to the stem

-CELLED suffix meaning cavity or individual unit

CENTRAL situated at, in, or near the center

CESPITOSE, CAESPITOSE tufted; several or many stems in a close tuft, as in bunchgrasses

CHAFF bracts subtending a flower (usually small, membranous, and dry) in the ASTERACEAE

CHANNELED deeply grooved

CHAPARRAL vegetation type with dense thickets of woody plants adapted to dry summers and wet winters

CHARTACEOUS having the texture of writing paper and usually not green

CILIATE fringed with hairs on the margin

CILIOLATE fringed with minute hairs on the margins

CLASPING an organ or tissue partially or totally wrapped around a second

CLAVATE, CLAVIFORM club-shaped, widening toward the apex

CLAW narrowed base of some sepals and petals

CLEFT divided into teeth or divisions that extend halfway or more to the midvein

CLEISTOGAMOUS applied to flowers or florets fertilized without opening, as in some species in the genera *Leptochloa, Nassella, Danthonia,* and others

CLUB-SHAPED widening toward the apex

CLUMP dense cluster

CLUSTER number of similar tissues or organs growing together; a bunch

COARSE composed of relatively large parts; not finely textured or finely structured

COBWEBBY tuft of tangled, fine hairs

COIL 1 or more loops

COLLAR area on the abaxial side of a leaf at the junction of the blade and sheath in grasses

COLONY group of plants of the same species growing in close association with each other; all members of the group may have originated from a single plant

COLUMN upper portion of the lemma of grasses located just below the divided awn

COMA tuft of hairs

COMBLIKE pectinate; with narrow, closely set, and divergent segments like the teeth in a comb

COMPACT having a small, dense structure

COMPOUND composed of several parts united into a single structure

COMPRESSED flattened laterally

COMPRISED made up of

CONCAVE hollowed inward, like the inside of a bowl

CONE cluster of scales on an axis, scales may be persistent or deciduous

CONFLUENT merging; coming together, blended together as one

CONGESTED overcrowded

CONIC, CONICAL cone-shaped with the point of attachment at the broad end

CONNATE fusion of like parts, such as petals, to form a corolla tube

CONSPICUOUS obvious; easy to notice

CONSTRICTED drawn together; appearing to be tightly held

CONTAMINATE the introduction of unwanted materials causing a reduction of value or use

CONTORTED bent; twisted

CONTRACTED panicle inflorescences that are narrow or dense, frequently spikelike

CONVEX rounded on the surface, like the bottom or exterior of a bowl

CONVOLUTE rolled longitudinally

COOL-SEASON a category of plants that grow during the cool portions of the year

COPIOUS abundant

CORDATE heart-shaped, with rounded lobes and a sinus at the base

CORIACEOUS leathery in texture

CORM the bulblike base of a stem, usually fleshy and underground

COROLLA all of the petals considered collectively

CORYMB simple racemose inflorescence that is flat-topped; an indeterminate inflorescence

CORYMBIFORM having the form, but not necessarily the structure, of a corymb; corymblike

COTTONY having the texture of cotton

COTYLEDON leaf of the embryo of a seed; a seed leaf

COUNTERIRRITANT substance or act that overcomes an irritation

CRAVING to want greatly

CREEPING continually spreading; a shoot or horizontal stem (usually a rhizome or stolon) that roots at the nodes

CRENATE having rounded teeth; scalloped margins

CRENULATE diminutive of crenate

CREOSOTE aromatic mixture of phenolic compounds obtained from the distillation of wood tar, commonly used to preserve wood

CRESCENT shape having a convex edge and a concave edge; shaped like the moon in its first quarter

CRESTED with an elevated ridge or appendage on the top or back

CRINKLY wrinkled or rippled surface

CROSS SECTION cut at a right angle to the main axis; transverse

CROSS-SEPTATE partitioned by cross-walls

CROWDED pressed close together; a number of structures in a small space

CROWN persistent base of a herbaceous perennial; the shape of the foliage of a shrub or tree; the tuft of hairs at the summit of the lemma in some grasses, as in the genera *Nassella, Hesperostipa,* and others

CRUCIFORM shaped like a cross

CULM hollow or pithy jointed stem or stalk of a grass, sedge, or rush

CUMULATIVE increasing by successive additions

CUNEATE wedge-shaped; narrowly triangular with the narrow end at the point of attachment

CUP, CUPULE cuplike structure at the base of a fruit such as an acorn in the genus *Quercus*

CUPULATE shaped like a cup

CURE drying, as in standing herbage or hay; relief or recovery from a disease

CURLED formed in the shape of curves or spirals

CUSPIDATE bearing a sharp, firm, and elongated point at the apex

CYANOGENETIC capable of producing cyanide

CYLINDRIC, CYLINDRICAL shaped like a cylinder

CYME convex or flat-topped flower cluster with the central flower the first to open; a determinate inflorescence

CYMIFORM having the shape, but not necessarily the structure, of a cyme; cymelike

CYMOSE resembling a cyme or bearing cymes

CYPSELA, CYPSELAE fruit in the ASTERACEAE resembling an achene; from multiple carpeled pistils

DECIDUOUS not persistent, but falling away in less than 1 year

DECOCTION extract obtained by boiling the plant material

DECUMBENT curved upward from a horizontal or inclined base, with only the end ascending

DECURRENT extending downward from the point of attachment

DEHISCENT opening at maturity along a definite suture

DELICATE fine structure or texture

DELTATE, DELTOID triangular; shaped like the Greek letter delta

DENSE crowded

DENTATE with pointed, coarse teeth spreading at right angles to the margin

DENTICULATE diminutive of dentate

DEPRESSED flattened from above; pressed down

DERMATITIS inflammation of the skin

DICHOTOMOUS branching repeatedly in pairs

DIFFUSE open and much-branched, loosely branching

DIGITATE several members arising from 1 point at the summit of a support, like the fingers arising from the hand as a point of origin

DILATED enlarged; expanded; widened

DIMORPHIC 2 types (forms) of leaves, flowers, or other structures on the same plant

DIOECIOUS unisexual flowers on separate plants; pistillate and staminate flowers on separate plants (see *Bouteloua dactyloides*)

DISARTICULATING, DISARTICULATION separating at maturity at a node or joint

DISCOID resembling a disk; in the ASTERACEAE, with all the flowers of a head tubular and perfect

DISHED shaped like a dish

DISK outgrowth of the receptacle that surrounds the base of the ovary or ovaries

DISK FLORETS regular flower of the ASTERACEAE

DISSECTED deeply divided into numerous parts

DISSEMINATE to disperse or spread

DISTAL remote from the place of attachment; toward the apex

DISTANT separated by space

DISTICHOUS conspicuously 2-ranked leaves, leaflets, flowers, or spikelets

DISTINCT clearly evident; separate; apart

DISTURBANCE alteration or destruction of the vegetative cover

DISTURBED sites in which the vegetative cover has been altered or destroyed

DIURETIC substance causing an increase in the flow of urine

DIVARICATE widely and stiffly divergent

DIVERGENT widely spreading

DIVIDED separated or cut into distinct parts by inclusions extending to near the base or midvein

DIVISION one of the parts of the whole

DOLABRIFORM hatchet- or T-shaped; attached in the middle (compare with basifixed)

DOMINANT species of plant that is one of the most frequently occurring in the vegetation; species that controls the character of the vegetation

DORMANCY an inactive state; period during which plants are not active, as in winter

DORSAL relating to the back of an organ; opposite the ventral side

DOTTED marked with small spots

DOUBLY SERRATE having teeth on the teeth, especially on the margins of leaves

DOWNY soft, fine pubescence

DROOP to hang downward; pendulous

DRUPE fleshy fruit, indehiscent, usually a single seed with a stony endocarp (e.g., a cherry)

DULL lacking brilliance or luster; not shiny

DUMBBELL structure that is narrow in the middle and large on both ends

DWARF less than normal size

DYE pigment or other substance used to color other items such as cloth

EDIBLE fit to be eaten; consumed as food

ELLIPSOID solid body circular in cross section and elliptic in long-section

ELLIPTIC, ELLIPTICAL shaped like an ellipse; narrowly pointed at the ends and widest in the middle

ELONGATE narrow, the length many times the width or thickness

EMARGINATE having a shallow notch at the tip

EMBEDDED enclosed in a supporting structure or organ; imbedded

EMETIC substance causing vomiting

ENTIRE whole; with a continuous margin

ENVELOPED enclosed within

EPHEDRINE crystalline alkaloid extracted from species of *Ephedra*

ERECT upright; not reclining or leaning

ERGOT fungus disease of grasses

EROSE irregularly notched at the apex; appearing gnawed or eroded

EVEN- prefix meaning number of structures divisible by 2

EVERGREEN woody plants that retain their leaves throughout the year

EVIDENT obvious; distinct; easily seen

EXCEED greater than; larger than

EXFOLIATE shedding in flakes or thin layers, such as bark of woody species

EXPANDED increased; extended

EXPECTORANT substance that promotes the discharge of mucus from the respiratory tract

EXPOSED open to view

EXSERTED protruding or projecting beyond; not included

EXTENSIVE having a wide or considerable range or spread

EXTRACT to separate or remove; material that has been separated

FAINT lacking distinctness

FALCATE curved or shaped like a sickle

FAMINE extreme scarcity of food

FAN-SHAPED shaped like a segment of a circle with the point of attachment at the narrow end

FASCICLE small bundle or cluster, such as needles of members of the genus *Pinus* in clusters of 2–5

FEATHERY having the texture or appearance of feathers; a central axis with pinnate subdivisions that are in close association

FELTY closely matted with intertwining hairs; having the texture or appearance of felt

FERN vascular plant with highly divided, delicate leaves which produces spores rather than seeds

FERTILE capable of producing fruit; does not refer to stamen presence or absence in grasses

FIBRILLOSE having small slender fibers or filaments

FIBROUS consisting of, or containing, mostly fibers; commonly used to describe branching root systems (compare with taproot)

FILAMENT stalk of a stamen supporting the anther; a threadlike structure

FILIFORM threadlike; long and very slender

FIRM hard; indurate

FISSURE deep groove

FLAKE thin, flattened piece or layer

FLANKED to be situated at either side of a structure

FLAT, FLATTENED having the major surfaces essentially parallel and distinctly greater than the minor surfaces; without a slope

FLESHY pulpy; succulent; having high water content within the plant material

FLEXIBLE capable of being easily bent

FLEXUOUS bent alternately in opposite directions; a wavy form

FLOCCOSE covered with long, soft, fine hairs that are loosely spreading and rub off easily

FLORET lemma and palea with included flower of the POACEAE; also ray and disk flowers of the ASTERACEAE

FLORIFEROUS flower-bearing

FLOWER part of a plant that produces the seed; usually contains petals, a pistil, and pollen-bearing stamens

FOLDED part or organ that is doubled over or laid over; V-shaped in cross section

FOLIAGE plant material that is mainly leaves

-FOLIATE suffix pertaining to or consisting of leaflets (i.e., trifoliate means that the leaves are made up of 3 leaflets)

FOLLICLE dry, dehiscent fruit splitting along 1 suture; a small closed or nearly closed cavity

FOOTHILLS region at the base of a mountain range

FORB herbaceous plants other than grasses and grasslike plants

FRAGRANT having a sweet or delicate odor

FREE not attached to other organs

FRINGED having a border consisting of hairs or other structures

FROND large, divided fern leaf

FRUIT ripened ovary; the seed-bearing organ

FUNNELFORM shaped like a funnel

FURROWED bearing longitudinal grooves or channels; sulcate

FUSED attached

FUSIFORM shaped like a spindle

GENICULATE bent abruptly, like a knee (awns or culms may be bent in this manner)

GLABRATE, GLABRESCENT nearly glabrous or becoming so with age

GLABROUS without hairs

GLAND protuberance or depression that appears to secrete a fluid

GLANDULAR supplied with glands

GLAUCOUS covered with a waxy coating that gives a bluish-green color; possessing a waxy surface that easily rubs off

GLOBOSE nearly spherical in shape

GLOMERULE dense cluster; a dense, headlike cyme

GLOSSY having a surface luster; shiny

GLUMES the pair of bracts at the base of a spikelet in grasses

GLUTINOUS with a firm, sticky substance covering the surface

GLYCOSIDE, GLUCOSIDE organic compounds that yield a sugar and another substance upon hydrolysis; may be found in plants and may be toxic to animals

GNARLED twisted and deformed; knotty

GRANULATE covered with small knobs or irregularly placed granules

GRASS monocotyledonous herbaceous plants (except bamboo)

GRASSLIKE herbaceous plants similar in appearance to grasses such as sedges and rushes

GRAZE to consume growing and/or standing grass or forb herbage; to place animals on pastures to enable them to consume the herbage

GROOVE long, narrow channel or depression; sulcus

GROWING POINT apical tissue; vegetative bud

GULLY trench eroded in the land surface by running water

HARSH texture disagreeable to the touch; rough; unpleasant

HEAD dense cluster of sessile or nearly sessile flowers on a short or common axis; an inflorescence type

HEMISPHERIC shaped like one-half of a sphere

HERB, HERBACEOUS not woody; dying each year or dying back to the crown

HERBAGE aboveground material produced by herbaceous plants; vegetation that is available for consumption by grazing animals

HERRINGBONE pattern made up of 2 rows of parallel lines with adjacent rows slanting in reverse directions

HIP fruit consisting of the fleshy floral tube surrounding the mature ovaries, as in the genus *Rosa*

HIRSUTE with straight, rather stiff hairs

HISPID with stiff or rigid hairs; bristly hairs

HOARY covered with fine gray or white pubescence

HOLLOW unfilled space; empty

HOOKED curved or bent like a hook

HORIZONTAL parallel to the plane of the earth

HORN exserted appendage

HYALINE thin and translucent or transparent

HYDROCYANIC ACID aqueous solution of hydrogen cyanide that is poisonous

HYPANTHIUM ring or cup around the ovary formed by a fusion of the bases of sepals, stamens, and petals

ILL-SCENTED unpleasant odor

IMBEDDED enclosed in a supporting structure or organ; embedded

IMBRICATE overlapping (like shingles on a roof)

IMPACT to fix firmly as if by packing or wedging

IMPENETRABLE area or object that cannot be entered; commonly applied to thick vegetation that grazing or browsing animals will not enter

IMPERFECT having either stamens or a pistil only

IMPROPER GRAZING animal utilization of herbaceous material and impact on the site that causes a significant decline in the plant community

INCISED cut sharply, irregularly, and more or less deeply

INCLUDED not exserted nor protruding

INCONSPICUOUS not easily seen; not evident

INCURVED curved toward the applicable center axis

INDEHISCENT not opening, staying closed at maturity; not splitting

INDISTINCT not easily seen; not sharply outlined or separable

INDURATE hard

INDUSIUM thin epidermal outgrowth on a fern leaf that covers the sorus

INFERIOR lower; below; less elevated in position

INFLATED swollen or expanded; puffed up; bladdery

INFLEXED turned in at the margins

INFLORESCENCE arrangement of flowers on an axis subtended by a leaf or portion thereof

INFRA- prefix meaning below

INFRASTIPULAR below the stipules

INFRUCTESCENCE the arrangement of fruits on an axis

INFUSION liquid resulting from soaking or steeping material in water without boiling

INROLLED curved or rolled adaxially toward the central axis of the structure; opposite of revolute

INSIPID lacking flavor

INTERMINGLED intermixed; mixed together

INTERNODE part of a stem between 2 successive nodes

INTERRUPT to break the uniformity; to come between 2 similar objects or structures

INTRICATE having many complex parts or elements

INTRODUCED not native to North America

INTRUDE to place, thrust, or force between

INVOLUCRAL pertaining to the involucre

INVOLUCRE whorl or circles of bracts below a flower or flower cluster (head)

INVOLUTE rolled inward from the edges, the upper surface within

IRREGULAR asymmetrical; not equal in similar parts

JAM, JELLY food made by boiling fruit and sugar to a thick consistency

JOINT section of a stem or culm from which a leaf or branch arises; node of a grass culm

JOINTED possessing nodes or articulations

JUNCTION place at which 2 structures or organs join

JUVENILE young; not mature or fully developed

KEEL sharp fold or ridge at the back of a compressed sheath, blade, glume, lemma, or palea of POACEAE; united lower petals of FABACEAE

KEELED ridged, like the keel of a boat

KNOT base of a woody branch enclosed in the stem from which it arises

LAC resinous substance secreted by an insect that was used as glue

LACERATE appearing torn at the edge or irregularly cleft several times

LACINATE deeply cut into narrow segments

LANATE woolly with long intertwined, curly hairs

LANCE-LINEAR shaped like a narrow lance head

LANCEOLATE rather narrow, pointed and tapering to both ends, widest below the middle

LATERAL belonging to or borne on the side

LAX loose; open and spreading

LEAFLESS without leaves

LEAFLET division of a compound leaf

LEAFY with many leaves

LEATHERY resembling leather in texture and appearance

LEDGE narrow, flat surface on the face of a mountain, hill, or canyon

LEGUME in the FABACEAE, fruit composed of a single carpel with 2 sutures and dehiscing at maturity along the sutures

LEMMA abaxial bract of the floret that subtends the grass flower and palea; the lower bract of a grass floret

LENTICEL slightly raised, lens-shaped area on the surface of a woody stem

LEPIDOTE covered with scales

LICHEN an association of fungi living symbiotically with algae on rocks or vascular plants

LIFE SPAN length of time a plant will live

LIGULATE in the ASTERACEAE, referring to flowering heads solely composed of the flat, strap-shaped florets on the margin (ray florets) of the disk; flowers of the head that are straplike, as in the genus *Taraxacum*

LIGULE in the POACEAE and several CYPERACEAE, the appendage, membrane, or ring of hairs on the adaxial side of a leaf at the junction of the sheath and blade; in the ASTERACEAE, the strap-shaped corolla of a ray floret

LIMB expanded portion of a sympetalous corolla; the strap-shaped corolla of a ray floret in the ASTERACEAE

LIMBER flexible; supple

LINEAR long and narrow with parallel sides

LOBE projecting part of an organ with divisions less than one-half the distance to the base or midvein, usually rounded or obtuse

LOCULAR having compartments divided by septae

LOCULE compartment of an ovary, fruit, or anther

LOMENT jointed, dry fruit, constricted and breaking apart between the seeds

LOOSE not arranged tightly together

LUSTROUS reflecting light evenly without glitter or sparkle

MARGIN edge; border

MARSH area of wet soil characterized by herbaceous vegetation, primarily grasses and grasslike plants

MAT tangled mass of plants growing close to the soil surface and generally rooting at the nodes

MATURE fully developed, usually includes production of seeds

MEADOW moist, level, lowland on which grasses dominate

MEALY surfaces flecked with a lighter color that may rub off

MEMBRANE thin, soft, and pliable tissue

MEMBRANOUS, MEMBRANACEOUS thin, opaque, not green; like a membrane

MERICARP 1-seeded portion of a schizocarp; a portion of a dry dehiscent fruit that splits away as it separates

-MEROUS suffix referring to the number of parts

MESA relatively flat-topped elevated land surface

MESIC characterized by moderately moist conditions; neither dry nor wet

MICRO- prefix meaning small

MIDVEIN central or principal vein of a leaf or bract

MINUTE small

MONOECIOUS plants with male and female flowers at different locations on the same plant; all flowers unisexual

MOTTLED marked with spots or blotches

MOUTH opening of any tube, canal, or cavity (e.g., the top of a leaf sheath in the POACEAE)

MUCILAGE gelatinous substance similar to a plant gum

MUCRO short, sharply pointed tip; a very short awn in the POACEAE

MUCRONATE tipped with a short, slender awn

MUCRONULATE tipped with a very small mucro

MUSH ground meal boiled in water

NAKED uncovered; lacking pubescence; lacking enveloping structures

NATIVE occurring in North America before settlement by Europeans

NAUSEOUS sickening; extremely unpleasant

NECK junction of 2 structures, such as the junction of the lemma and awn of plants in the genera *Nassella* and *Hesperostipa*

NECTAR GUIDES lines inside of flowers thought to serve as guides to pollinating insects

NEEDLE slender leaf, pointed or blunt, as in the genus *Pinus*

NEUTER lacking stamens and pistil; without functional sexual parts

NITRATES compound of nitrogen accumulated by some plants that can cause poisoning if consumed by animals

NODDING inclined somewhat from the vertical; drooping

NODE joints along a stem where leaves are borne; a joint in a stem or inflorescence

NODULOSE with minute knobs or nodules, frequently associated with roots

NOTCH gap; a V-shaped indentation

NUT indehiscent, dry, 1-seeded fruit

NUTLET small, usually 1-seeded hard fruit that is indehiscent; a small nut

OB- prefix meaning inversely

OBCONIC, OBCONICAL inversely cone-shaped with the attachment at the broad end rather than the narrow end

OBCORDATE inversely heart-shaped or cordate with the attachment at the point

OBDELTOID inversely triangular-shaped with the attachment at the point of the triangle rather than along the side

OBLANCEOLATE inversely lanceolate with the broadest portion nearer the apex

OBLATE flattened at the poles

OBLIQUE having the axis not perpendicular to the base; neither perpendicular nor parallel

OBLONG longer than broad, with sides nearly equal; rounded on both sides with parallel sides

OBOVATE opposite of ovate, with the widest part toward the far end; egg-shaped with the widest part above the middle

OBOVOID opposite of ovoid, with the attachment at the narrower end

OBSCURE inconspicuous; not easily seen

OBTURBINATE inversely turbinate; conical; shaped like a top

OBTUSE shape of an apex, with an angle greater than 90°

OBVIOUS easily seen

ODD- number not evenly divisible by 2

OPPOSITE borne across from one another at the same node

ORBICULATE, ORBICULAR nearly circular in outline

ORIGIN place where the species originally occurred

ORNAMENTAL plants cultivated for their beauty rather than for agronomic use

OVAL broadly elliptic with round ends; shorter than oblong

OVARY expanded basal part of the pistil that contains the ovules

OVATE, OVOID shaped like an egg, with the broadest portion closer to the base

OVERFLOW SITES land on which water flows across occasionally

OVERLAP to extend over and cover part of an adjacent structure

OVULATE bearing ovules

OVULE immature seed located in the ovary; the egg-containing part of the ovary

OXALATE group of chemical compounds that may cause poisoning in animals; salts of oxalic acid

PAIRED 2; together

PALATABLE acceptable in taste and texture for consumption

PALE dim; not bright; deficient in color

PALEA adaxial or upper bract of a grass floret; the upper bract subtending a disk floret in the ASTERACEAE

PALER lighter in color

PALMATE with 3 or more lobes, veins, or leaflets arising from a common point

PANICLE inflorescence with a main axis and rebranched branches

PANICULIFORM having the shape, but not necessarily the structure, of a panicle; paniclelike

PAPERY having the texture of writing paper

PAPILIONACEOUS flower type in the FABACEAE having a banner petal, 2 wing petals, and 2 partially fused keel petals

PAPILLA, PAPILLAE small, rounded bump or projection

PAPILLOSE having minute papillae

PAPPUS, PAPPI group of hairs, scales, or bristles that crown the summit of the cypsela in the ASTERACEAE flower (considered to be a modified calyx)

PARCH to toast with dry heat

PARKLAND vegetation type characterized by scattered clumps of trees in prairie

PECTINATE comblike; divided into numerous narrow segments

PEDICEL stalk of a spikelet or single flower in an inflorescence

PEDICELLATE having a pedicel

PEDICELLED borne on a pedicel

PEDUNCLE stalk of a flower cluster or spikelet cluster

PELLUCID translucent or transparent

PELTATE shield-shaped; attached to the petiole on the lower surface

PEMMICAN concentrated food made by some Native Americans consisting of powdered meat mixed with melted fat and occasionally dried fruit

PENDANT, PENDULOUS suspended or hanging downward; drooping

PERENNIAL lasting more than 2 years; applied to plants or plant parts that live more than 2 years

PERFECT applied to flowers having both functional stamens and pistil

PERIANTH floral envelope consisting of the calyx and corolla collectively

PERICARP fruit wall; wall of a ripened ovary

PERIGYNIUM inflated sac that encloses the achene in the genus *Carex*

PERISPORE membrane surrounding a spore

PERSISTENT remaining attached

PETAL part or member of the corolla, usually brightly colored

PETALOID resembling a petal; bracts of a calyx that resemble petals because of color

PETIOLAR growing from the petiole; pertaining to the petiole

PETIOLATE with a petiole

PETIOLE stalk of a leaf blade

PHOTOSENSITIZATION hypersensitivity of the skin to sunlight due to the ingestion of photodynamic compounds from certain plants

PHYLLARY bract of the involucre at the outside of the head of florets in the ASTERACEAE

PIGMENTED colored with a substance

PILOSE with long soft, straight hairs

PILOSULOUS minutely pilose

PINNA primary division of a pinnate leaf

PINNATE having 2 rows of lateral divisions along a main axis (like barbs of a feather)

PINNATIFID deeply cut in a pinnate manner, but not cut entirely to the main axis

PINOLE finely ground flour made from parched corn; any of various flours resembling pinole and ground from seeds of other plants

PIPE-SHAPED shaped like a pipe used to smoke tobacco

PISTIL combination of the stigma, style, and ovary; the female reproductive organ of a flower

PISTILLATE applied to flowers bearing pistils only; unisexual flowers

PIT small depression in a surface

PITH soft, spongy material located in the center of a stem or culm

PLACENTA structure by which the ovule is attached to the wall of the ovary

PLAINS extensive area of flat to rolling land usually covered with vegetation dominated by grasses

PLANOCONVEX flat on 1 side and convex on the other

PLATEAU area of land with a relatively flat surface raised sharply above the adjacent land at least on 1 side; tableland

PLUMBEOUS lead-colored; greenish-gray

PLUME arrangement of hairs that resembles a feather

PLUMOSE feathery, with long pubescence or pinnately arranged hairs or bristles

PLUMP rounded, full

POD dry, dehiscent fruit splitting along 2 sutures, 1 carpel

POLLINATION process of the transfer of pollen from an anther to a stigma

POLYGAMODIOECIOUS monoecious plants having some perfect flowers

POLYGAMOUS with bisexual and unisexual flowers on the same plant

POME inferior, fleshy, many-seeded fruit in which the receptacle forms the outer, fleshy portion (e.g., an apple)

POTHERB plants that are boiled before being eaten

PRAIRIE extensive tract of level to rolling land with vegetation comprised of dominate grasses together with forbs, shrubs, and grasslike plants

PRICKLE small, sharp outgrowth of the bark or epidermis

PRIMARY first

PRIMARY BRANCH branch arising directly from the main inflorescence axis or stem axis

PROCUMBENT prostrate; lying flat on the ground; trailing but not taking root

PROMINENT readily noticeable; projecting out beyond the surface

PROSTRATE lying flat on the ground; procumbent

PROW bow of a ship; the projecting front part; shape of the leaf apices of some species of the genus *Poa*

PSEUDOLATERAL appearing to originate from the side of a structure, but arising from the apex

PUBERULENT diminutive of pubescent

PUBESCENT covered with short, soft hairs

PULP soft, succulent portion of a fruit

PUNCTATE having dots, usually with small glandular pits

PUNGENT sharp and penetrating odor; firm- or sharp-pointed

PUSTULAR, PUSTULATE having small eruptions or blisters

PYRAMIDAL shaped like a pyramid

PYRIFORM pear-shaped

QUININE a bitter crystalline alkaloid used as a tonic or to treat malaria

RACEME inflorescence in which all of the spikelets or flowers are pediceled on a rachis

RACEMOSE racemelike branch of the inflorescence

RACHILLA small axis (applied especially to the axis of a spikelet)

RACHIS, RACHISES axis of a spike, spicate raceme, or raceme inflorescence or pinnately compound leaf

RADIATE term used to describe the ASTERACEAE flower arrangement with the marginal florets ligulate and the disk or central florets tubular; spreading from a common center

RADIATING spreading from a common center

RAME branch of a panicle inflorescence, as in the ANDROPOGONEAE, with paired spikelets at each node

RANKED arranged in 1 or more vertical rows

-RANKED suffix indicating rows or series

RAVINE narrow, steep-sided valley typically eroded by running water

RAY ligulate florets in the ASTERACEAE; the branch of an umbel

RECEPTACLE upper end of the stem of a plant to which the flowering parts are attached

RECURVED curved away from the apex

REDUCED smaller than normal; not functional

REFLEXED bent or turned downward abruptly

REGULAR having structures of the flower, especially the corolla, of similar shape and equally spaced about the center of the flower; flowers with 2 or more planes of symmetry

REMOTE widely spaced

RENIFORM kidney-shaped

REPEL to force away

RESINOUS producing any of numerous viscous substances such as resin or amber

RESTORATION returning the contour of the land and the vegetation to its original condition

RETICULATE in the form of a network; netted as many leaf veins in dicots

RETRORSE pointing backward toward the base

RETUSE with a slight notch at a rounded apex

REVEGETATION replacing current vegetation or starting vegetation on denuded land

REVOLUTE margins rolled downward toward the abaxial surface; opposite of inrolled

RHIZOMATOUS having rhizomes

RHIZOME underground stem with nodes, scalelike leaves, and short internodes

RHOMBIC having the shape of a 4-sided figure with opposite sides parallel and equal but with 2 of the angles oblique; diamond-shaped

RIB prominent vein

RIGHT-OF-WAY usually vegetated land along roads, highways, railroad tracks, pipelines, or transmission lines

RIGID firm; not flexible

ROBUST healthy; full-sized

ROOTSTOCK underground stem; rhizome

ROSETTE basal whorl of leaves, usually crowded

ROSULATE in the form of or resembling a rosette

ROUGH not smooth; surface marked by inequalities

RUDIMENT imperfectly developed organ or part, usually nonfunctional

RUDIMENTARY underdeveloped

RUGOSE wrinkled surface

RUMEN large first compartment of the stomach of a ruminant animal

SAC pouch or baglike structure

SAGITTATE arrowhead-shaped with the lobes turned downward, such as the leaf bases in the genus *Balsamorhiza*

SALICIN crystalline glucoside found in the bark and leaves of some trees and used as a medicine; similar to the active ingredient in aspirin

SALINE nonsodic soil containing sufficient soluble salts to impair its productivity

SAMARA dry, indehiscent 1-seeded fruit with a prominent wing

SAP fluid contained in a plant

SAPONIN any of various glucosides found in plants and marked by the property of producing soapy lather

SCABERULOUS slightly roughened

SCABRIDULOUS minutely roughened

SCABROUS rough to the touch; short, angled hairs requiring magnification for observation

SCALES reduced leaves at the base of a shoot or a rhizome; a thin chafflike portion of the bark of woody plants; a thin, flat structure

SCALY having scales

SCAPE leafless peduncle arising from the ground or basal whorl of leaves and bearing 1 head or solitary flower, as in the genus *Taraxacum*

SCAPIFORM scapelike, but not entirely leafless

SCAPOSE bearing a flower or flowers on a scape or resembling a scape

SCAR mark on the stem where a leaf, bud, flower, or fruit was formerly attached

SCARIOUS thin, dry, membranous, not green

SCHIZOCARP dry, dehiscent fruit consisting usually of 2 or more carpels that, when mature, split apart forming 1-seeded parts (mericarps)

SCLEROTIUM in fungi, such as ergot, a hardened compact mass of mycelium that gives rise to the fruiting bodies; replaces the caryopsis in ergot-infested POACEAE

SCOURS diarrhea in livestock

SCURFY covered with minute scales or specialized mealy hairs

SEASON-LONG throughout 1 season; often used to describe grazing during the growing season

SECRETION materials such as resins, mucilages, gums, oils, and nectar on the exterior of plant parts

SEED ripened ovule

SEEP place where water oozes slowly to the land surface

SEGMENT part of a structure which may be separated from the other parts

SELENIUM nonmetallic element frequently extracted from the soil by plants and may be poisonous to animals in relatively large quantities

SENESCENCE process of plant aging; decline following maturity

SEPAL member of the calyx bracts, usually green and subtending the petals

SEPTATE divided by 1 or more wall or membrane

SEPTUM dividing wall or membrane

SERIES group with an order of arrangement; in the ASTERACEAE the number of rows of bracts in the involucre

SERRATE saw-toothed margins, with teeth pointing toward the apex

SERRULATE minutely serrate

SESSILE without a pedicel or stalk

SETACEOUS bristlelike hairs

SETOSE covered with hairs; hispid

SHATTER to break apart; fall and scatter as seeds, leaves, or fruits

SHEATH lower part of a leaf that encloses the stem in POACEAE

SHEATHING structure that encloses or wraps around another structure, such as the base of a grass leaf when it surrounds the culm

SHEDDING casting off parts

SHINY lustrous; possessing a sheen

SHOOT young stem or branch

SHOWY attractive, such as a large colorful flower; striking appearance

SHRED long, narrow strip; fibrous exfoliating bark

SHREDDING detaching in long, narrow strips

SHRUB low-growing woody plant; bush with 1 or more trunks

SILICLE short capsule of 2 carpels, about at long as broad; a type of capsule in the BRASSICACEAE

SILIQUE long, slender capsular fruit of 2 carpels; a type of capsule in the BRASSICACEAE

SILKY fine, lustrous, long hair; resembling silk in appearance or texture

SILVERY lustrous and gray or white; having the luster of silver

SIMPLE not branched; not compound; single

SINUATE strong, wavy margins

SINUS indentation between 2 lobes or teeth segments, such as on leaf blade margins

SLOUGH place in which shallow water stands for most or all of the year

SNUFF to draw forcibly into the nostrils

SOD-FORMING creating a dense mat, usually associated with rhizomatous plants

SOLITARY single; alone; 1 by itself

SORUS, SORI cluster of sporangia on the surface of a fern frond

SPARINGLY meager; not dense

SPARSE scattered; opposite of dense

SPATHE modified sheathing bract of the inflorescence; a modified leaf sheath

SPATULATE shaped like a spatula, being broader above than below

SPHERICAL, SPHEROID having the shape of a sphere or ball; a globular body

SPICATE spikelike

SPICIFORM having the shape, but not necessarily the structure, or a spike; spikelike

SPIKE unbranched inflorescence in which the spikelets or flowers are sessile on the central axis

SPIKELIKE having the appearance of a spike

SPIKELET unit of inflorescence in grasses usually consisting of 2 glumes, 1 or more florets, and a rachilla

SPINE stiff, pointed outgrowth that is usually woody

SPINESCENT bearing a spine; terminating in a spine

SPINOSE having spines

SPLIT divided lengthwise

SPLOTCH blotch or spot; blending of a spot and blotch

SPORANGIUM, SPORANGIA spore-bearing sac or case

SPORE reproductive cell from meiotic division in a sporangium

SPOT GRAZING heavy, repeated grazing of localized areas

SPRIG vegetative shoot that is planted to establish a new plant

SPROUT shoot of a plant, especially the first from a root or a germinating seed

SPUR any slender, hollow projection of a flower, such as the flowers of the genus *Delphinium*

SQUARROSE having spreading and recurved parts, such as phyllary bracts surrounding an inflorescence, as in the genus *Grindelia*

STALK supporting structure of an organ

STAMEN pollen-producing structure of a flower; typically an anther borne at the apex of a filament

STAMINATE flower containing only stamens, unisexual flowers

STASIS slowing down or cessation of flow or a process

STEEP to soak in water at a temperature under the boiling point

STELLATE star-shaped, usually referring to hairs with many branches from the base; type of hair

STEM portion of the plant bearing nodes, internodes, leaves, and buds

STERILE without functional pistils, may or may not bear stamens

STICKY covered with an adhesive substance

STIFF not easily bent; rigid

STIGMA apical portion of the pistil that receives the pollen

STIPE stalk or stem that supports an organ

STIPITATE borne on a stalk or stipe

STIPULES appendages, usually leaflike, occurring in pairs, 1 on either side of the petiole base; may be leaflike, glands, or modified to spines

STOLON horizontal, aboveground, modified propagating stem with nodes, internodes, and leaves

STOLONIFEROUS bearing stolons

STONE the hard, inner portion of a drupe that contains the seed

STOUT sturdy; strong; rigid

STRAGGLY scattered irregularly

STRAMINEOUS straw-colored

STRIATE marked with slender, longitudinal grooves or lines; appearing striped

STRICT narrow, with close, upright branches

STRIGOSE rough, with short, stiff hairs or bristles

STRIGULOSE minutely strigose

STRIKING attractive; showy

STRINGY resembling or consisting of fibrous material

STRYCHNINE poisonous alkaloid

STYLAR pertaining to a style

STYLE slender, elongated portion of the pistil which bears the stigma at its apex

SUB- prefix that denotes somewhat, slightly, or in less degree

SUBSHRUB suffrutescent perennial plant

SUBTEND to underlie; located below

SUBTERRANEAN below ground

SUBULATE shaped like an awl

SUCCULENT fleshy and juicy

SUFFRUTESCENT slightly shrubby; having a slightly woody or shrubby base

SUFFRUTICOSE plants herbaceous above and woody at the base

SULCATE having grooves or furrows

SULCUS groove or furrow

SUMMIT top; apex

SUPERIOR higher; above; more elevated in position

SURMOUNT directly on top of

SURPASS exceed

SUTURE line or seam marking the union of 2 parts; the line of dehiscence of a fruit or capsule

SWALE low-lying depression in the land

SWAMP seasonally flooded area with more woody plants than a marsh

SWOLLEN enlarged

SYMMETRICAL divisible into 2 similar parts by more than 1 plane passing through the center

SYMPETALOUS petals united at or near the base

TAPERING regularly narrowing toward one end

TAPROOT primary root of a plant that grows directly downward and gives rise to lateral branches

TAR brown or black viscous, odorous organic substance

TARDILY late

TAWNY pale brown or dirty yellow

TEETH pointed divisions, as on leaf blade margins

TEPAL segment of a perianth that is not clearly differentiated into a calyx and corolla; a sepal or petal

TERETE cylindric and slender; circular in cross section

TERMINAL borne at or belonging to the extremity or summit

TETRAHEDRAL geometric form with 4 faces

THICKET dense growth of shrubs or small trees

THROAT opening; the orifice of a corolla or calyx

THYRSE flower cluster of racemosely arranged cymes organized into an elongate panicle

THYRSOID, THRYSIFORM resembling a thyrse

TILLER shoot from an adventitious bud at the base of a plant

TINGED slightly colored

TIP apex

TOMENTOSE surface covered with matted, tangled, densely woolly hairs

TOMENTULOSE finely or slightly tomentose

TOMENTUM covering of dense, woolly hairs

TONIC beverage that refreshes, stimulates, restores, or invigorates

TOOTH pointed projection or division

TRAILING prostrate and creeping, but not rooting

TRANSLUCENT semitransparent; transmitting light rays only partially

TRANSVERSE at 90° angles to the long axis; crosswise; cross section

TRI- prefix meaning 3

TRIANGULAR having 3 edges and 3 angles

TRICHOME epidermal hair

TRIDENTATE 3-toothed, such as *Artemisia tridentata* leaves

TRIFID divided into 3 parts

TRIFOLIATE having 3 leaflets, such as *Medicago polymorpha*

TRIGONOUS having 3 angles, such as achenes of many species of *Carex*

TRUNCATE ending abruptly; appearing to be cut off at the end

TRUNK main stem of a tree or shrub

TUBERCLE small projection from the surface of an organ or structure

TUBERCULATE furnished with small projections

TUBULAR having the shape of a tube, such as the corolla of some flowers

TUFT cluster; bunch; not rhizomatous

TUMBLE to roll over and over as when blown by the wind

TURBINATE top-shaped; inversely conical

TWIG small branch of a tree or shrub

ULTIMATE smallest subdivisions

UMBEL simple flat-topped or rounded inflorescence with pedicels or rays radiating from a common point

UMBELLATE resembling an umbel

UMBELLIFORM in the shape of an umbel

UNARMED without thorns, prickles, or spines

UNDULATE strongly wavy in a perpendicular plane

UNILATERAL arranged on or directed toward one side

UNISEXUAL flowers containing only stamens or only pistils

UNITED two or more wholly or partially fused parts

UNPALATABLE not desirable for food; not readily eaten

URCEOLATE shaped like an urn

URN hollow and contracted near the mouth; urceolate

UTRICLE small 1-seeded fruit with a thin wall, dehiscing by the breakdown of the thin wall; similar to an achene

VALVE a single portion of a compound ovary; part of a pod or capsule

VARIETY category or taxonomic rank below the species and subspecies level

VEIN single branch of the vascular system of a plant

VELUTINOUS velvety vestiture

VELVETY soft and smooth like velvet; velutinous

VENATION pattern formed by the veins in a leaf or other plant organ

VERTICIL whorl of 2 or more levels of branching

VERTICILLATE whorls; arranged in verticils

VESTIGE rudimentary structure

VESTITURE any covering on a surface making it other than glabrous

VILLOUS with long, soft macrohairs; similar to pilose, but with a higher density of hairs

VISCID sticky or clammy

WARM-SEASON category of plants that grow during the warmer portions of the year

WART growth or large blister on the epidermis, resembling a wart on an animal

WASHES areas of active water erosion

WAVY margin with small, regular lobes; undulating surface or margin

WEAK frail; not stout nor rigid; partially or incompletely

WEBBED bearing fine, tangled hairs

WHORL cluster of several branches or leaves around the axis arising from a common node

WING thin projection or border

WIRY being thin and resilient

WITHERED appearing shriveled and shrunken

WOOLLY covered with long, entangled soft hairs

WRINKLE small ridges and/or furrows on a surface

XEROPHYTE plant adapted to a dry habitat

ZIGZAG series of short, sharp bends

ZYGOMORPHIC irregular; divisible into equal parts in only 1 plane (contrast with actinomorphic)

Abbreviations for Nomenclature Authorities

ABRAMS LeRoy Abrams (1874–1956), Stanford University botanist

AELLEN Paul Aellen (1896–1973), botanist in Basel, Switzerland

ALL. Carlo Allioni (1728–1804), Italian physician and University of Turin professor of botany

AITON William Aiton (1731–1793), botanist and Royal Gardener with the Royal Botanic Gardens, Kew, England

ALLRED Kelly W. Allred (b. 1949), New Mexico State University agrostologist

ALLRED & GOULD Kelly W. Allred (b. 1949), New Mexico State University agrostologist, and Frank Walton Gould (1913–1981), Texas A&M University agrostologist

ALEXANDER Edward Johnston Alexander (1905–1985), botanist and assistant curator with the New York Botanical Garden

ANDERSON, L.C. Loran Crittendon Anderson (b. 1936), Florida State University botanist

ANDERSSON Nils Johan Andersson (1821–1880), Swedish botanist and director of the botanical department of the Swedish Museum of Natural History, Stockholm

ASCH. & SCHWEINF. Paul Friedrich August Ascherson (1834–1913), German botanist, and Georg August Schweinfurth (1836–1925), German botanist

BARKWORTH Mary Elizabeth Barkworth (b. 1941), Utah State University agrostologist

BARKWORTH & D.R. DEWEY Mary Elizabeth Barkworth (b. 1941), Utah State University agrostologist, and Douglas R. Dewey (1929–1993), U.S. Department of Agriculture and Utah State University cytogeneticist

BAUM, B.R. & FINDLAY Bernard René Baum (b. 1937), French-born Canadian botanist, and Judy N. Findlay (b. 1941), Canadian research technician

BEAL William James Beal (1833–1924), Michigan State University agrostologist

BEAUV., P. Ambroise Marie Francois Joseph Palisot de Beauvois (1752–1820), French naturalist

BEETLE Alan Ackerman Beetle (1913–2003), University of Wyoming taxonomist

BENTH. George Bentham (1800–1884), taxonomist with the British Museum of Natural History, London

BERTOL. Antonio Bertoloni (1775–1869), Italian professor of botany

BESSER Wilbert Swibert Joseph Gottlieb von Besser (1784–1842), Austrian-born botanist who worked in the Ukraine

BESSEY Charles Edwin Bessey (1845–1915), University of Nebraska botanist and administrator

BIEB., M. Baron Friedrich August Marschall von Bieberstein (1768–1826), German explorer (Russia and the Caucasus)

BIEHLER Johann Friedrich Theodor Biehler (1785–1850), German botanist

BIERNER Mark William Bierner (b. 1946), Southwest Texas State University botanist

BIGELOW Jacob Bigelow (1787–1879), Harvard Medical School physician and botanist

BISCH. Gottlieb Wilhelm Bischoff (1797–1854), German professor of botany

BLAKE, S.F. Sidney Fay Blake (1892–1959), U.S. Department of Agriculture scientist

BOL. Henry Bolander (1831–1897), German-born state botanist for California

BONPL. Aimé Jacques Alexader Bonpland (1773–1858), French-born botanist who worked extensively in South America

BOOTT Francis Boott (1805–1887), American botanist and authority on *Carex*

BORY Jean Baptiste Georges Geneviève Marcellin Bory de Saint-Vincent (1778–1846), French botanist

BOTSCH. Victor Petrovič Botschantzev (1910–1990), Russian botanist

BOWDEN Wray Merrill Bowden (1914–1996), cytogeneticist with the Department of Agriculture, Canada

BRIQ. John Isaac Briquet (1870–1931), botanist and director of the Conservatoire Botanique, Geneva, Switzerland

BR., R. Robert Brown (1777–1858), botanist and first keeper of botany with the British Museum of Natural History, London

BRITTON Nathaniel Lord Britton (1859–1934), director-in-chief of the New York Botanical Garden

BRITTON & ROSE Nathaniel Lord Britton (1859–1934), director-in-chief of the New York Botanical Garden, and Joseph Nelson Rose (1862–1928), botanist with the U.S. Department of Agriculture and U.S. National Herbarium

BRITTON & RUSBY Nathaniel Lord Britton (1859–1934), director-in-chief of the New York Botanical Garden, and Henry Hurd Rusby (1855–1940), dean of the New York College of Pharmacy and botanical collector, especially in South America

BUCKLEY Samuel Botsford Buckley (1809–1884), naturalist and state geologist of Texas

BURTT DAVEY Joseph Burtt Davy (1870–1940), English botanist and agrostologist

BUSH Benjamin Franklin Bush (1858–1937), amateur Missouri botanist

CARO & E.A. SÁNCHEZ José Aristide (Alfredo) Caro (1919–1985), Argentinian agrostologist, and Evangelina A. Sánchez (b. 1934), Argentinian naturalist and taxonomist

CAV. Antonio José Cavanilles (1745–1804), Spanish botanist

CHASE Mary Agnes Merrill Chase (1869–1963), agrostologist and custodian of grasses with the U.S. National Herbarium

CHODAT Robert Hippolyte Chodat (1865–1934), Swiss botanist

CHU, G.L. & S.C. SAND. Ge Ling Chu (b. 1934), Chinese botanist, and Stewart Cottam Sanderson (b. 1942), American botanist

CLARKE, C.B. Charles Baron Clarke (1832–1906), botanist who worked in India and with the Royal Botanic Gardens, Kew, England

COCKERELL Theodore Dru Alison Cockerell (1866–1948), professor of biological sciences in New Mexico and Colorado

COLLOTZI Albert William Collotzi (1936–1997), botanist who worked in Utah

COLUMBUS James Travis Columbus (b. 1962), agrostologist with the Rancho Santa Ana Botanic Garden, Claremont, California

CORY Victor Louis Cory (1880–1964), Southern Methodist University botanist

COULT., J.M. John Merle Coulter (1851–1928), University of Chicago professor of botany

COVILLE Fredrick Vernon Coville (1867–1937), curator of the U.S. National Herbarium

CZEREP. Serge Kirillovich Czerepanov (b. 1921), Russian botanist

DANDY James Edgar Dandy (1903–1976), English botanist and keeper of botany with the British Museum of Natural History, London

DC. Augustin Pyramus de Candolle (1778–1841), Swiss botanist and professor of botany

DC., A. Alphonse Louis Pierre Pyramus de Candolle (1806–1893), French-Swiss botanist and son of Augustin Pyramus de Candolle

DESF. René Louiche Desfontaines (1750–1833), French botanist

DESR. Louis Auguste Joseph Desrousseau (1753–1838), French botanist

DESV. Nicaise Auguste Desvaux (1784–1856), French botanist and director of the botanical gardens in Angers, France

DEWEY Reverend Chester Dewey (1784–1867), University of Rochester *Carex* specialist

DEWEY, D.R. Douglas R. Dewey (1929–1993), U.S. Department of Agriculture and Utah State University cytogeneticist

DIETR., D. David Nathanael Friedrich Dietrich (1799–1888), German botanist

DOLL, R. Reinhard Doll (b. 1941), German botanist

DON, D. David Don (1799–1841), English botanist

DOUGLAS David Douglas (1798–1834), Scottish botanical collector in northwestern United States and Canada

DUMORT. Barthélemy Charles Joseph Dumortier (1797–1878), French agronomist and horticulturist

DUNAL Michel Felix Dunal (1789–1856), French botanist

DU ROI John Philipp Du Roi (1741–1785), German physician and specialist in the taxonomy of American trees

EASTW. Alice Eastwood (1859–1953), Canadian-born curator of botany with the California Academy of Sciences

EATON Amos Eaton (1776–1842), American botanist, produced first botanical manual in United States with descriptions in English

EATON, D.C. Daniel Cady Eaton (1843–1895), Yale University botanist and herbarium curator

ELMER Adolph Daniel Edward Elmer (1870–1942), American botanist

ENDL. Stephan Friedrich Ladislaus Endlicher (1804–1849), Austrian botanist

ENGELM. George Engelmann (1809–1884), physician and botanist in St. Louis, Missouri

FERNALD Merritt Lyndon Fernald (1873–1950), plant geographer and systematist, director of the Gray Herbarium, Harvard University

FERNALD & J.F. MACBR. Merritt Lyndon Fernald (1873–1950), plant geographer and systematist, director of the Gray Herbarium, Harvard University, and James Francis Macbride (1897–1976), taxonomist with the Gray Herbarium

FIORI & PAOL. Andriano Fiori (1865–1950), Italian botanist, and Giulio Paoletti (1865–1941), Italian botanist

FISCH. Friedrich Ernst Ludwig von Fischer (1782–1854), director of the Botanical Garden, St. Petersburg, Russia

FISCH. & TRAUTV. Friedrich Ernst Ludwig von Fischer (1782–1854), director of Botanical Garden, St. Petersburg, Russia, and Ernst Rudolph von Trautvetter (1809–1889), Russian botanist

FLOD. Björn Gustaf Oscar Floderus (1867–1941), Swedish physician, botanist, and authority on *Salix*

FOURN., E. Eugene Pierre Nicolas Fournier (1834–1884), physician and amateur botanist in Paris

GAERTN. Joseph Gaertner (1732–1791), German botanist

GAUDIN Jean François Aimé Théophile Philippe Gaudin (1766–1833), Swiss botanist

GEYER Carl Andreas Geyer (1809–1853), Austrian botanist who collected in the northwestern United States

GOODEN. Samuel Goodenough (1743–1827), English botanist and bishop of Carlisle, England

GOULD Frank Walton Gould (1913–1981), Texas A&M University agrostologist

GRAY, A. Asa Gray (1810–1888), Harvard University professor of botany

GREENE Edward Lee Greene (1842–1915), University of California botanist

GRIFFITHS David Griffiths (1867–1935), British-born agronomist with the U.S. Department of Agriculture

GRISEB. August Heinrich Rudolf Grisebach (1814–1879), German professor of botany in Göttingen

GROSSH. Alexander Alfonsovich Grossheim (1888–1948), Ukrainian botanist and authority on ferns

HACK. Eduard Hackel (1850–1926), Austrian agrostologist

HARTM. Carl Johann Hartman (1790–1849), Swedish physician and botanist

HARV. & A. GRAY William Henry Harvey (1811–1866), Irish botanist, and Asa Gray (1810–1888), Harvard University professor of botany

HAUSSKN. Heinrich Carl Haussknecht (1838–1903), botanist and professor in Weimar, Germany

HELLER, A. Amos Arthur Heller (1867–1944), botanist in Pennsylvania

HENDRICKSON James Solberg Hendrickson (b. 1940), botanist with the Plant Resources Center, University of Texas

HENRARD Johannes Theodoor Henrard (1881–1974), Dutch pharmacist and conservator of Rijksherbarium, Netherlands

HERTER Wilhelm Gustav Herter (1884–1958), German botanist who resided in Uruguay after 1924

HITCHC. Albert Spear Hitchcock (1865–1935), agrostologist with the U.S. Department of Agriculture

HITCHC. & CHASE Albert Spear Hitchcock (1865–1935), agrostologist with the U.S. Department of Agriculture, and Mary Agnes Merrill Chase (1869–1963), agrostologist and custodian of grasses with the U.S. National Herbarium

HOFFM., O. Karl August Otto Hoffmann (1853–1909), German botanist

HOLUB Josef Ludwig Holub (1930–1999), Czech botanist

HONDA Masaji (Masazi) Honda (1897–1984), Japanese botanist

HOOK. Sir William Jackson Hooker (1785–1865), director of the Royal Botanic Gardens, Kew, England

HOOK. F. Joseph Dalton Hooker (1817–1911), son of Sir William Jackson Hooker

HOOK. & ARN. Sir William Jackson Hooker (1785–1865), director of the Royal Botanic Gardens, Kew, England, and George Arnott Walker-Arnott (1799–1868), Scottish botanist

HOWELL Thomas Jefferson Howell (1842–1912), botanist in Oregon

HOWELL, J.T. John Thomas Howell (1903–1994), botanist in California

HOST Nicolaus Thomas Host (1761–1834), Austrian physician and botanist

HUDS. William Hudson (1730–1793), British botanist and apothecary

HULTÉN Oskar Eric Gunnar Hultén (1894–1981), Swedish botanist and plant explorer

HUMB. & BONPL. Friedrich Wilhelm Heinrich Alexander von Humboldt (1769–1859), Prussian naturalist and explorer, and Aimé Jacques Alexandre Bonpland (1773–1858), French botanist and explorer

HYL. Nils Hylander (1904–1970), Swedish botanist and mycologist

IIJIN Modest Mikhaĭlovich Iijin (1889–1967), Russian botanist

ISELY Duane Isely (1918–2000), Iowa State University taxonomist

JACQ. Nikolaus Joseph Baron von Jacquin (1727–1817), Austrian taxonomist

JONES, M.E. Marcus Eugene Jones (1852–1934), botanist and early explorer in the western United States

JORD. Claude Thomas Alexis Jordan (1814–1897), French botanist

KEARNEY Thomas Henry Kearney (1874–1956), botanist and agronomist with the United States Department of Agriculture

KELLOGG Albert Kellogg (1813–1887), physician and botanist, San Francisco

KITAG. Masao Kitagawa (1909–1995), Japanese botanist

KLOTZSCH Johann Friedrich Klotzsch (1805–1860), German pharmacist, botanist, and director of the Royal Herbarium and Botanical Garden, Berlin

KOEHNE Bernhard Adalbert Emil Koehne (1848–1914), German botanist

KRAUSE, E.H.L. Ernst Hans Ludwig Krause (1859–1942), German botanist

KROCK. Anton Johann Krocker (1744–1823), Czechoslovakian medical botanist

KUHN Friedrich Adalbert Maximilian Kuhn (1842–1894), German fern taxonomist

KUNTH Carl Sigismund Kunth (1788–1850), German botanist

KUNTZE Carl Ernst Otto Kuntze (1843–1907), German botanist

L. Carolus Linnaeus (1707–1778), Swedish botanist and author of *Species Plantarum*, recognized as the starting point for botanical nomenclature

L. F. Carolus Linnaeus, the son, (1741–1783), successor to his father as professor of botany, Uppsala, Sweden

LAG. Mariano Lagasca y Segura (1776–1839), Spanish professor

LAM. Jean Baptiste Antoine Pierre Monnet de Lamarck (1744–1829), French botanist

LANGE Johan Martin Christian Lange (1818–1889), professor of botany, Copenhagen, Denmark

LAWSON, P. & C. LAWSON Peter Lawson (1762–1820), and Sir Charles Lawson (1794–1873), father and son, Scottish nurserymen

LEDEB. Carl Friedrich von Ledebour (1785–1851), professor and author, Dorpat, Estonia

LEMMON John Gill Lemmon (1832–1908), botanist who worked in California and Arizona

LESS. Christian Friedrich Lessing (1809–1862), German physician and ASTERACEAE specialist

LEWIS, H.F. Harrison Flint Lewis (1893–1974), amateur botanist and chief of the Canadian Wildlife Service

LEYSS. Freidrich Wilhelm von Leysser (1731–1815), German botanist and author of the *Flora of Halle*

L'HÉR. Charles Lous L'Héritier de Brutelle (1746–1800), French magistrate and botanist

LINDB., H. Harald Lindberg (1871–1963), Swedish-born botanist and first custodian with the botanical museum in Helsinki, Finland

LINDL. John Lindley (1799–1865), English professor of botany

LINK Johann Heinrich Friedrich Link (1767–1851), German professor of natural science, Berlin

LOUIS-MARIE Pére Louis-Marie (1896–1978), Canadian botanist

LÖVE, Á. Áskell Löve (1916–1994), Icelandic botanist who worked in North America

LÖVE, Á. & D. LÖVE Áskell Löve (1916–1994), Icelandic botanist who worked in North America, and Doris Benta Maria (Mrs. Áskell) Löve (1918–2000)

LÖVE. Á. & D. LÖVE & KAPOOR Áskell Löve (1916–1994), Icelandic botanist who worked in North America, Doris Benta Maria (Mrs. Áskell) Löve (1918–2000), and Brij Mohan Kapoor (1936–2003), Indian-born botanist

LUNELL Jöel Lunell (1850–1920), Swedish-born botanist who worked in North Dakota

MACBR, J.F. James Francis Macbride (1892–1976), American botanist who studied the flora of Peru

MACMILLAN Conway MacMillan (1867–1929), University of Minnesota botanist

MACOUN John Macoun (1831–1920), Irish-born Canadian botanist

MALTE Oscar Malte (1880–1933), chief botanist with the National Herbarium of Canada, Ottawa

MARSHALL Humphry Marshall (1722–1801), American botanist

MARTENS, M. Martin Martens (1797–1863), Netherlands-born, Belgian botanist and chemist

MARTENS, M. & GALEOTTI Martin Martens (1797–1863), Netherlands-born, Belgian botanist and chemist, and Henri Guillaume Galeotti (1814–1858), French-Belgian botanist and director of the botanical gardens in Brussels

MCCLURE Floyd Alonzo McClure (1897–1970), botanist with the U.S. Department of Agriculture and research associate with the Smithsonian Institution

MCMINN Howard Earnest McMinn (1891–1963), professor of botany in California

MEDIK. Friedrich Kasimir Medikus (1736–1808), physician, botanist, and curator of the botanical garden in Mannheim, Germany

MEEUSE, A. & A. SMIT Adrianus Dirk Jacob Meeuse (b. 1914), Netherlands-born botanist who worked in South Africa, and A. Smit (b. 1934), Danish botanist

MELDRIS Aleksandre Meldris (1909–1986), European taxonomist and author

MERR. Elmer Drew Merrill (1876–1956), director of the New York Botanical Garden and Arnold Arboretum, Harvard University

MEY., C.A. Carl Anton von Meyer (1795–1855), director of the Botanical Garden, St. Petersburg, Russia

MEY., E. Ernest Heinrich Friedrich Meyer (1791–1858), professor of botany and director of the botanical garden with the University of Königsberg, Russia

MICHX. Andre Michaux (1746–1802), French botanist and explorer of North America

MILL. Philip Miller (1691–1771), English botanist

MOENCH Conrad Moench (1744–1802), German botanist

MOLINA Friar Giovanni Ignazio Molina (1737–1829), Chilean priest, botanist, and ornithologist

MOQ. Christian Horace Benedict Alfred Moquin-Tandon (1804–1863), French botanist

MUHL. Gotthilf Heinrich Ernest Muhlenberg (1753–1815), German-educated Lutheran minister and pioneer botanist in Pennsylvania

MÜLL., C.H. Cornelius Herman Müller (1909–1997), University of California, Santa Barbara, botanist and herbarium curator

NASH George Valentine Nash (1864–1921), agrostologist and head gardener with the New York Botanical Garden

NEES Christian Gottfried Daniel Nees von Esenbeck (1776–1858), German botanist

NELSON, A. Aven Nelson (1859–1952), University of Wyoming professor of botany and president

NELSON, A. & J.F. MACBR. Aven Nelson (1859–1952), University of Wyoming professor of botany and president, and James Francis Macbride (1892–1976), American botanist who studied the flora of Peru

NESOM, G.L. & G.I. BAIRD Guy L. Nesom (b. 1945), Botanical Research Institute of Texas botanist, and Gary I. Baird (b. 1955), Brigham Young University-Idaho taxonomist

NEVSKI Sergei Arsenjevic Nevski (1908–1938), Russian agrostologist

NIEUWL. Julius Arthur Nieuwland (1878–1936), Belgian-born, priest and professor of botany and chemistry with the University of Notre Dame

NUTT. Thomas Nuttall (1786–1859), English-American naturalist who collected in the western United States

OPIZ, P. Philipp Maxmilian Opiz (1787–1858), Bohemian botanist

OSTENF. Carl Hansen Ostenfeld (1873–1931), Danish botanist

PAINTER, J.H. Joseph Hannum Painter (1879–1908), botanist and assistant curator in the Division of Plants, United States National Museum

PALL. Peter Simon Pallas (1741–1811), German botanist

PALLA Eduard Palla (1864–1922), Austrian botanist and authority on sedges

PALMER, E.J. Ernest Jesse Palmer (1875–1962), collector for the Arnold Arboretum, Harvard University, and Missouri Botanical Garden

PARRY Charles Christopher Parry (1823–1890), botanist with the Mexican Border Survey

PAVON José Antonio Pavon Jiménez (1754–1844), Spanish botanist who studied the flora of Chile and Peru

PAYNE, W.W. Willard William Payne (b. 1937), University of Illinois botanist

PERS. Christiaan Hendrick Persoon (1761–1836), Dutch-born South African and French botanist

PETERSON, P.M. Paul M. Peterson (b. 1954), agrostologist with the National Museum of Natural History

PETERSON, P.M. & N. SNOW Paul M. Peterson (b. 1954), agrostologist with the National Museum of Natural History, and Neil Wilton Snow (b. 1960), botanist and musician in Colorado and Kansas

PETERSON, P.M. & SAARELA Paul M. Peterson (b. 1954), agrostologist with the National Museum of Natural History, and Jeffery Michael Saarela (b. 1978), research botanist with the Canadian Museum of Nature, Ottawa

PHIL. Rudolf Amandus Philippi (1808–1904), Chilean botanist

PILG. Robert Knud Friedrich Pilger (1876–1953), German botanist, Berlin

PIPER Charles Vancouver Piper (1867–1926), Washington State University professor of botany and zoology

PLANCH. Jules Émile Planchon (1823–1888), botanist and head of the department of botanical sciences with the University of Montpellier, France

POHL, R.W. Richard Walter Pohl (1916–1993), Iowa State University taxonomist specializing in POACEAE

POIR. Jean Louis Marie Poiret (1755–1834), French botanist

POLJAKOV Peter Petrovich Poljakov (1902–1974), Russian botanist

POPL. Henrietta Ipplitovna Poplavskaja (1886–1916), Russian botanist

PORSILD Morton Pedersen Porsild (1872–1956), Danish botanist

PORTER & J.M. COULT. Thomas Conrad Porter (1822–1901), professor of botany with Lafayette College, Pennsylvania, and John Merle Coulter (1851–1928), University of Chicago professor of botany

POTT Johann Friedrich Pott (1738–1805), German botanist and physician

PRANTL Karl Anton Eugen Prantl (1849–1893), professor and director of the botanical garden Breslau University, Germany

PRESL, J. Jan Svatopluk Presl (1791–1849), natural scientist and professor in Prague

PURSH Fredrick Traugott Pursh (1774–1820), German author and botanical collector in North America

RAF. Constantin Samuel Rafinesque (1783–1840), Constantinople-born pioneer naturalist in Kentucky

RAUP Hugh Miller Raup (1901–1995), research associate with the Arnold Arboretum, Harvard University

RAUSCHERT Stephan Rauschert (1931–1986), German botanist

RICH. Louis Claude Marie Richard (1754–1821), French botanist and collector in South America and the West Indies

RICHARDS, A.J. Adrian John Richards (b. 1943), botanist with the University of Newcastle upon Tyne, England

RICHARDSON John Richardson (1787–1865), Scottish physician and botanist

RICHT., K. Karl Richter (1855–1891), Austrian botanist

RICKER Percy Leroy Ricker (1878–1973), agronomist who specialized in POACEAE and FABACEAE with the U.S. Department of Agriculture

ROEM. Johann Jacob Roemer (1763–1819), Swiss botanist

ROEM. & SCHULT. Johann Jacob Roemer (1763–1819), Swiss botanist, and Joseph August Schultes (1773–1831), Austrian botanist

ROEM., M. Max Joseph Roemer (1791–1849), German botanist

ROLLINS Reed Clark Rollins (1911–1998), Harvard University botanist and professor

ROOF, J. James Bernard Roof (1910–1983), botanist in California

ROSE & PAINTER Joseph Nelson Rose (1862–1928), botanist with the U.S. Department of Agriculture and National Herbarium, and William Hunt Painter (1835–1910), British clergyman and botanist

ROSHEV. Roman Julievich Roshevitz (1882–1949), Russian botanist

ROTH Albrecht Wilhelm Roth (1757–1834), German physician and botanist

ROWLEE Gordon Douglas Rowlee (b. 1921), English botanist

RYDB. Per Axel Rydberg (1860–1931), Swedish-born botanist with the University of Nebraska and the New York Botanical Garden

RZED. Jerzy Rzedowski (b. 1926), Polish-born botanist who worked in Mexico

SARG. Charles Sprague Sargent (1841–1927), Harvard University botanist

SCHEELE Georg Heinrich Adolf Scheele (1808–1864), German botanist who described plants collected in Texas

SCHRAD. Heinrich Adolph Schrader (1767–1836), German botanist and professor

SCHREB. Johann Christian Daniel von Schreber (1739–1810), German botanist and professor in Erlangen

SCHULT. Joseph August Schultes (1773–1831), Austrian botanist and professor

SCHUM., K. Karl Moritz Schumann (1851–1904), German botanist

SCHRANK Franz von Paula von Schrank (1747–1835), German-born priest, botanist, and entomologist

SCHWARZ, O. Otto Karl Anton Schwarz (1900–1983), botanist with Friedrich Schiller University and Jena Botanical Garden, Germany

SCOP. Joannes Antonius Scopoli (1723–1788), Austrian botanist and physician

SCOTT, A.J. Andrew John Scott (b. 1950), European botanist

SCRIBN. Frank Lamson Scribner (1851–1938), agrostologist with the U.S. Department of Agriculture

SCRIBN. & MERR. Frank Lamson Scribner (1851–1938), agrostologist with the U.S. Department of Agriculture, and Elmer Drew Merrill (1876–1956), director of the New York Botanical Garden and Arnold Arboretum, Harvard University

SCRIBN. & J.G. SM. Frank Lamson Scribner (1851–1938), agrostologist with the U.S. Department of Agriculture, and Jared Gage Smith (1866–1925), agrostologist with the U.S. Department of Agriculture

SCRIBN. & T.A. WILLIAMS Frank Lamson Scribner (1851–1938), agrostologist with the U.S. Department of Agriculture, and Thomas Albert Williams (1865–1900), agrostologist with the U.S. Department of Agriculture

SEGLER & EBINGER David Stanley Seigler (b. 1940), University of Illinois botanist, and John Edward Ebinger (b. 1933), Eastern Illinois University botanist

SENNEN & PAU Padre Sennen (Etienne Marcellin Granie-Blanc) (1861–1937), French-born clergyman, botanical collector, and taxonomist in Ecuador, and Carlos Pau y Espanola (1857–1937), Spanish botanist

SESSÉ & MOC. Martin de Sessé y Lacasta (1751–1808), Spanish botanist and director of the botanical garden in Mexico City, and José Mariano Mociño (1757–1820), Mexican physician

SHINNERS Lloyd Herbert Shinners (1918–1971), Canadian-born botanist, professor of botany with Southern Methodist University

SIMONK. Lajos von Simonkai (Simkovics) (1851–1910), Hungarian phytogeographer

SIMS John Sims (1749–1831), English botanical editor

SM., J.G. Jared Gage Smith (1866–1925), agrostologist with the U.S. Department of Agriculture

SMALL John Kunkel Small (1869–1938), botanist and curator of museums with the New York Botanical Garden

SPRENG. Curt Polykarp Joachim Sprengel (1766–1833), professor of botany at Halle, Germany

STENT Sydney Margaret Stent (1875–1942), South African botanist

STEUD. Ernst Gottlieb Steudel (1783–1856), German physician and authority on the POACEAE

ST. JOHN, H. Harold St. John (1892–1991), University of Hawaii botanist

ST. JOHN, H. & HARDIN Harold St. John (1892–1991), University of Hawaii botanist, and James Walker Hardin (b. 1929), North Carolina State University taxonomist

ST.-LAG. Jean Baptiste Saint-Lager (1825–1912), French physician and botanist in Lyon, France

STOKES Susan Gabriella Stokes (1868–1954), Rocky Mountain naturalist and San Diego high school teacher

STUNTZ Stephen Conrad Stuntz (1875–1918), American botanist

SUSKD. Wilhelm Nikolas Suksdorf (1850–1932), German-born American botanist

SW. Olof Peter Swartz (1760–1818), Swedish botanist

SWEET Robert Sweet (1783–1835), English botanist

SWEZEY Goodwin Deloss Swezey (1851–1934), University of Nebraska botanical collector, botanist, and professor of astronomy

THARP & F.A. BARKLEY Benjamin Carroll Tharp (1885–1964), University of Texas botanist, and Fred Alexander Barkley (1908–1989), American botanist

THURB. George Thurber (1821–1890), American botanist with Mexican Boundary Survey

TIDESTR. Ivar Frederick Tidestrom (1864–1956), Swedish-born botanist who worked for the United States Department of Agriculture and later for Catholic University

TORR. John Torrey (1796–1873), American physician and botanist

TORR. & FRÉM. John Torrey (1796–1873), American physician and botanist, and John Charles Frémont (1813–1890), soldier, explorer, and presidential candidate

TORR. & A. GRAY John Torrey (1796–1873), American physician and botanist, and Asa Gray (1810–1888), Harvard University professor of botany

TREL. William Trelease (1857–1945), University of Illinois professor of botany

TRIN. Carl Bernhard von Trinius (1778–1844), Russian physician, poet, and authority on grasses

TRIN. & RUPR. Carl Bernhard von Trinius (1778–1844), Russian physician, poet, and authority on grasses, and Franz Joseph Ruprecht (1814–1870), Austrian-born Russian physician and botanist

TUTIN Thomas Gaskell Tutin (1908–1987), professor of botany with the University of Leicester, England

VAIL Anna Murray Vail (1863–1955), botanist and first librarian with the New York Botanical Garden

VASEY Geroge Vasey (1822–1893), English-born botanist and curator of the U.S. National Herbarium

VENT. Etienne Pierre Ventenat (1757–1808), French professor of botany

VITMAN Fulgenzio Vitman (1728–1806), Italian clergyman and botanist

WAHLENB. Georg (Göran) Wahlenberg (1780–1851), Swedish naturalist

WALP. Wilhelm Gerhard Walpers (1816–1853), German botanist

WALTER Thomas Walter (1740–1789), British-American botanist

WANGENH. Friedrich Adam Julius von Wangenheim (1749–1800), German forester

WARD, G.H. George Henry Ward (1917–2003), botanist with Knox College, Galesburg, Illinois

WARNOCK Barton Holland Warnock (1911–1998), Sul Ross State University botanist

WATSON, S. Sereno Watson (1826–1892), curator of the Gray Herbarium, Harvard University

WEBER Frédéric Albert Constantin Weber (1830–1903), French botanist and member of a French expedition in Mexico

WEBER, W.A. William Alfred Weber (b. 1918), University of Colorado botanist and curator of the herbarium

WELSH, S.L. Stanley Larson Welsh (b. 1928), Brigham Young University botanist

WIGG., F.H. Friedrich Heinrich Wiggers (1746–1811), taxonomist and author of the *Flora of Holstein* (1780)

WIGHT & ARN. Robert Wight (1796–1872), Scottish physician and botanist, and George Arnott Walker-Arnott (1799–1868), Scottish botanist

WILLD. Carl Ludwig Willdenow (1765–1812), German botanist

WILLIAMS, T.A. Thomas Albert Williams (1865–1900), agrostologist with the United States Department of Agriculture

WIMM. & GRAB. Christian Friedrich Heinrich Wimmer (1803–1868), German botanist in Breslau, and Heinrich Emanuel Grabowski (1792–1842), Polish-born botanist and pharmacist who worked in Germany

WITH. William Withering (1741–1799), English botanist, geologist, and physician

WOOD, ALPH. Alphonso W. Wood (1810–1881), botanist and author of first American book to employ dichotomous keys

WOOTON & STANDL. Elmer Otis Wooton (1865–1945), New Mexico State University professor of biology, and Paul Carpenter Standley (1884–1963), U.S. National Herbarium curator

ZUCC. Joseph Gerhard Zuccarini (1797–1848), professor of botany with the University of Munich, Germany

ZULOAGA & MORRONE Fernando Omar Zulouga (b. 1951), Argentinean botanist, and Osvaldo Morrone (1957–2011), Argentinean botanist

Checklist of Wildland Plants

GRASSES

TRIBE/FAMILY	GENUS/SPECIFIC EPITHET	LONGEVITY[1]	ORIGIN[2]	COMMON NAME
ANDROPOGONEAE	*Andropogon gerardi*	P	N	Big bluestem
ANDROPOGONEAE	*Andropogon virginicus*	P	N	Broomsedge bluestem
ANDROPOGONEAE	*Bothriochloa laguroides*	P	N	Silver bluestem
ANDROPOGONEAE	*Heteropogon contortus*	P	N	Tanglehead
ANDROPOGONEAE	*Schizachyrium scoparium*	P	N	Little bluestem
ANDROPOGONEAE	*Sorghastrum nutans*	P	N	Indiangrass
ANDROPOGONEAE	*Tripsacum dactyloides*	P	N	Eastern gamagrass
ARISTIDEAE	*Aristida oligantha*	A	N	Prairie threeawn
ARISTIDEAE	*Aristida purpurea*	P	N	Purple threeawn
AVENEAE	*Agrostis stolonifera*	P	I	Redtop
AVENEAE	*Avena barbata*	A	I	Slender oats
AVENEAE	*Avena fatua*	A	I	Wild oats
AVENEAE	*Avenula hookeri*	P	N	Spikeoats
AVENEAE	*Calamagrostis canadensis*	P	N	Bluejoint
AVENEAE	*Calamagrostis rubescens*	P	N	Pine reedgrass
AVENEAE	*Deschampsia cespitosa*	P	N	Tufted hairgrass
AVENEAE	*Koeleria macrantha*	P	N	Prairie junegrass
AVENEAE	*Phalaris arundinacea*	P	N	Reed canarygrass
AVENEAE	*Phleum alpinum*	P	N	Alpine timothy
AVENEAE	*Phleum pratense*	P	I	Timothy
AVENEAE	*Trisetum spicatum*	P	N	Spike trisetum
BROMEAE	*Bromus carinatus*	P	N	Mountain brome
BROMEAE	*Bromus diandrus*	A	I	Ripgut brome
BROMEAE	*Bromus hordeaceus*	A	I	Soft brome
BROMEAE	*Bromus inermis*	P	I	Smooth brome
BROMEAE	*Bromus tectorum*	A	I	Cheatgrass
CYNODONTEAE	*Bouteloua curtipendula*	P	N	Sideoats grama
CYNODONTEAE	*Bouteloua dactyloides*	P	N	Buffalograss
CYNODONTEAE	*Bouteloua eriopoda*	P	N	Black grama
CYNODONTEAE	*Bouteloua gracilis*	P	N	Blue grama

CYNODONTEAE	*Bouteloua hirsuta*	P	N	Hairy grama
CYNODONTEAE	*Bouteloua repens*	P	N	Slender grama
CYNODONTEAE	*Chloris cucullata*	P	N	Hooded windmillgrass
CYNODONTEAE	*Cynodon dactylon*	P	I	Bermudagrass
CYNODONTEAE	*Hilaria belangeri*	P	N	Curly mesquite
CYNODONTEAE	*Hilaria jamesii*	P	N	Galleta
CYNODONTEAE	*Hilaria mutica*	P	N	Tobosa
CYNODONTEAE	*Schedonnardus paniculatus*	P	N	Tumblegrass
CYNODONTEAE	*Spartina gracilis*	P	N	Alkali cordgrass
CYNODONTEAE	*Spartina pectinata*	P	N	Prairie cordgrass
DANTHONIEAE	*Danthonia californica*	P	N	California oatgrass
DANTHONIEAE	*Danthonia intermedia*	P	N	Timber oatgrass
DANTHONIEAE	*Danthonia parryi*	P	N	Parry oatgrass
ERAGROSTEAE	*Blepharoneuron tricholepis*	P	N	Pine dropseed
ERAGROSTEAE	*Calamovilfa longifolia*	P	N	Prairie sandreed
ERAGROSTEAE	*Distichlis spicata*	P	N	Saltgrass
ERAGROSTEAE	*Eragrostis curvula*	P	I	Weeping lovegrass
ERAGROSTEAE	*Eragrostis trichodes*	P	N	Sand lovegrass
ERAGROSTEAE	*Leptochloa dubia*	P	N	Green sprangletop
ERAGROSTEAE	*Muhlenbergia montana*	P	N	Mountain muhly
ERAGROSTEAE	*Muhlenbergia porteri*	P	N	Bush muhly
ERAGROSTEAE	*Muhlenbergia torreyi*	P	N	Ring muhly
ERAGROSTEAE	*Redfieldia flexuosa*	P	N	Blowoutgrass
ERAGROSTEAE	*Scleropogon brevifolius*	P	N	Burrograss
ERAGROSTEAE	*Sporobolus airoides*	P	N	Alkali sacaton
ERAGROSTEAE	*Sporobolus compositus*	P	N	Tall dropseed
ERAGROSTEAE	*Sporobolus cryptandrus*	P	N	Sand dropseed
MELICEAE	*Melica bulbosa*	P	N	Oniongrass
PANICEAE	*Digitaria californica*	P	N	Arizona cottontop
PANICEAE	*Panicum hallii*	P	N	Hall panicum
PANICEAE	*Panicum obtusum*	P	N	Vinemesquite
PANICEAE	*Panicum virgatum*	P	N	Switchgrass
PANICEAE	*Paspalum distichum*	P	N	Knotgrass
PANICEAE	*Pennisetum ciliare*	P	I	Buffelgrass
PANICEAE	*Setaria leucopila*	P	N	Plains bristlegrass
POEAE	*Dactylis glomerata*	P	I	Orchardgrass
POEAE	*Festuca campestris*	P	N	Rough fescue
POEAE	*Festuca idahoensis*	P	N	Idaho fescue
POEAE	*Poa fendleriana*	P	N	Muttongrass

POEAE	*Poa pratensis*	P	I	Kentucky bluegrass
POEAE	*Poa secunda*	P	N	Sandberg bluegrass
POEAE	*Vulpia octoflora*	A	N	Sixweeksgrass
STIPEAE	*Achnatherum hymenoides*	P	N	Indian ricegrass
STIPEAE	*Achnatherum nelsonii*	P	N	Columbia needlegrass
STIPEAE	*Hesperostipa comata*	P	N	Needleandthread
STIPEAE	*Nassella leucotricha*	P	N	Texas wintergrass
STIPEAE	*Nassella pulchra*	P	N	Purple needlegrass
TRITICEAE	*Agropyron cristatum*	P	I	Crested wheatgrass
TRITICEAE	*Elymus canadensis*	P	N	Canada wildrye
TRITICEAE	*Elymus elymoides*	P	N	Squirreltail
TRITICEAE	*Elymus trachycaulus*	P	N	Slender wheatgrass
TRITICEAE	*Hordeum jubatum*	P	N	Foxtail barley
TRITICEAE	*Hordeum pusillum*	A	N	Little barley
TRITICEAE	*Leymus cinereus*	P	N	Basin wildrye
TRITICEAE	*Pascopyrum smithii*	P	N	Western wheatgrass
TRITICEAE	*Pseudoroegneria spicata*	P	N	Bluebunch wheatgrass
TRITICEAE	*Taeniatherum caput-medusae*	A	I	Medusahead rye
TRITICEAE	*Thinopyrum intermedium*	P	I	Intermediate wheatgrass

GRASSLIKE PLANTS

TRIBE/FAMILY	GENUS/SPECIFIC EPITHET	LONGEVITY[1]	ORIGIN[2]	COMMON NAME
CYPERACEAE	*Carex filifolia*	P	N	Threadleaf sedge
CYPERACEAE	*Carex geyeri*	P	N	Elk sedge
CYPERACEAE	*Carex nebrascensis*	P	N	Nebraska sedge
CYPERACEAE	*Carex utriculata*	P	N	Beaked sedge
CYPERACEAE	*Schoenoplectus acutus*	P	N	Hardstem bulrush
JUNCACEAE	*Juncus balticus*	P	N	Wire rush

FORBS AND WOODY PLANTS

TRIBE/FAMILY	GENUS/SPECIFIC EPITHET	LONGEVITY[1]	ORIGIN[2]	COMMON NAME
ANACARDIACEAE	*Rhus aromatica*	P	N	Skunkbrush
APIACEAE	*Conium maculatum*	P	I	Poison hemlock
ANTHEMIDEAE	*Achillea millefolium*	P	N	Yarrow
ANTHEMIDEAE	*Artemisia cana*	P	N	Silver sagebrush
ANTHEMIDEAE	*Artemisia filifolia*	P	N	Sand sagebrush
ANTHEMIDEAE	*Artemisia frigida*	P	N	Fringed sagebrush
ANTHEMIDEAE	*Artemisia ludoviciana*	P	N	Cudweed sagewort
ANTHEMIDEAE	*Artemisia nova*	P	N	Black sagebrush

ANTHEMIDEAE	*Artemisia spinescens*	P		N	Budsage
ANTHEMIDEAE	*Artemisia tridentata*	P		N	Big sagebrush
ASTEREAE	*Amphiachyris dracunculoides*	A		N	Annual broomweed
ASTEREAE	*Chrysothamnus viscidiflorus*	P		N	Little rabbitbrush
ASTEREAE	*Ericameria nauseosa*	P		N	Rubber rabbitbrush
ASTEREAE	*Grindelia squarrosa*	P		N	Curlycup gumweed
ASTEREAE	*Gutierrezia sarothrae*	P		N	Broom snakeweed
ASTEREAE	*Heterotheca villosa*	P		N	Hairy goldaster
ASTEREAE	*Solidago missouriensis*	P		N	Prairie goldenrod
CICHORIEAE	*Agoseris glauca*	P		N	False dandelion
CICHORIEAE	*Crepis acuminata*	P		N	Tapertip hawksbeard
CICHORIEAE	*Taraxacum officinale*	P		I	Dandelion
EUPATORIEAE	*Liatris punctata*	P		N	Dotted gayfeather
HELIANTHEAE	*Ambrosia deltoidea*	P		N	Triangleleaf bursage
HELIANTHEAE	*Ambrosia dumosa*	P		N	White bursage
HELIANTHEAE	*Ambrosia psilostachya*	P		N	Western ragweed
HELIANTHEAE	*Baileya multiradiata*	P		N	Desert marigold
HELIANTHEAE	*Balsamorhiza sagittata*	P		N	Arrowleaf balsamroot
HELIANTHEAE	*Flourensia cernua*	P		N	Tarbush
HELIANTHEAE	*Hymenoxys hoopesii*	P		N	Orange sneezeweed
HELIANTHEAE	*Hymenoxys odorata*	A		N	Bitterweed
HELIANTHEAE	*Ratibida columnifera*	P		N	Prairie coneflower
HELIANTHEAE	*Wyethia amplexicaulis*	P		N	Mulesears
HELIANTHEAE	*Wyethia mollis*	P		N	Woolly mulesears
SENECIONEAE	*Senecio flaccidus*	P		N	Threadleaf groundsel
SENECIONEAE	*Senecio serra*	P		N	Sawtooth butterweed
SENECIONEAE	*Tetradymia canescens*	P		N	Gray horsebrush
BETULACEAE	*Alnus incana*	P		N	Mountain alder
BRASSICACEAE	*Descurainia pinnata*	A		N	Tansymustard
BRASSICACEAE	*Stanleya pinnata*	P		N	Desert princesplume
CANNABACEAE	*Celtis pallida*	P		N	Spiny hackberry
CAPRIFOLIACEAE	*Symphoricarpos albus*	P		N	Snowberry
CAPRIFOLIACEAE	*Symphoricarpos occidentalis*	P		N	Western snowberry
CHENOPODIACEAE	*Atriplex canescens*	P		N	Fourwing saltbush
CHENOPODIACEAE	*Atriplex confertifolia*	P		N	Shadscale saltbush
CHENOPODIACEAE	*Atriplex gardneri*	P		N	Saltbush
CHENOPODIACEAE	*Grayia spinosa*	P		N	Spiny hopsage
CHENOPODIACEAE	*Halogeton glomeratus*	A		I	Halogeton

CHENOPODIACEAE	*Kochia americana*	P		N	Greenmolly summercypress
CHENOPODIACEAE	*Kochia scoparia*	A		I	Kochia
CHENOPODIACEAE	*Krascheninnikovia lanata*	P		N	Winterfat
CHENOPODIACEAE	*Salsola tragus*	A		I	Russian thistle
CHENOPODIACEAE	*Sarcobatus vermiculatus*	P		N	Greasewood
CLUSIACEAE	*Hypericum perforatum*	P		I	St. Johnswort
CORNACEAE	*Cornus sericea*	P		N	Redosier dogwood
CUPRESSACEAE	*Juniperus monosperma*	P		N	Oneseed juniper
CUPRESSACEAE	*Juniperus scopulorum*	P		N	Rocky Mountain juniper
DENNSTAEDTIACEAE	*Pteridium aquilinum*	P		N	Bracken fern
ELAEAGNACEAE	*Shepherdia canadensis*	P		N	Russet buffaloberry
EPHEDRACEAE	*Ephedra trifurca*	P		N	Longleaf ephedra
ERICACEAE	*Arctostaphylos pungens*	P		N	Pointleaf manzanita
ERICACEAE	*Arctostaphylos uva-ursi*	P		N	Bearberry
FABACEAE	*Acacia berlandieri*	P		N	Guajillo
FABACEAE	*Acacia farnesiana*	P		N	Huisache
FABACEAE	*Acacia greggii*	P		N	Catclaw acacia
FABACEAE	*Acacia rigidula*	P		N	Blackbrush acacia
FABACEAE	*Amorpha canescens*	P		N	Leadplant
FABACEAE	*Astragalus mollissimus*	P		N	Woolly loco
FABACEAE	*Dalea purpurea*	P		N	Purple prairieclover
FABACEAE	*Lupinus caudatus*	P		N	Tailcup lupine
FABACEAE	*Medicago polymorpha*	A		I	Burclover
FABACEAE	*Oxytropis lambertii*	P		N	Purple locoweed
FABACEAE	*Prosopis glandulosa*	P		N	Honey mesquite
FABACEAE	*Psoralidium tenuiflorum*	P		N	Slimflower scurfpea
FAGACEAE	*Quercus gambelii*	P		N	Gambel oak
FAGACEAE	*Quercus stellata*	P		N	Post oak
GERANIACEAE	*Erodium botrys*	A		I	Broadleaf filaree
GERANIACEAE	*Erodium cicutarium*	A		I	Redstem filaree
GERANIACEAE	*Geranium richardsonii*	P		N	Wildwhite geranium
GROSSULARIACEAE	*Ribes cereum*	P		N	Wax currant
KRAMERIACEAE	*Krameria erecta*	P		N	Range ratany
MALVACEAE	*Sphaeralcea coccinea*	P		N	Scarlet globemallow
ONAGRACEAE	*Chamaenerion angustifolium*	P		N	Fireweed
PINACEAE	*Pinus edulis*	P		N	Pinyon pine

PINACEAE	*Pinus ponderosa*	P	N	Ponderosa pine
PLANTAGINACEAE	*Penstemon glaber*	P	N	Blue penstemon
PLANTAGINACEAE	*Plantago patagonica*	A	N	Woolly plantain
RANUNCULACEAE	*Delphinium bicolor*	P	N	Low larkspur
RANUNCULACEAE	*Delphinium occidentale*	P	N	Tall larkspur
RHAMNACEAE	*Ceanothus cuneatus*	P	N	Wedgeleaf buckbrush
RHAMNACEAE	*Ceanothus fendleri*	P	N	Fendler ceanothus
RHAMNACEAE	*Ceanothus integerrimus*	P	N	Deerbrush
RHAMNACEAE	*Ceanothus velutinus*	P	N	Buckbrush
ROSACEAE	*Adenostoma fasciculatum*	P	N	Chamise
ROSACEAE	*Amelanchier alnifolia*	P	N	Servicberry
ROSACEAE	*Cercocarpus ledifolius*	P	N	Curlleaf mountainmahogany
ROSACEAE	*Cercocarpus montanus*	P	N	Birchleaf mountainmahogany
ROSACEAE	*Coleogyne ramosissima*	P	N	Blackbrush
ROSACEAE	*Dasiphora fruticosa*	P	N	Shrubby cinquefoil
ROSACEAE	*Fallugia paradoxa*	P	N	Apache plume
ROSACEAE	*Prunus virginiana*	P	N	Chokecherry
ROSACEAE	*Purshia stansburyana*	P	N	Mexican cliffrose
ROSACEAE	*Purshia tridentata*	P	N	Antelope bitterbrush
ROSACEAE	*Rosa woodsii*	P	N	Wild rose
SALICACEAE	*Populus tremuloides*	P	N	Quaking aspen
SALICACEAE	*Salix bebbiana*	P	N	Beaked willow
SALICACEAE	*Salix exigua*	P	N	Sandbar willow
ZYGOPHYLLACEAE	*Larrea tridentata*	P	N	Creosotebush

[1]A=annual, P=perennial; [2]N=native to North America, I=introduced into North America

Selected References

Abrams, Leroy. 1951. Illustrated flora of the Pacific States. Volume 3. Stanford University Press, Stanford, California.

Abrams, Leroy, and Roxana Stinchfield Ferris. 1960. Illustrated flora of the Pacific States. Volume 4. Stanford University Press, Stanford, California.

Agricultural Research Service. 1970. Selected weeds of the United States. Agricultural Research Service, U.S. Department of Agriculture, Washington, D.C.

Agricultural Research Service, United States Department of Agriculture. Germplasm Resources Information Network (GRIN) taxonomy for plants. Accessed January 11, 2016. http://www.ars-grin.gov/cgi-bin/npgs/html/index.pl.

Albee, B. J., L. M. Shultz, and S. Goodrich. 1988. Atlas of the vascular plants of Utah. Utah Museum of Natural History, Salt Lake City.

Alley, Harold P., and Gary A. Lee. 1969. Weeds of Wyoming. Bulletin 498. Agricultural Experiment Station, University of Wyoming, Laramie.

Allred, Kelley W. 1982. Describing the grass inflorescence. Journal of Range Management 35:672–675.

———. 2005. A field guide to the grasses of New Mexico. New Mexico Agricultural Experiment Station, New Mexico State University, Las Cruces.

Andersen, Berniece A. Undated. Desert plants of Utah. Extension Circular 376. Cooperative Extension Service, Utah State University, Logan.

Andersen, Berniece A., and Arthur H. Holmgren. 1985. Mountain plants of northeastern Utah. Circular 319. Cooperative Extension Service, Utah State University, Logan.

Babcock, Ernest Brown. 1947. The genus *Crepis*. Part 2. Systematic treatment. University of California Press, Los Angeles.

Bare, Janet E. 1979. Wildflowers and weeds of Kansas. Regents Press of Kansas, Lawrence.

Barkworth, Mary E., Laurel K. Anderton, Kathy A. Capels, Sandy Long, and Michael B. Piep. 2007. Manual of the grasses of North America. Intermountain Herbarium and Utah State University Press, Logan.

Barnard, Iralee. 2014. Field guide to the common grasses of Oklahoma, Kansas, and Nebraska. University Press of Kansas, Lawrence.

Barth, Roland, and Neal Ratzlaff. 2004. Field guide to wildflowers. Fontenelle Nature Association, Bellevue, Nebraska.

Beetle, Alan Ackerman. 1960. A study of sagebrush: The section *Tridentatae* of *Artemisia*. Bulletin 368. Agricultural Experiment Station, University of Wyoming, Laramie.

———. 1977. Noteworthy grasses from Mexico. Phytologia 37:317–407.

Beetle, Alan Ackerman, and Donald J. Gordon. 1991. Gramineas de Sonora. COTECOCA, Hermosillo, Sonora.

Beetle, Alan Ackerman, and Kendall L. Johnson. 1982. Sagebrush in Wyoming. Bulletin 779. Agricultural Experiment Station, University of Wyoming, Laramie.

Beetle, Alan Ackerman, and Morton May. 1971. Grasses of Wyoming. Research Journal 39. Agricultural Experiment Station, University of Wyoming, Laramie.

Beetle, Alan Ackerman, Rafael Guzmán Mejía, Víctor Jaramillo Luque, Matilde P. Guerrero Sánchez, Elizabeth Manrique Forceck, Aurora Chimal Hernández, Celia Shariff Bujdud, and Irama Núñez Tancredi. 1983. Las Gramineas de Mexico. Tomo 1. COTECOCA, Mexico City.

Beetle, Alan Ackerman, Elizabeth Manrique Forceck, Víctor Jaramillo Luque, M. Patricia Guerrero Sánchez, J. Alejandro Miranda Sánchez, Irama Núñez Tancredi, and Aurora Chimal Hernández. 1987. Las Gramineas de Mexico. Tomo 2. COTECOCA, Mexico City.

Beetle, Alan Ackerman, Elizabeth Manrique Forceck, Javier Alejandro Miranda Sánchez, Víctor Jaramillo Luque, Aurora Chimal Hernández, Angélica Maria Rodriguez Rodreguez. 1991. Las Gramineas de Mexico. Tomo 3. COTECOCA, Mexico City.

Beetle, Alan Ackerman, Javier Alejandro Miranda Sánchez, Víctor Jaramillo Luque, Angélica Maria Rodriguez Rodreguez, Laura Aragón Melchor, Martha Aurora Vergara Batalla, Aurora Chimal Hernández, and Oscar Dominguez Sepúlverda. 1995. Las Gramineas de Mexico. Tomo 4. COTECOCA, Mexico City.

Benson, L. D., and R. A. Darrow. 1981. The trees and shrubs of southwestern deserts. University of New Mexico Press, Albuquerque.

Bentley, Henry Lewis. 1898. Grasses and forage plants of central Texas. Bulletin 10. Division of Agrostology. U.S. Department of Agriculture, Washington, D.C.

Best, Keith F., Jan Looman, and J. Baden Campbell. 1971. Prairie grasses. Publication 1413. Canada Department of Agriculture, Saskatchewan.

Biota of North America Project. North American plants atlas. Accessed January 11, 2016. http://www.bonap.org/.

Blackwell, Will H. 1990. Poisonous and medicinal plants. Prentice-Hall, Englewood Cliffs, New Jersey.

Booth, W. E., and J. C. Wright. 1959. Flora of Montana. Part 2. Dicotyledons. Department of Botany and Microbiology, Montana State University, Bozeman.

Brummitt, Richard K., and C. Emma Powell (eds.). 1992. Authorities of plant names. Royal Botanic Gardens, Kew.

Budd, A. C. 1957. Wild plants of the Canadian Prairies. Publication 983. Canada Department of Agriculture, Saskatchewan.

Burbidge, Nancy T., and Surrey W. L. Jacobs. 1984. Australian grasses. Angus and Robertson Publishers, Sydney.

Campbell, J. Baden, Robert W. Lodge, and A. C. Budd. 1956. Poisonous plants of the Canadian Prairies. Publication 900. Canada Department of Agriculture, Ottawa.

Chase, Agnes. 1922. First book of grasses. The Macmillan Company, New York.

Cheeke, Peter R., and Lee R. Shull. 1985. Natural toxicants in feeds and poisonous plants. AVI Publishing, Westport, Connecticut.

Christiansen, Paul, and Mark Müller. 1999. An illustrated guide to Iowa prairie plants. University of Iowa Press, Iowa City.

Clark, Lewis J. 1973. Wild flowers of British Columbia. Gray's Publications, Sidney, British Columbia.

Clausen, J., David D. Keck, and William Hiesey. 1940. Experimental studies on the nature of species, 1. Effect of varied environments on western North American plants, the *Achillea millefolium* complex. Carnegie Institute Publication 520:296–324.

Coffey, C. R., and R. L. Stevens. 2004. Grasses of southern Oklahoma and north Texas: A pictorial guide. Samuel Roberts Nobel Foundation, Ardmore, Oklahoma.

Coon, Nelson. 1979. Using plants for healing. Rodale Press, Emmaus, Pennsylvania.

Copple, R. F., and A. E. Aldous. 1932. The identification of certain native and naturalized grasses by their vegetative characters. Technical Bulletin 32. Agricultural Experiment Station, Kansas State College, Manhattan.

Copple, R. F., and C. P. Pase. 1967. A vegetative key to some common Arizona range grasses. Research Paper RM-27. Rocky Mountain Forest and Range Experiment Station, Forest Service, U.S. Department of Agriculture, Washington, D.C.

Correll, Donovan S., and Marshall C. Johnston. 1970. Manual of the vascular plants of Texas. Texas Research Foundation, Renner.

Cronquist, Arthur. 1980. Vascular flora of the southeastern United States. Volume 1. Asteraceae. University of North Carolina Press, Chapel Hill.

Cronquist, Arthur, Arthur H. Holmgren, Noel H. Holmgren, James R. Reveal, and Patricia K. Holmgren. 1977. Intermountain flora. Volume 6. The monocotyledons. Columbia University Press, New York.

———. 1984. Intermountain flora. Volume 4. Subclass Asteridae. Columbia University Press, New York.

———. 1989. Intermountain flora. Volume 3, Part B. Fabales (by Rupert C. Barneby). Columbia University Press, New York.

———. 1994. Intermountain flora. Volume 5. Asterales. Columbia University Press, New York.

Cronquist, Arthur, Noel H. Holmgren, and Patricia K. Holmgren. 1997. Intermountain flora. Volume 3, Part A. Subclass Rosidae. Columbia University Press, New York.

Davis, Ray J. 1952. Flora of Idaho. William C. Brown Company, Dubuque, Iowa.

Dayton, William A. 1931. Important western browse plants. Miscellaneous Publication 101. U.S. Department of Agriculture, Washington, D.C.

Densmore, Frances. 1974. How Indians use wild plants. Dover Publications, New York.

Diggs, George M., Jr., Barney L. Lipscomb, and Robert J. O'Kennon. 1999. Illustrated flora of north central Texas. Botanical Research Institute of Texas, Fort Worth.

Dodge, Natt N. 1976. Flowers of the Southwest deserts. 1976. Southwest Parks and Monuments Association, Globe, Arizona.

Dollahite, J. W., G. T. Householder, and B. J. Camp. 1966. Oak poisoning in livestock. Bulletin 1049. Agricultural Experiment Station, Texas A&M University, College Station.

Dorn, Robert D. 2001. Vascular plants of Wyoming. Mountain West Publishing, Cheyenne.

Duft, Joseph F., Robert K. Mosley. 1989. Alpine Wildflowers of the Rocky Mountains, Mountain Press Publishing Company, Missoula, Montana.

Durrell, L. W. 1951. Halogeton—a new stock-poisoning weed. Circular 170-A. Agricultural Experiment Station, Colorado State College, Fort Collins.

Durrell, L. W., and I. E. Newsom. 1939. Colorado's poisonous and injurious plants. Bulletin 445. Agricultural Experiment Station, Colorado State College, Fort Collins.

Elias, Thomas S. 1980. Trees of North America. Van Nostrand Reinhold, New York.

Elmore, Francis H. 1976. Shrubs and trees of the Southwest uplands. Southwest Parks and Monuments Association, Tucson, Arizona.

Emboden, William. 1979. Narcotic plants. Macmillan, New York.

Eppas, Alan C. 1976. Wild edible and poisonous plants of Alaska. Publication 28. Cooperative Extension Service, University of Alaska, Fairbanks.

Evers, Robert A., and Roger P. Link. 1972. Poisonous plants of the Midwest and their effects on livestock. Special Publication 24. College of Agriculture, University of Illinois, Urbana-Champaign.

Farrar, John Laird. 1995. Trees of the northern United States and Canada. Iowa State University Press, Ames.

Farrar, Jon. 2011. Wildflowers of Nebraska and the Great Plains. Nebraska Game and Parks Commission, Lincoln.

Featherly, H. I. 1938. Grasses of Oklahoma. Technical Bulletin 3. Agricultural Experiment Station, Oklahoma Agricultural and Mechanical College, Stillwater.

Fernald, Marritt Lyndon. 1950. Gray's manual of botany. American Book Company, New York.

Flora of North America Editorial Committee. 1993. Volume 2. Pteridophytes and Gymnosperms. Oxford University Press, New York.

———. 1993. Volume 3. Magnoliophyta: Magnoliidae and Hamamelidae. Oxford University Press, New York.

———. 2000. Volume 22. Magnoliophyta: Alismatidae, Arecidae, Commelinidae (in part), and Zingiberidae. Oxford University Press, New York.

———. 2002. Volume 23. Magnoliophyta: Commelinidae (in part): Cyperacea, part 2. Oxford University Press, New York.

———. 2003. Volume 4. Magnoliophyta: Caryophyllidae, part 1. Oxford University Press, New York.

———. 2003. Volume 25. Magnoliophyta: Commelinidae (in part): Poaceae, part 2. Oxford University Press, New York.

———. 2005. Volume 5. Magnoliophyta: Caryophyllidae, part 2. Oxford University Press, New York.

————. 2006. Volume 19. Magnoliophyta: Asterideae (in part): Asteraceae, part 1. Oxford University Press, New York.

————. 2006. Volume 20. Magnoliophyta: Asterideae (in part): Asteraceae, part 2. Oxford University Press, New York.

————. 2006. Volume 21. Magnoliophyta: Asterideae (in part): Asteraceae, part 3. Oxford University Press, New York.

————. 2007. Volume 24. Magnoliophyta: Commelinidae (in part): Poaceae, part 1. Oxford University Press, New York.

————. 2009. Volume 8. Magnoliophyta: Paeoniaceae to Ericaceae. Oxford University Press, New York.

————. 2010. Volume 7. Magnoliophyta: Salicaceae to Brassicaceae. Oxford University Press, New York.

————. 2014. Volume 9. Magnoliophyta: Picramnaceae to Rosaceae. Oxford University Press, New York.

————. 2015. Volume 6. Magnoliophyta: Cucurbitaceae to Droseraceae. Oxford University Press, New York.

Flores, Glafiro J. Alanís, Marcela González Alvarez, Marco Antonio Guzmán Lucio, and Gerónimo Cano y Cano. 1995. Flora representativa de Chipinque: Arboles y arbustos. Universidad Autonoma de Nuevo Leon, Monterrey.

Forest Service. 1937. Range plant handbook. Forest Service, U.S. Department of Agriculture, Washington, D.C.

Frankton, Clarence. 1955. Weeds of Canada. Canada Department of Agriculture, Ottawa.

Gates, Frank C. 1937. Grasses in Kansas. Volume 55, Number 220-A. Kansas State Board of Agriculture, Topeka.

Gay, Charles W., Don Dwyer, Chris Allison, Stephan Hatch, and Jerry Schickedanz. 1980. New Mexico range plants. Circular 374. Cooperative Extension Service, New Mexico State University, Las Cruces.

Gleason, Henry A. 1952. Illustrated flora of the northeastern United States and adjacent Canada. Volume 1. The Pteridophyta, Gymnospermae, and Monocotyledoneae. Lancaster Press, Lancaster, Pennsylvania.

————. 1952. Illustrated flora of the northeastern United States and adjacent Canada. Volume 2. The chloripetalous Dycotyledoneae. Lancaster Press, Lancaster, Pennsylvania.

————. 1952. Illustrated flora of the northeastern United States and adjacent Canada. Volume 3. The sympetalous Dicotyledoneae. Lancaster Press, Lancaster, Pennsylvania.

Gould, Frank W. 1951. Grasses of southwestern United States. Biological Sciences Bulletin Number 7. University of Arizona, Tucson.

————. 1975. Texas plants—a checklist and ecological summary. Miscellaneous Publication 585. Agricultural Experiment Station, Texas A&M University Press, College Station.

————. 1978. Common Texas grasses. Texas A&M University Press, College Station.

Gould, Frank W., and R. Moran. 1981. The grasses of Baja California, Mexico. Memoir 12. San Diego Society of Natural History.

Gould, Frank W., and Robert B. Shaw. 1983. Grass systematics. Texas A&M University Press, College Station.

Great Plains Flora Association. 1977. T. M. Barkley (ed.). Atlas of the flora of the Great Plains. Iowa State University Press, Ames.

———. 1986. T. M. Barkley (ed.). Flora of the Great Plains. University Press of Kansas, Lawrence.

Grinnell, George Bird. 1962. The Cheyenne Indians—their history and ways of life. Volume 2. Cooper Square Publisher, New York.

Haddock, Michael John. 2005. Wildflowers and grasses of Kansas. University of Kansas Press, Lawrence.

Hafenrichter, A. L., Lowell A. Mullen, and Robert L. Brown. 1949. Grasses and legumes for soil conservation in the Pacific Northwest. Miscellaneous Publication 678. U.S. Department of Agriculture, Washington, D.C.

Hallsten, Gregory P., Quentin D. Skinner, and Alan A. Beetle. 1987. Grasses of Wyoming. Research Journal 202. Agricultural Experiment Station, University of Wyoming, Laramie.

Harrington, H. D. 1954. Manual of the plants of Colorado. Sage Books, Denver.

———. 1957. How to identify plants. Swallow Press, Chicago.

———. 1967. Edible native plants of the Rocky Mountains. University of New Mexico Press, Albuquerque.

Harrington, H. D., and L. W. Durrell. 1944. Key to some Colorado grasses in vegetative condition. Technical Bulletin 33. Agricultural Experiment Station, Colorado State College, Fort Collins.

Harris, James G., and Melinda Woolf Harris. 2001. Plant identification terminology. Spring Lake Publishing, Spring Lake, Utah.

Hart, Jeff, and Jacqueline Moore. 1976. Montana native plants and native people. Montana Historical Society, Helena.

Hatch, Stephan L. 2009. Gould's grasses of Texas. Department of Ecosystem Science and Management, Texas A&M University, College Station.

Hatch, Stephan L., K. N. Gandhi, and L. E. Brown. 1990. Checklist of the vascular plants of Texas. MP 1655. Agricultural Experiment Station, Texas A&M University, College Station.

Hatch, Stephan L., and Jennifer Pluhar. 1993. Texas range plants. Texas A&M University Press, College Station.

Hatch, Stephan L., Joseph L. Schuster, and D. Lynn Drawe. 1999. Grasses of the Texas gulf prairies and marshes. Texas A&M University Press, College Station.

Hayes, Doris W., and George A. Garrison. 1960. Key to important woody plants of eastern Oregon and Washington. Agriculture Handbook 148. Forest Service, U.S. Department of Agriculture, Washington, D.C.

Hermann, F. J. 1966. Notes on western range forbs. Agriculture Handbook 293. Forest Service, U.S. Department of Agriculture, Washington, D.C.

————. 1970. Manual of the carices of the Rocky Mountains and Colorado Basin. Agriculture Handbook 374. Forest Service, U.S. Department of Agriculture, Washington, D.C.

Hignight, K. W., J. K. Wipff, and Stephan L. Hatch. 1988. Grasses (Poaceae) of the Texas Cross Timbers and Prairies. Miscellaneous Publication 1657. Agricultural Experiment Station, Texas A&M University, College Station.

Hilken, Thomas O., and Richard F. Miller. 1980. Medusahead (*Taeniatherum asperum* Nevski): A review and annotated bibliography. Station Bulletin 644. Agricultural Experiment Station, Oregon State University, Corvallis.

Hitchcock, A. S. 1951. Manual of the grasses of the United States. Revised by Agnes Chase. Miscellaneous Publication 200. U.S. Department of Agriculture, Washington, D.C.

Hitchcock, C. Leo. 1937. A key to the grasses of Montana. John S. Swift Company, St. Louis.

Hitchcock, C. Leo, and Arthur Cronquist. 1973. Flora of the Pacific Northwest. University of Washington Press, Seattle.

Holmgren, Arthur H. 1965. Handbook of vascular plants of the northern Wasatch. Nature Press, Palo Alto, California.

Holmgren, Arthur H., and James L. Reveal. 1966. Checklist of the vascular plants of the Intermountain Region. Research Paper INT-32. Forest Service, U.S. Department of Agriculture, Washington, D.C.

Hosie, R. C. 1973. Native trees of Canada. Canadian Forest Service, Department of Environment, Ottawa.

Hulten, Eric. 1968. Flora of Alaska and neighboring territories. Stanford University Press, California.

Humphrey, Robert R. 1958. Arizona range grasses. Bulletin 298. Agricultural Experiment Station, University of Arizona, Tucson.

Instituto Nacional de Estadistica. 1995. Catálogo de Herbario INEGI. Tomo 1. Geografia e Informatica, Mexico City.

————. 1995. Catálogo de Herbario INEGI. Tomo 2. Geografia e Informatica, Mexico City.

————. 1995. Catálogo de Herbario INEGI. Tomo 3. Geografia e Informatica, Mexico City.

Isely, Duane. 1990. Vascular flora of the southeastern United States. Volume 3, Part 2. Leguminosae (Fabaceae). University of North Carolina Press, Chapel Hill.

Jepson, Willis L. 1951. Flowering plants of California. University of California Press, Berkeley.

Johnson, James R., and Gary E. Larson. 1999. Grassland plants of South Dakota and the Northern Great Plains. Bulletin 566. Agricultural Experiment Station, South Dakota State University, Brookings.

Johnson, W. M. 1964. Field key to the sedges of Wyoming. Bulletin 419. Agricultural Experiment Station, University of Wyoming, Laramie.

Jones, F. B. 1975. Flora of the Texas Coastal Bend. Mission Press, Corpus Christi, Texas.

Judd, B. Ira. 1962. Principal forage plants of Southwestern ranges. Station Paper 69. Rocky Mountain Forest and Range Experiment Station, Forest Service, U.S. Department of Agriculture, Washington, D.C.

Judziewicz, Emmit J., Robert W. Freckmann, Lynn G. Clark, and Merel R. Black. 2014. Field guide to Wisconsin grasses. University of Wisconsin Press, Madison.

Kannowski, Paul. B. 1989. Wildflowers of North Dakota. University of North Dakota Press, Grand Forks.

Kaul, Robert B., David Sutherland, and Steven Rolfsmeier. 2011. The flora of Nebraska. Conservation and Survey Division, University of Nebraska, Lincoln.

Kearney, Thomas H., and Robert H. Peebles. 1960. Arizona flora. University of California Press, Berkeley.

Keeler, Richard F., Kent R. Van Kampen, and Lynn F. James. 1978. Effects of poisonous plants on livestock. Academic Press, New York.

Keim, F. D., G. W. Beadle, and A. L. Frolik. 1932. The identification of the more important prairie grasses of Nebraska by their vegetative characteristics. Research Bulletin 65. Agricultural Experiment Station, University of Nebraska, Lincoln.

Kinch, Raymond C., Leon Wrage, and Raymond A. Moore. 1975. South Dakota weeds. Cooperative Extension Service, South Dakota State University, Brookings.

Kingsbury, John M. 1964. Poisonous plants of the United States and Canada. Prentice-Hall, Englewood Cliffs, New York.

Knobel, Edward. 1980. Field guide to the grasses, sedges and rushes of the United States. Dover Publications, New York.

Langman, I. K. 1964. A selected guide to the literature on the flowering plants of Mexico. University of Pennsylvania Press, Philadelphia.

Larson, Gary. 1993. Aquatic and wetland vascular plants of the northern Great Plains. General Technical Report RM-238. Forest Service, U.S. Department of Agriculture, Washington, D.C.

Larson, Gary E., and James R. Johnson. 2007. Plants of the Black Hills and Bear Lodge Mountains. B732. Agricultural Experiment Station, South Dakota State University, Brookings.

Leithead, Horace L., Lewis L. Yarlett, and Thomas N. Shiflet. 1971. 100 native forage grasses in 11 southern states. Agriculture Handbook 389. Soil Conservation Service, U.S. Department of Agriculture, Washington, D.C.

LeSueur, H. 1945. The ecology of the vegetation of Chihuahua, Mexico, north of Parallel 28. Publication 4521. University of Texas Press, Austin.

Lewis, Walter H. 1977. Medical botany. John Wiley & Sons, New York.

Little, Elbert L., Jr. 1971. Atlas of United States trees. Volume 1. Conifers and important hardwoods. Miscellaneous Publication 1146. U.S. Department of Agriculture, Washington, D.C.

———. 1976. Atlas of United States trees. Volume 3. Minor western hardwoods. Miscellaneous Publication 1314. U.S. Department of Agriculture, Washington, D.C.

Lommasson, Robert C. 1973. Nebraska wild flowers. University of Nebraska Press, Lincoln.

Looman, J., and K. F. Best. 1979. Budd's flora of the Canadian prairie provinces. Publication 1662. Research Branch Agriculture, Canada.

Marsh, C. D. 1929. Stock-poisoning plants of the range. Bulletin 1245. U.S. Department of Agriculture, Washington, D.C.

Martin, S. Clark. 1975. Ecology and management of southwestern semi-desert grass-shrub ranges: The status of our knowledge. Research Paper RM-156. Forest Service, U.S. Department of Agriculture, Washington, D.C.

Martin, William C., and Charles R. Hutchins. 1980. A flora of New Mexico. Volume 1. J. Cramer, Germany.

———. 1981. A flora of New Mexico. Volume 2. J. Cramer, Germany.

May, Morton. 1960. Key to the major grasses of the Big Horn Mountains—based on vegetative characters. Bulletin 371. Agricultural Experiment Station, University of Wyoming, Laramie.

McArthur, E. Durant, and Bruce L. Welch. 1984. Proceedings—symposium on the biology of *Artemisia* and *Chrysothamnus*. General Technical Report INT-200. Forest Service, U.S. Department of Agriculture, Washington, D.C.

McKean, William T. 1976. Winter guide to central Rocky Mountain shrubs. Colorado Department of Natural Resources, Denver.

McKell, Cyrus M., James P. Blaisdell, and Joe R. Goodin. 1972. Wildland shrubs—their biology and utilization. General Technical Report INT-1. Forest Service, U.S. Department of Agriculture, Washington, D.C.

Missouri Botanical Garden. Tropicos. Accessed January 11, 2016. http://www.tropicos.org/.

Moreno, N. P. 1984. Glosario botanico ilustrado. Instituto National de Investigaciones sobre Recursos Bióticos, Xalapa, Veracruz, Mexico.

Morris, H. E., W. E. Booth, G. F. Payne, and R. E. Stitt. 1950. Important grasses on Montana ranges. Bulletin 470. Agricultural Experiment Station, Montana State College, Bozeman.

Morton, Julia F. 1974. Folk remedies of the low country. E. A. Seemann Publishing, Miami, Florida.

Moser, L. E., D. R. Buxton, and M. D. Casler (eds.). 1996. Cool-season forage grasses. Agronomy Monograph 34. American Society of Agronomy, Madison, Wisconsin.

Mozingo, Hugh H. 1987. Shrubs of the Great Basin. University of Nevada Press, Reno.

Muenscher, Walter Conrad. 1939. Poisonous plants of the United States. Macmillan, New York.

Munson, T. V. 1883. Forest and forest trees of Texas. American Journal of Forestry 1:433–451.

Munz, Philip A., and David D. Keck. 1959. A California flora. University of California Press, Berkeley.

Natural Resources Conservation Service, United States Department of Agriculture. Plants database. Accessed January 11, 2016. http://plants.usda.gov/java/.

Nelson, Ruth Ashton. 1968. Wild flowers of Wyoming. Bulletin 490. Cooperative Extension Service, University of Wyoming, Laramie.

Ode, David J. 2006. Dakota flora. South Dakota State Historical Society Press, Pierre.

Owensby, Clinton E. 1980. Kansas prairie wildflowers. Iowa State University Press, Ames.

Pammel, L. H., Carleton R. Ball, and F. Lamson-Scribner. 1904. The grasses of Iowa. Part 2. Iowa Geological Survey, Des Moines.

Parker, Karl G., Lamar R. Mason, and John F. Vallentine. Undated. Utah grasses. Extension Circular 384. Cooperative Extension Service, Utah State University, Logan.

Parker, Kittie F. 1972. Arizona weeds. University of Arizona Press, Tucson.

Parks, H. B. 1937. Valuable plants native to Texas. Bulletin 551. Agricultural Experiment Station, Texas A&M University, College Station.

Pavlick, Leon. 1995. *Bromus* L. of North America. Royal British Columbia Museum, Victoria, British Columbia.

Perryman, Barry L., and Quentin D. Skinner. 2007. A field guide to Nevada grasses. Indigenous Rangeland Management Press, Lander, Wyoming.

Phillips, Jan. 1979. Wild edibles of Missouri. Missouri Department of Conservation, Jefferson City.

Phillips Petroleum Company. 1963. Pasture and range plants. Phillips Petroleum Company, Bartlesville, Oklahoma.

Pool, Raymond J. 1971. Handbook of Nebraska trees. Bulletin 32. Conservation Survey Division, University of Nebraska, Lincoln.

Porter, C. L. 1960. Wyoming trees. Circular 164R. Cooperative Extension Service, University of Wyoming, Laramie.

Powell, A. Michael. 1988. Trees and shrubs of Trans-Pecos, Texas. Big Bend Natural History Association, Big Bend.

——. 1994. Grasses of the Trans-Pecos and adjacent areas. University of Texas Press, Austin.

Preston, Richard J., Jr. 1976. North American trees. Iowa State University Press, Ames.

Ratzlaff, Neal S., and Roland E. Barth. 2007. Field guide to trees, shrubs, woody vines, grasses, sedges and rushes. Fontenelle Nature Association, Bellevue, Nebraska.

Reed, P. B., Jr. 1988. National list of plant species that occur in wetlands: Texas. U.S. Fish and Wildlife Service, U.S. Department of Interior, Washington, D.C.

Richardson, Alfred. 1995. Plants of the Rio Grande Delta. University of Texas Press, Austin.

Rickett, Harold William. 1966. Wild flowers of the United States. McGraw-Hill, New York.

Runkel, Sylvan T., and Dean M. Roosa. 1989. Wildflowers of the tallgrass prairie. Iowa State University Press, Ames.

Rydberg, Per Axel. 1932. Flora of the prairies and plains of central North America. Hafner Publishing, New York.

——. 1954. Flora of the Rocky Mountains and adjacent plains. Hafner Publishing, New York.

Rzedowski, J. 1978. Vegetacion de Mexico. Editorial Limusa, Mexico City.

Sampson, Arthur W. 1924. Native American forage plants. John Wiley & Sons, New York.

Sampson, Arthur W., and Agnes Chase. 1927. Range grasses of California. Bulletin 430. University of California, Berkeley.

Sampson, Arthur W., Agnes Chase, and Donald W. Hedrick. 1951. California grasslands and range forage grasses. Bulletin 724. Agricultural Experiment Station, University of California, Berkeley.

Saunders, Charles Francis. 1976. Edible and useful wild plants. Dover Publications, New York.

Scoggan, H. J. 1979. The flora of Canada. Part 2, Pteriodophyta, Gymnospermae and Monocotyledoneae. Museums of Canada, Ottawa.

———. 1979. The flora of Canada. Part 3, Dicotyledoneae (Saururaceae to Violaceae). Museums of Canada, Ottawa.

———. 1979. The flora of Canada. Part 4, Dicotyledoneae (Loasaceae to Compositae). Museums of Canada, Ottawa.

Shantz, H. L., and R. Zon. 1936. The natural vegetation of the United States. Pages 1–29. In Atlas of American agriculture, Part 1, Section E. U.S. Department of Agriculture, Washington, D.C.

Shaw, Richard J. 1989. Vascular plants of northern Utah: An identification manual. Utah State University Press, Logan.

———. 1995. Utah Wildflowers: A field guide to northern and central mountains and valleys. Utah State University Press, Logan.

Shaw, Robert B. 2008. Grasses of Colorado. University Press of Colorado, Boulder.

———. Guide to Texas grasses. 2012. Texas A&M University Press, College Station.

Shaw, Robert B., and J. D. Dodd. 1976. Vegetative key to the Compositae of the Rio Grande Plains of Texas. Miscellaneous Publication 1274. Agricultural Experiment Station, Texas A&M University, College Station.

Silveus, W. A. 1933. Texas grasses. Clegg Company, San Antonio.

Simpson, Benny J. 1996. A field guide to Texas trees. Gulf Publishing Company, Houston.

Smeins, Fred E., and Robert B. Shaw. 1978. Natural vegetation of Texas and adjacent areas, 1675–1975: A bibliography. Miscellaneous Publication 1399. Agricultural Experiment Station, Texas A&M University, College Station.

Soil Conservation Service. 1965. Important native grasses for range conservation in Florida. Soil Conservation Service, U.S. Department of Agriculture, Washington, D.C.

Soreng, R. J., G. Davidse, P. M. Peterson, F. O. Zuloaga, E. J. Judziewicz, T. S. Filgueiras, and O. N. Marrone. 2003 (and onward). Online taxonomic novelties and updates, distributional additions and corrections, and editorial changes since the four published volumes of the Catalogue of New World Grasses (Poaceae) published in Contributions to United States National Herbarium, vol. 39, 41, 46, and 48. Accessed January 11, 2016. http://tropicos.org/Project/CNWG.

Spellenberg, Richard. 1994. National Audubon Society field guide to North American wildflowers, Western Region, Alfred A. Knopf, New York.

Sperry, O. E., J. W. Dollahite, G. O. Hoffman, and B. J. Camp. 1977. Texas plants poisonous to livestock. Cooperative Extension Service, Texas A&M University, College Station.

Stechman, John V. 1977. Common western range plants. Vocational Education Productions, California Polytechnic State University, San Luis Obispo.

Stephens, H. A. 1973. Woody plants of the north central plains. University Press of Kansas, Lawrence.

Stevens, O. A. 1963. Handbook of North Dakota plants. North Dakota Institute of Regional Studies, Fargo.

Steyermark, Julian A. 1963. Flora of Missouri. Iowa State University Press, Ames.

Stoddart, L. A., A. H. Holmgren, and C. W. Cook. 1949. Important poisonous plants of Utah. Special Report 2. Agricultural Experiment Station, Utah State University, Logan.

Stubbendieck, James. 1994. Rangeland plants. Pages 559–574. In Encyclopedia of agricultural science. Academic Press, New York.

Stubbendieck, James, Mitchell J. Coffin, and L. M. Landholt. 2003. Weeds of the Great Plains. Nebraska Department of Agriculture, Lincoln.

Stubbendieck, James, and Elverne C. Conard. 1989. Common legumes of the Great Plains. University of Nebraska Press, Lincoln.

Stubbendieck, James, and Thomas A. Jones. 1996. Other cool-season grasses. Pages 765–780. In Cool-season forage grasses. Agronomy Monograph 34. American Society of Agronomy, Madison, Wisconsin.

Stubbendieck, James, and Kay L. Kottas. 2005. Common grasses of Nebraska. Extension Circular 170. Cooperative Extension Service, University of Nebraska, Lincoln.

———. 2007. Common forbs and shrubs of Nebraska. Extension Circular 118. Cooperative Extension Service. University of Nebraska, Lincoln.

Sutherland, David M. 1975. A vegetative key to Nebraska grasses. Pages 283–316. In Prairie: A multiple view. University of North Dakota Press, Grand Forks.

Taylor, Ronald J. 1992. Sagebrush country: A wildflower sanctuary. Mountain Press Publishing Company Missoula, Montana.

Texas Forest Service. 1963. Forest trees of Texas. Texas Forest Service, College Station.

Thilenius, John F. 1975. Alpine range management in the western United States—principles, practices, and problems: The status of our knowledge. Forest Service Research Paper RM-157. U.S. Department of Agriculture, Washington, D.C.

Tidestrom, I. 1925. Flora of Utah and Nevada. U.S. Natural Herbarium Contributions 25:1–665.

Tsvelev, N. N. 1984. Grasses of the Soviet Union. Russian Translation Service 8. A. A. Balkema, Rotterdam.

Turner, Billie Lee. 1959. The legumes of Texas. University of Texas Press, Austin.

Turner, Billie Lee, Holly Nichols, Goeffrey C. Denny, and Oded Doron. 2003. Atlas of the vascular plants of Texas. Botany Research Institute of Texas Press, Fort Worth.

Tyrl, R. J., T. G. Bidwell, R. E. Masters, and R. D. Elmore. 2008. Field guide to Oklahoma plants: Commonly encountered prairie, shrubland, and forest species. Oklahoma State University, Stillwater.

United States Government. Integrated Taxonomic Information System (ITIS). Accessed January 11, 2016. www.itis.gov.

Utah State University. Manual on the web. Accessed January 11, 2016. http://herbarium.usu.edu/webmanual/.

Valdés-Renya, J. 2015. Gramineas de Coahuila. Ultradigital Press, S.A. de C.V., Mexico.

Van Bruggen, Theodore. 1976. The vascular plants of South Dakota. Iowa State University Press, Ames.

Vasey, George. 1891. Illustrations of North American grasses. Volume 1. Grasses of the Southwest. U.S. Department of Agriculture, Washington, D.C.

Villarreal-Q., J. A. 1983. Malezas de Buenavista Coahuila. Universidad Autonoma Agraria "Antonio Narro." Buenavista, Saltillo, Mexico.

Vines, Robert A. 1960. Trees, shrubs and woody vines of the southwest. University of Texas Press, Austin.

Wagner, Warren L., and Earl F. Aldon. 1978. Manual of the saltbushes (*Atriplex* spp.) in New Mexico. General Technical Report RM-57. Forest Service, U.S. Department of Agriculture, Washington, D.C.

Waterfall, U. T. 1962. Keys to the flora of Oklahoma. Oklahoma State University Press, Stillwater.

Watson, L., and M. J. Dallwitz. 1988. Grass genera of the world. The Australian National University, Canberra.

————. 1992. Grass genera of the world. CAB International, University Press, Cambridge.

Weiner, Michael A. 1980. Earth medicine—earth food: Plant remedies, drugs, and natural foods of the North American Indians. Macmillan, New York.

Welsh, Stanley L., Duane Atwood, Sherel Goodrich, and Larry C. Higgins. 2007. Utah flora. Brigham Young University Press, Provo.

Western Regional Technical Committee W-90. 1972. Galleta: Taxonomy, ecology, and management of *Hilaria jamesii* on western rangelands. Bulletin 487. Agricultural Experiment Station, Utah State University, Logan.

Williams, Kim. 1977. Eating wild plants. Mountain Press, Missoula, Montana.

Winward, A. H. 1980. Taxonomy and ecology of sagebrush in Oregon. Station Bulletin 642. Agricultural Experiment Station, Oregon State University, Corvallis.

Yatskievych, George. 1999. Flora of Missouri. Missouri Botanical Garden Press, St. Louis.

Index

Accepted scientific names appear in bold-italic type

496